Oceanic and Anthropogenic Controls of Life in the Pacific Ocean

The GeoJournal Library

Volume 21

The titles published in this series are listed at the end of this volume.

Oceanic and Anthropogenic Controls of Life in the Pacific Ocean

Proceedings of the 2nd Pacific Symposium on Marine Sciences, Nadhodka, Russia, August 11–19, 1988

edited by

V. I. ILYICHEV and V. V. ANIKIEV

Institute of Oceanology,
Academy of Sciences, Vladivostok, Russia

KLUWER ACADEMIC PUBLISHERS
DORDRECHT / BOSTON / LONDON

Library of Congress Cataloging-in-Publication Data

Pacific Symposium on Marine Sciences (2nd : 1988 : Nakhodka,
Primorskiĭ kraĭ, R.S.F.S.R.)
Oceanic and anthropogenic controls of life in the Pacific Ocean :
proceedings of the 2nd Pacific Symposium on Marine Sciences,
Nakhodka, Russia, August 11-19, 1988 / edited by V.I. Ilyichev and
V.V. Anikiev.
p. cm. -- (GeoJournal library ; v. 21)
ISBN 0-7923-1854-4 (alk. paper)
1. Oceanography--North Pacific Ocean--Congresses. 2. Marine
resources--North Pacific Ocean--Congresses. 3. Marine ecology-
-North Pacific Ocean--Congresses. I. Il'ichev, V. I. II. Anikiev,
V. V. III. Title. IV. Series.
GC771.P33 1992
574.5'2636'091644--dc20 92-17798

ISBN 0-7923-1854-4

Published by Kluwer Academic Publishers,
P.O. Box 17, 3300 AA Dordrecht, The Netherlands.

Kluwer Academic Publishers incorporates
the publishing programmes of
D. Reidel, Martinus Nijhoff, Dr W. Junk and MTP Press.

Sold and distributed in the U.S.A. and Canada
by Kluwer Academic Publishers,
101 Philip Drive, Norwell, MA 02061, U.S.A.

In all other countries, sold and distributed
by Kluwer Academic Publishers Group,
P.O. Box 322, 3300 AH Dordrecht, The Netherlands.

Printed on acid-free paper

Printed in the Netherlands

CONTENTS

PART V METHODS OF OBSERVATIONS AND CALCULATIONS

PART I
INTRODUCTION

INTRODUCTION

S.M. Haq

The Intergovernmental Oceanographic Commission (IOC) is a central body within the United Nations System, which, in accordance with its Statute, is responsible for recommending, developing and coordinating of programmes in ocean sciences and ocean services to be implemented through concerted action of its Member States. The Commission also performs its tasks as a joint specialized mechanism under the umbrella of the Intersecretariat Committee on Scientific Programmes Relating to Oceanography (ICSPRO) of which the United Nations, UNESCO, FAO, WMO and IMO are the member organizations.

Following the large scale oceanographic expeditions organized and co-ordinated by IOC during the first fifteen years of its existence (e.g., International Indian Ocean Expeditions, IIOE), the Co-operative Investigations of the Caribbean (CICAR), the Co-operative Investigations of the Kuroshio in the Pacific (CSK) and others. The Commission's programmes and activities progressively evolved under the impact of two major influences. Firstly, the need to reshape its major programs in ocean science and ocean services on a continuing basis in keeping with rapid advances in ocean research. Secondly, the growing demand from the vast majority of its Member States in different regions of the Globe for developing programmes and activities in the area of common interest and in accordance with their priority needs. In order to fulfil its global objective and responsibilities, the Commission reorganized its programmes and structure to respond to those needs. The IOC established technical subsidiary bodies, responsible for major global programmes in ocean sciences and ocean services, and regional subsidiary bodies in different regions. In ocean sciences the technical bodies deal with major programmes which include the Global Investigations of Pollution in the Marine Environment (GIPME); Ocean Science in relation to Living Resources (OSLR) - jointly with FAO; Ocean Science in Relation to Non-Living Resources (OSNLR) - jointly with UN; Ocean Processes and Climate (OPC); a joint SCOR-IOC Committee on Climatic Changes and the Ocean (CCCO) and a joint IOC-IHO Guiding Committee for the General Bathymetric Chart of the Ocean (GEBCO). In ocean services the major programmes of the Commission include the International Oceanographic Data Exchange (IODE) and its Marine Information Management (MIM); the Integrated Global Ocean Services System (IGOSS) - jointly with WMO.

Under CCCO, IOC co-operates with the Joint Scientific Committee (JSC) of ICSU, and WMO for the World Climate Research Programme (WCRP) in outer programmes such as the Tropical Global Atmosphere (TOGA) and the World Ocean Circulation Experiment (WOCE).

One of the major activities of the Commission is Training, Education and Mutual Assistance (TEMA) in marine science through which not only assistance is provided to developing countries, but also co-operation between developed and developing countries is fostered in areas relating to the ongoing programmes of the Commission. The UNESCO/IOC Comprehensive Plan for a Major Assistance to Enhance Marine Science Capabilities of Developing Countries (hereafter called Comprehensive Plan) sets out strategies required for the development of effective co-ordination of all marine research and related activities including formulation of a marine science policy at the national level in a given country, so as to ensure economic utilization of all available national resources and effective liaison with international organizations concerned with marine affairs. As part of the strategy, IOC is involved in the preparation of Marine Science Country Profiles defining the scope and the basis for the development of marine scientific research capability in the context of ocean development at the national level, as well as the formulation of a technical assistance project, at the request of Member State(s), for possible international funding.

The regional subsidiary bodies on the other hand were established by the Commission in response to the request of the Member States. These include: IOCARIBE (in the Caribbean); WESTPAC (in the Western Pacific); IOCINCWIO (in the North, Central and Western Indian Ocean); IOCEA (in the Eastern Atlantic); IOCINDIO (in the North-Central Indian Ocean). It also co-operates with other bodies in regional programmes such as the Joint IOC-WMO-CPPS Investigations of El Nino (in the South-Eastern Pacific), the Joint CCOP(SOPAC)-IOC Working Group on Post-IODE Studies on East Asian Tectonics and Resources (SEATAR) and joint CCOP(SOPAC)-IOC Working Group on South Pacific and Resources (STAR). The programmes and activities of each of the regional subsidiary body is recommended, developed and implemented by the Member States concerned with IOC, wherever applicable with other bodies, playing a co-ordinating role.

PROGRAMME AND ACTIVITIES OF WESTPAC

The IOC Regional Committee for WESTPAC was established in 1978 soon after the termination of the IOC Co-operative Investigations of the Kuroshio. Although during the first three sessions of WESTPAC there had been modest progress in the implementation of the approved programme and activities, it was at the Fourth Session held in Bankgkok, Thailand, 22-26 June 1987, that the Regional Committee under the Chairmanship of Dr. John Bunt of Australia, took several concrete actions to identify specific co-operative projects for implementation by all interested Member States of WESTPAC. It laid particular emphasis on:

- establishing priority programmes that are of specific interest to the region as regional components of the IOC major programmes;

- effective implementation of the priority programmes through creation of mechanisms that should ensure development of co-operative networks relative to each approved project amongst the interested countries and their scientist/institutions in the region.

The <u>nine projects</u> approved by the Regional Committee for WESTPAC are as follows:

OCEAN SCIENCES

<u>Regional Component of OSLR in the Western Pacific</u>

(i) Recruitment of Penaeid Prawns in the Indo-Western Pacific;

(ii) Toxic and Anoxic phenomena associated with the algal blooms in the Western Pacific region.

Regional Component of OSNLR in the Western Pacific

(iii) WESTPAC Palaeographic Map
(iv) Margins of Active Plates (MAP).

Regional Component of Ocean Dynamics and Climate (ODC)

(v) Co-operative Research Study of Banding of Porites Corals as a component of ocean climate studies;

(vi) Co-operative Research Study of the Continental Shelf Circulation in the Western Pacific;

(vii) Co-operative Research Study of the Ocean Dynamics in the North-West Pacific.

Regional Component of Marine Pollution Research and Monitoring (GIPME)/(MARPPOLMON)

(viii) Assessment of River Inputs to the Sea in the WESTPAC region;

(ix) Monitoring Heavy Metals and Organochlorine Pesticides using the Musselwatch approach.

It is important to note that the Regional Committee endorsed the concept of a co-operative network to be developed in the long run which should be based on the principle of partnership, self help and concerted action amongst the interested institutions/scientists, for the purpose of planning, promoting, co-ordinating and implementing agreed projects of common interest to them. For a network to be functional and successful, the operational plan for a given project was considered as a first step which should determine the precise nature of participation by each co-operating institution/scientist and identification of training or assistance needs, to ensure effective implementation by interested institutions of a given project. In this context, it was considered essential that a mechanism, namely, the Project Steering Group, be established to serve a "motor" for the implementation of the project. Such a group is composed of active scientists from participating institutions and led by a Project Leader to be appointed by the Chairman of WESTPAC in consultation with the Secretary IOC.

Although the Regional Committee decided to implement preferentially the projects (i); (ii); (iii); (v); (viii) and (ix) referred to above, it also encouraged the implementation by interested Member States of the remaining projects, i.e., (iv); (vi) and (vii).

OCEAN SERVICES

In the field of ocean services no specific project of the kind referred to above was considered. However, the Regional Committee approved several actions with a view to strengthening the following programmes and activities:

encouraged Member States to increase their participation on IGOSS Ships of Opportunity Programme, particularly by increasing their TESAC data, and their efforts to identify sources of real time data, mainly from research programmes in the the IGOSS System;

expressed support for the regional component of IOC Global Sea Level Observing System (GLOSS) and requested for the development of a tide gauge network in the region;

endorsed strengthening participation of Member States in the IODE System in the region;

welcomed the preparation of national inventories of marine scientific institutions and marine scientists so as to make available to all Member States of the region and, if possible, to incorporate it into the relevant registers of Aquatic Fisheries Information System (ASFIS);

supported the establishment of regional marine scientific information networks to be included in the proposed Marine Information Management Strategic Plan and, in this respect, recommended co-operation between IOC-FAO and SEAFDEC.

LIAISON WITH OTHER PROGRAMMES/ACTIVITIES IN THE REGION

One of the important decisions of the Committee was the maintenance of liaison with other programmes and activities in progress in the region which are the results of initiatives taken by the countries of the region. It is the decision of the Committee to place such activities under the category of serving the objectives of WESTPAC. No criteria was established for such categorization, the decision was left to the judgement of the Chairman of WESTPAC and the Secretary of IOC.

ESTABLISHMENT OF A SUB-COMMISSION FOR WESTPAC

Considering the rapid advances made by the countries in the region in the field of ocean science and ocean services, the Member States of the region decided that a Sub-Commission for WESTPAC should be established. In this context the Secretary of IOC is negotiating with the Government of Thailand, which offered to host the secretariat for the proposed Sub-Commission and a final decision in this regard is expected to be taken by the Fifteenth Session of the IOC Assembly scheduled to be held in the first half of 1989.

TEMA COMPONENT OF WESTPAC

Training, Education and Mutual Assistance in Marine Sciences are an integral part of all programmes and projects approved by WESTPAC. The Project Steering Group responsible for each of the projects approved by the Regional Committee was therefore advised to define TEMA needs while preparing an operational plan relevant to their activities. Two basic types of assistance under TEMA have been foreseen: one relates to the training of scientists, as well as equipment relative to each of the agreed projects, aimed at strengthening the required capabilities of interested institutions of developing countries of the region so that they are able to participate effectively in these projects; and the other is the formulation of a regional technical assistance project under the UNESCO/IOC Comprehensive Plan, referred to above, for possible funding from international sources.

A number of Member States of the region offered assistance including: Australia, China, Japan and the USSR, to developing countries of the region which are in the process of being negotiated and/or implemented by IOC.

The Regional Committee unanimously elected Professor Takahisa Nemoto, Director of the Ocean Research Institute, Tokyo, and Dr. Choompol Swasdiyakorn, Secretary General, National Research Council of Thailand as Vice-Chairman.

SUMMARY OF INTERSESSIONAL ACTIVIES

During the intersessional period a number of activities endorsed by the Regional Committee have either been implemented or are in the process of being implemented. These include:

- Appointment of project leaders for the project approved by the Regional Committee.

- Operational plan for implementation of the project on Toxic and Anoxic Phenomena Associated with Algal Blooms in the WESTPAC Region was developed at the IOC Workshop held, following the International Red Tide Symposium organized in Takamatsu in Japan in November 1987.

- Operational plan for implementation of the project on Recruitment of Shrimp in the Indo-West Pacific Region, at the IOC Workshop held at the CSIRO Marine Laboratories, Cleveland, Australia, 25-30 July 1988.

- Operational Plan for the Project on Banding of Porites Corals as a Component of Ocean Climate Studies, which is expected to be developed at the IOC training workshop now being held, following the International Coral Reef Symposium, scheduled at the Australian Institute of Marine Sciences, Townsville, 15-17 August 1988.

- Operational plan for WESTPAC Palaeographic Map which is scheduled to be developed at the IOC Workshop, following the International Conference on Asian Marine Geology to be held in Tongjin, Shanghai, early September 1988.

- Submission of a project to UNDP for assistance on Recruitment of Shrimps in the WESTPAC region, to be executed jointly by FAO/IOC.

- Submission of a project aid UNDP for funding on Strengthening Marine Science and Services for Ocean Development in the Region.

- Planning of a training course jointly with UNESCO on the Use of Micro-computer in Oceanographic Data Centre in Support of the IODE Programme in the region to be held at the Asian Institute of Technology, Bangkok, Thailand, in early 1989.

In addition to these, the IOC has organized several shipboard training on board Japanese research vessels and a training course in oceanographic data management scheduled in 1988 at the Japanese Oceanographic Data Centre, Tokyo. IOC is also considering an offer from the USSR to provide training for scientists from the region on board of the USSR research vessels in the South Pacific in late 1988 and early 1989.

While the above mentioned activies highlight the objectives, programmes and activities being undertaken by the countries of the region within the context of WESTPAC, it is important to bear in mind the limitation of these activities. The programmes/projects referred to above are specific in nature and target oriented and the aim of these activies is to develop understanding of the marine scientific problems that are of common interest to the countries of the region so as to provide adequate scientific bases for rational development and management of the marine resources and the environment. These activities by no means encompass numerous other marine scientific problems of the Pacific region. That is why the Regional Committee for WESTPAC considered development of parallel scientific research activities which are the result of the initiatives taken out individually and/or collectively by the countries of the region in the multidisciplinary and interdisciplinary fields of marine science, as a very significant development contributing towards an enhancement of regional capacity to deal with ocean affairs. Activities such as these are, doubtless, complementary to the objective of WESTPAC. It is therefore logical the IOC, as referred to earlier, would be happy to encourage increased collaboration with other national and region efforts in the region of the Pacific.

S.M.Haq
Intergovernmental Oceanographic Commission
(UNESCO)

PART II
CHANGEABILITY OF NATURAL FIELDS AND FLUXES OF MATTER AND ENERGY

1 GENERATION MECHANISM OF THERMOHALINE FRONT IN SHELF SEA

Tetsuo Yanagi and Atsuhiko Isobe

Abstract

A remarkable thermohaline front was observed at the mouth of Tokyo Bay, Japan, in winter. Water temperature changed from 14.5°C to 17.0°C and salinity from 34.0 ppt. to 34.8 ppt. across the front, but little density difference was observed. The generation mechanism of such a thermohaline front is investigated by use of a three-dimensional numerical experiment. The thermohaline front is formed at the leading edge of the density-driven current due to the salinity efflux from the inner boundary of the Bay. The dependence of sharpness and position of the thermohaline front on the external parameters such as surface cooling rate, river discharge, and boundary conditions of water temperature and salinity is also examined numerically.

1.1 Introduction

The quasi-steady thermohaline front is generated in a transition zone between colder coastal water and warmer oceanic water in winter. The qualitative generation mechanism of such a thermohaline front is as follows. The colder coastal water does not become heavy enough to sink owing to the fresh water discharge. Similarly, oceanic saline water cannot become heavy enough to sink because of the mixing with warm open sea water. Water in the central region of the shelf area becomes the heaviest due to the cooling through the sea surface and it sinks there. As a consequence, there is little density difference between both sides of the thermohaline front because the salinity compensates the density difference, which might otherwise be caused by the temperature difference.

Such thermohaline fronts have been observed in the coastal areas of the NW Pacific Ocean in winter, e.g. in the Seto Inland Sea (Yoshioka, 1971, Yanagi, 1980), in the W area of Cheju Island (Lie, 1985) and at the mouth of Tokyo Bay (Yangagi et al., 1988).

On the other hand, the quantitative generation mechanism of thermohaline have been investigated numerically. Endoh (1977) first succeeded in reproducing a sharp thermohaline front by the coupling effects of cooling through the sea surface and the inflow of fresh water within a vertical two-dimensional numerical model without the Coriolis effect. A sharply edged shelf causing the marked difference of the heat capacity of the water columns over the shelf and in the deeper open ocean region is responsible for the formation of a thermohaline front over the shelf edge in his model. Harashima et al. (1978) also reproduced a sharp thermohaline front within a vertical two-dimensional numerical model with a flat bottom. The difference of heat capacity due to the variation of water depth in the model of Endoh (1977) was replaced by the boundary condition of warm water temperature at the open sea boundary in the model of Harashima et al. (1978). Oonishi et al. (1978) interpreted such numerical studies with a simple analytical model and they obtained the simple functional dependencies of the position and the width of the front on the cooling rate and the fresh water discharge. However, their studies does not consider the

V. I. Ilyichev and V. V. Anikiev (eds.), Oceanic and Anthropogenic Controls of Life in the Pacific Ocean, 11–33.

Coriolis effect, which may play an important role in the formation of a thermohaline front in the field.

Harashima and Oonishi (1981) studied the Coriolis effect on frontogeneses in steady buoyancy-driven circulation using a three dimensional numerical model. They clarified that the characteristics of the thermohaline front depend on two non-dimensional parameters, that is, the flux Rossby number and the flux Reynolds number. However, they could not reproduce the sharp thermohaline front in the three-dimensional numerical model including the Coriolis effect with the realistic parameters in the field.

We will first show the characteristics of the thermohaline front from the observation at the mouth of Tokyo Bay in winter. Next we will try to reproduce the sharp thermohaline front in the three-dimensional numerical model including the Coriolis effect with the realistic parameters in the field. Moreover we will reveal the characteristics of the buoyancy-driven current in the coastal sea in winter.

Figure 1.1: Observation stations (Stns. 29 to 33) by R/V Tansei-Maru at the mouth of Tokyo Bay. Thin line represents the depth contour and numbers show the depth in meters.

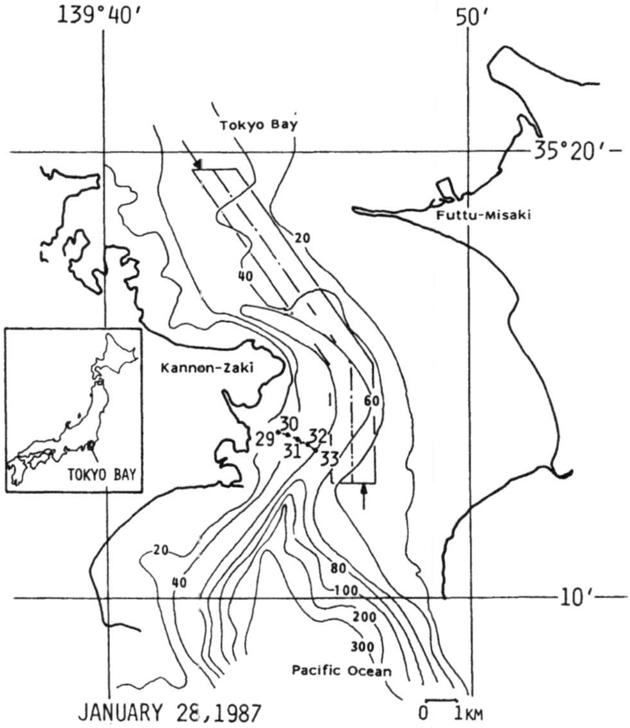

1.2 Observations

The field observations were carried out at the mouth of Tokyo Bay (Figure 1.1) by R/V Tansei-Maru of the Ocean Research Institute, University of Tokyo from 25 January to 28 January, 1987. The details of the field observation are written in another paper (Yanagi et al., 1988). The observed horizontal distributions of surface (-5 m) water temperature and salinity on 28 January, 1987, are shown in Figure 1.2.

Figure 1.2: Horizontal distributions of surface water temperature and salinity at the mouth of Tokyo Bay on 28 January 1987. Broken line shows the track of R/V Tansei-Maru.

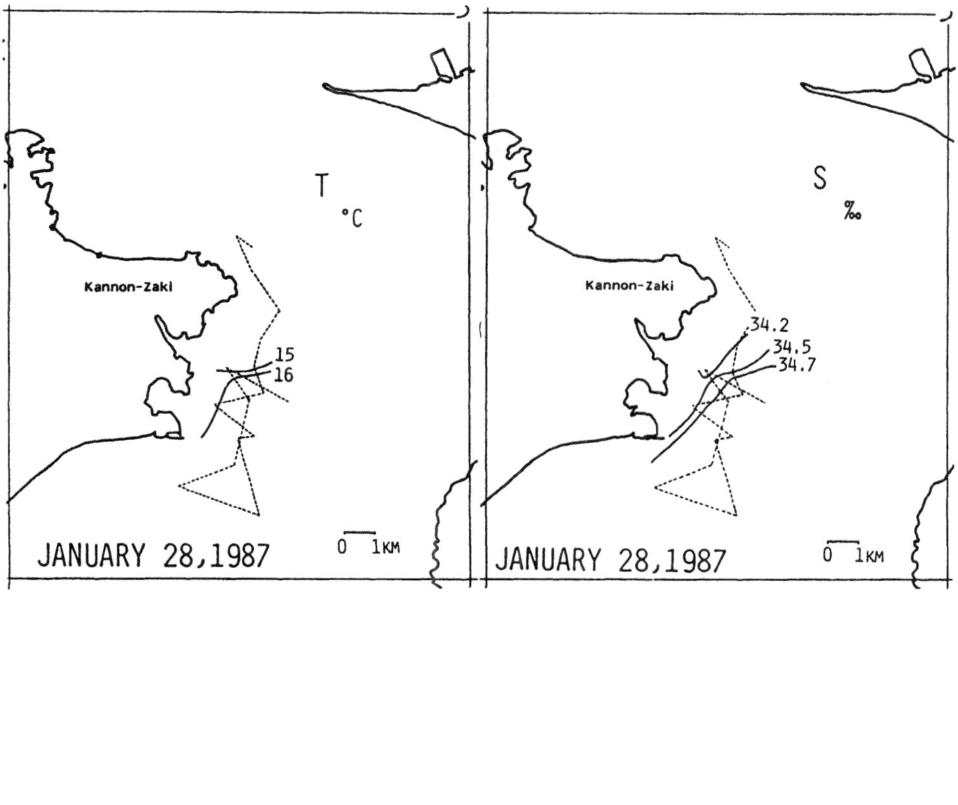

The fresh and cold coastal water exists near Kan-Non-Zaki along the Miura Peninsula and the salty and warm offshore water exists southeastward across the frontal structure. Figure 1.3 is a NOAA i.r.image, collected at 03:26 JMT on 28 January,1987, which shows the position of the thermohaline front at the mouth of Tokyo Bay.

Figure 1.3: NOAA 9 i.r. image at 03:26 JMT on 28 January,1987 showing the position of the thermohaline front at the mouth of Tokyo Bay. The grid size on the satellite image is 1.38 x 1.38 km. The light area on the image indicate warmer water. North is at the top of the image.

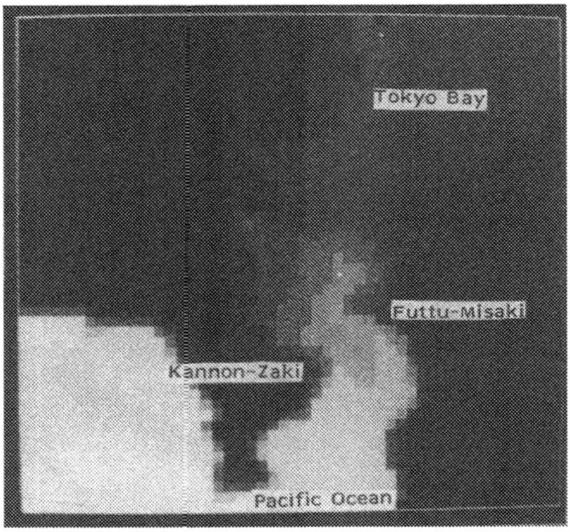

The thermohaline front, represented by the discontinuity line between cooler coastal water and warmer offshore water, runs across the mouth of Tokyo Bay. The front in the W part of the Bay mouth is sharper than that in the E part. Such asymmetry is observed in the Kii Channel (Yoshioka, 1971) and in Iyo-Nada (Yanagi, 1980).

Figure 4 shows the horizontal variations in water temperature, salinity and density at 5 m below the sea surface from Stn. 29 to Stn. 33 across the thermohaline front which exists near Stn. 30. Water temperature changes from 14.5 °C to 17.0 °C and salinity from 34.0 ppt. to 34.8 ppt. within about a hundred meters across the front. Conversely, density does not change across the front showing a small local maximum at the front. Figure 5 shows the vertical distributions of water temperature, salinity and density in the section along Stn. 29 to Stn. 33. The coastal water which has a little larger density sinks below the offshore water. Such a thermohaline front exists at the nearly same position throughout winter season (Yanagi and Koike, 1987, Yanagi et al., 1988).

Figure 1.4: Horizontal variations in water temperature, salinity and density at 5m below the sea surface from Stn. 29 to Stn. 33 across the thermohaline front at the mouth of Tokyo Bay on 28 January, 1987.

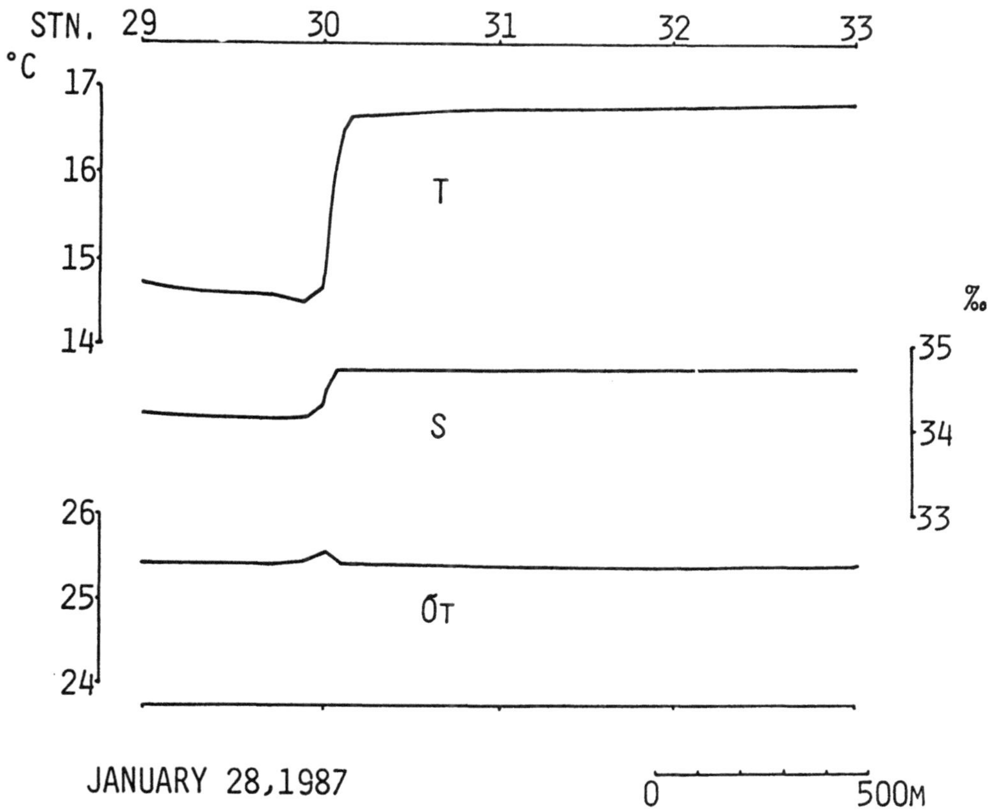

The characteristics of thermohaline front are summarized in the followings: 1) The thermohaline front is formed between the cold, fresh coastal water and the warm, saline offshore water in winter. 2) The thermohaline front is stationary from late autumn until early spring. 3) The differences of water temperature and salinity across the thermohaline front are very large but that of density is little. 4) The thermohaline front in the W part of the Bay is sharper than that in the eastern part. We will reproduce the thermohaline front with characteristics from 1) to 4) with the use of the three-dimensional numerical model including the Coriolis effect.

Figure 1.5: Vertical distributions of water temperature, salinity and density across the thermohaline front at the mouth of Tokyo Bay on 28 January, 1987.

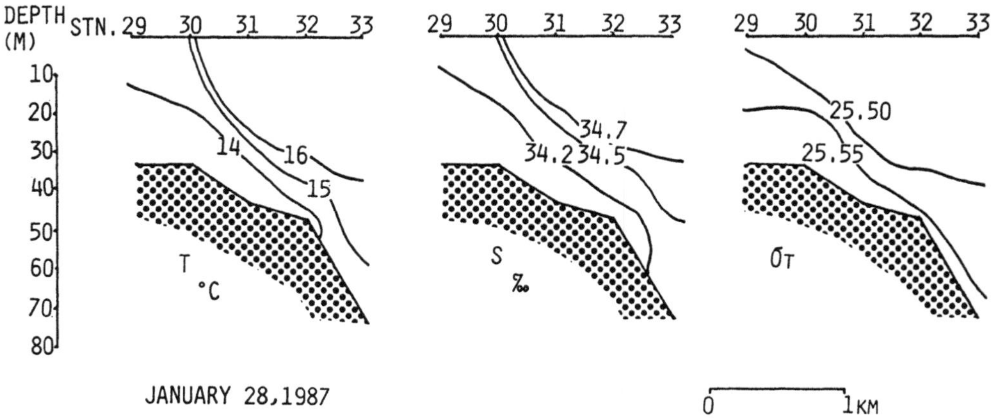

1.3 Numerical experiment

1.3.1 Formulation

Let us consider a model basin of depth H (H = 40 m), transverse scale B (B = 25 km), longitudinal scale L (L = 50 km), surrounded by longitudinal walls, the ocean side boundary and the Bay side boundary as shown in Figure 1.6.

Figure 1.6: Schematic diagram of the model basin.

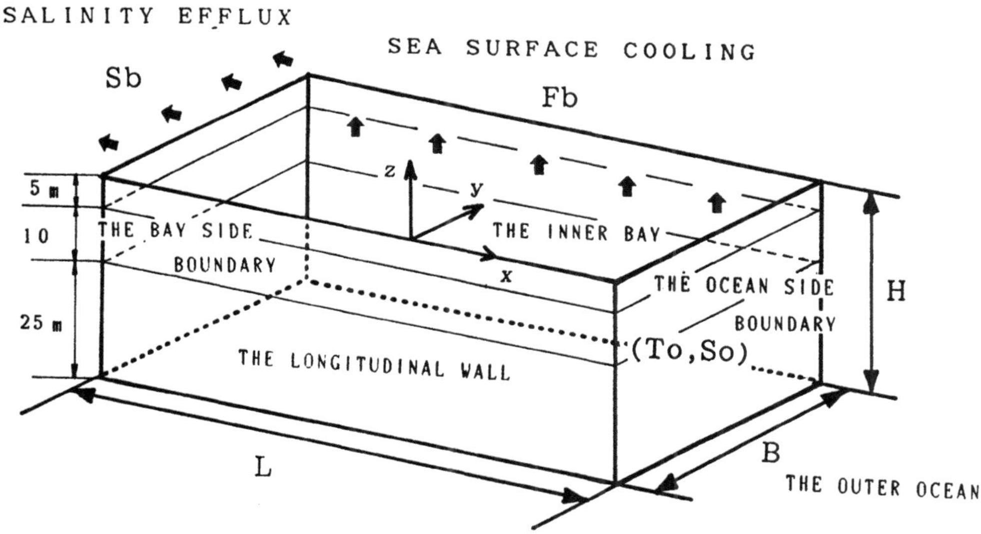

Buoyancy efflux at the sea surface F_b is given so as to correspond to cooling from above and buoyancy influx at the Bay boundary S_b to fresh water inflow. At the ocean side, temperature and salinity of the sea water are fixed to (T_0, S_0). Assumptions and simplifications are made as follows,

1) Boussinesq approximation.
2) Vertically hydrostatic approximation.
3) Fluid is assumed to be incompressible.
4) Convective adjustment for density in the vertical water column where unstable density stratification occurs.

Then the momentum and continuity equations are written as follows:

$$\frac{\partial u}{\partial t}+u\frac{\partial u}{\partial x}+v\frac{\partial u}{\partial y}+w\frac{\partial u}{\partial z}-fv=-\frac{1}{\rho_0}\frac{\partial p}{\partial x}+\nu_h\Delta u+\nu_V\frac{\partial^2 u}{\partial z^2} \tag{1}$$

$$\frac{\partial v}{\partial t}+u\frac{\partial v}{\partial x}+v\frac{\partial v}{\partial y}+w\frac{\partial v}{\partial z}+fu=-\frac{1}{\rho_0}\frac{\partial p}{\partial y}+\nu_h\Delta v+\nu_V\frac{\partial^2 v}{\partial z^2} \tag{2}$$

$$p=\rho_0 g\eta-\rho_0\int_0^{-z} b\,dz \tag{3}$$

$$\frac{\partial u}{\partial x}+\frac{\partial v}{\partial y}+\frac{\partial w}{\partial z}+fu=0 \tag{4}$$

$$\frac{\partial u}{\partial x}+\frac{\partial v}{\partial y}+\frac{\partial w}{\partial z}=0 \tag{5}$$

Here u and v are the horizontal velocity in x and y directions, respectively, w is the vertical velocity in z direction (Figure 1.6), f the Coriolis parameter, ρ_0 the overall mean density of the sea water, p the pressure, ρ the density, b the buoyancy, ν_h and ν_V the horizontal and vertical eddy viscosities, η the elevation from the mean sea surface, g the gravitational constant, Δ a horizontal Laplacian operator.

The state and advection-diffusion equations are changed according to each case and they will be shown later.
The boundary conditions for the water motion are as follows
i) at the sea surface ($z=0$)

$$\frac{\partial u}{\partial z}=\frac{\partial v}{\partial z}=0 \tag{6}$$

ii) at the sea bottom ($z=$-H)

$$u=v=w=0, \tag{7}$$

iii) at the longitudinal wall ($y=0,B$)

$$u=v=0, \tag{8}$$

iv) at the open side boundary ($x=L$)

$$u = 0, \frac{\partial v}{\partial x} = 0, \tag{9}$$

v) at the Bay side boundary ($x=0$)

$$u\text{-}v=0. \tag{10}$$

The boundary conditions for the material flux will be shown later.
The initial conditions for u,v, and w are set to be 0 and those for water temperature and salinity to be equal, that is, the whole basin is initially filled with water of uniform density at rest.

Table 1.1: Parameters used in the numerical experiment.

horizontal grid scale	Δx	...	2	km
time step	Δt	...	1000	sec
horizontal eddy viscosity	ν_h	...	10^5	$cm^2\ sec^{-1}$
vertical eddy viscosity	ν_z	...	10	$cm^2\ sec^{-1}$
horizontal eddy diffusivity	K_h	...	10^5	$cm^2\ sec^{-1}$
vertical eddy diffusivity	K_z	...	10	$cm^2\ sec^{-1}$
gravitational constant	g	...	980	$cm^2\ sec^{-1}$
Coriolis parameter	f	...	$7.3\text{x}10^{-5}$	$cm\ sec^{-1}$
overall mean density	ρ_0	...	1.0255	$g\ cm^{-3}$
heat flux	Q_s	...	-300	$cal\ cm^{-2}\ day^{-1}$
river discharge	R	...	70	$cm^3\ sec^{-1}\ cm^{-1}$
buoyancy of river water	b_R	...	24.5	$cm\ sec^{-2}$
buoyancy of open ocean	b_0	...	0	$cm\ sec^{-2}$
specific heat	c_p	...	0.932	$cal\ deg^{-1}\ g^{-1}$
heat expansion coefficient	α	...	$2\text{x}10^{-4}$	deg^{-1}

The finite difference scheme is made by adding the time integration of the advection-diffusion equations for materials to the scheme of Oonishi (1978). The model parameters are shown in Table 1.1.

The quasi-steady state is obtained about three months after the beginning of the calculation. The result of such a quasi-steady state will be shown in the following.

1.3.2 Results

Case 1)

Following Harashima and Oonishi (1981), transformations of the temperature and the salinity into buoyancy b and a dynamically passive tracer C is carried out. They combined the advection-diffusion equations of water temperature and salinity and used those of passive tracer C and buoyancy b, because they made the assumption the effects of salinity and water temperature to the change of density are linear. The difference of inner water and outer water is distinguished by the distribution of passive tracer C. The advection-diffusion equations of b and C are as follows,

$$\frac{\partial b}{\partial t}+u\frac{\partial b}{\partial x}+v\frac{\partial b}{\partial y}+w\frac{\partial b}{\partial z}=K_h \Delta b+K_v\frac{\partial^2 b}{\partial z^2} , \quad (11)$$

$$\frac{\partial C}{\partial t}+u\frac{\partial C}{\partial x}+v\frac{\partial C}{\partial y}+w\frac{\partial C}{\partial z} = K_h \Delta C+K_v\frac{\partial^2 u}{\partial z^2} . \quad (12)$$

Here K_h and K_v are the horizontal and vertical diffusivities, respectively. The boundary conditions for b and C are as follows.

i) at the sea surface ($z=0$)

$$K_v\frac{\partial b}{\partial z}=F_b,\ F_b=\frac{g\alpha Q_s}{\rho C_p},\ \frac{\partial C}{\partial z}=0 , \quad (13)$$

ii) at the sea bottom ($z=$-H)

$$\frac{\partial b}{\partial z}=\frac{\partial C}{\partial z}=0 , \quad (14)$$

iii) at the longitudinal wall ($y=0$,B)

$$\frac{\partial b}{\partial n}=\frac{\partial C}{\partial n}=0 \quad , \tag{15}$$

iv) at the open side boundary (x=L)

$$b=0, C=1, \tag{16}$$

v) at the Bay side boundary (x=0)

$$Q_R=(b_R - b)\ R,\ C=-1. \tag{17}$$

Here α denotes the expansion coefficient of sea water, Q_s the heat flux through the sea surface, C_p the specific heat, n the variation in the direction normal to the boundary, b_R the buoyancy of river water and R the river discharge per unit coast line. Harashima and Oonishi (1981) set the boundary condition for b at the inner side to be 0 instead of Eq. (17), that is, they assumed that the buoyancy of inner water was fixed. However, their boundary condition is not correct because the capacity of inner water is not so large as the open ocean water where the buoyancy is fixed. Therefore we set the boundary condition for b at the inner side as the flux type like Eq. (17).

The result as shown in Figure 1.7. the distribution of passive tracer C is essentially as the same as that of Harashima and Oonishi (1981), that is, it is point-symmetry at the centre point of the basin and the sharp front is not formed in this case.

Case 2)

Transformation of temperature and salinity into b and C in Case 1) is abandoned in Case 2) and distributions of water temperature T and salinity S are calculated separately as follows,

$$\frac{\partial T}{\partial t}+u\frac{\partial T}{\partial x}+v\frac{\partial T}{\partial y}+w\frac{\partial T}{\partial z} = K_h \,\Delta\, T+K_v\frac{\partial^2 T}{\partial z^2} \quad , \tag{18}$$

$$\frac{\partial S}{\partial t}+u\frac{\partial S}{\partial x}+v\frac{\partial S}{\partial y}+w\frac{\partial S}{\partial z} = K_h \,\Delta\, S+K_v\frac{\partial^2 S}{\partial z^2} \quad , \tag{19}$$

The linear relation among density, water temperature and salinity is as follows,

$$\rho = 1.0 - 8.41x10^{-4}T + 2.29x10^{-4}S. \tag{20}$$

The boundary conditions for water temperature and salinity are as follows,
i) at the sea surface (z=0)

Figure 1.7: Horizontal distributions in passive tracer (a) and horizontal velocity (b) in the surface layer in case (1).

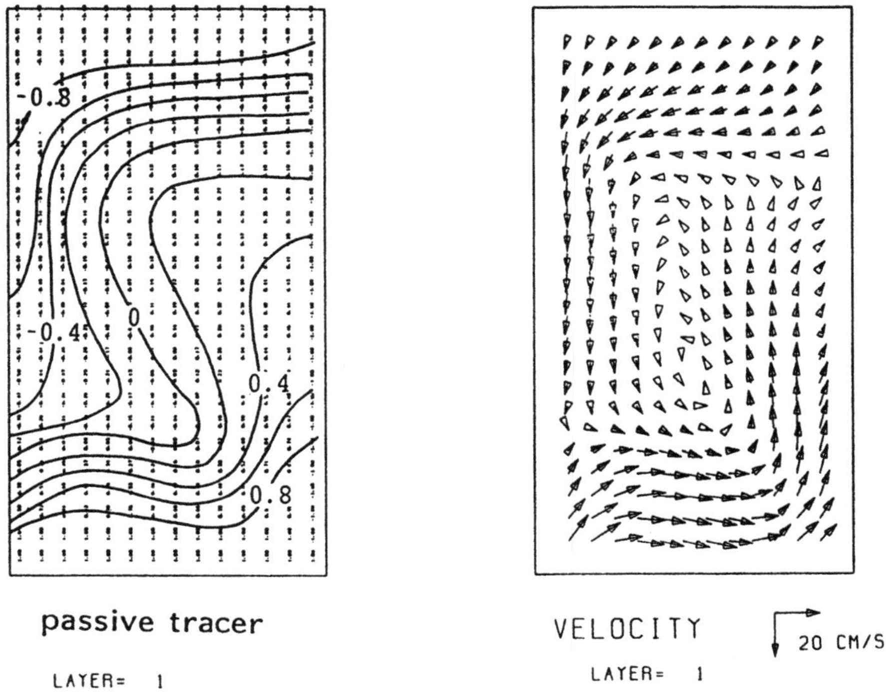

(a) (b)

$$K_v \frac{\partial T}{\partial z} = F_b,\ F_b = \frac{Q_s}{\rho C_p},\ \frac{\partial S}{\partial z} = 0 \quad , \tag{21}$$

ii) at the sea bottom (z=-H)

$$\frac{\partial T}{\partial z} = \frac{\partial S}{\partial z} = 0 \quad , \tag{22}$$

iii) at the longitudinal wall (y=0,B)

$$\frac{\partial T}{\partial n} = \frac{\partial S}{\partial n} = 0 \quad , \tag{23}$$

iv) at the open side boundary ($x=L$)

$$T=T_0, S=S_0 \qquad (24)$$

v) at the Bay side boundary ($x=0$)

$$Q_R = {}_{-S}R. \qquad (25)$$

The calculated horizontal distributions of water temperature, salinity density and velocity in the surface layer are shown in Figure 1.8. Patterns in this case are not symmetric like in Case 1) and sharp thermohaline front is formed at the left lower part of the basin. Such asymmetry patterns of temperature and salinity in the numerical model reproduces well the characteristics of the thermohaline front in the field. Water temperature and salinity change by 3 °C and 0.6 ppt between two meshes at the front, though density changes not significantly.

The reasons why the results of Cases 1) and 2) are different are investigated. The boundary conditions of buoyancy at both side boundaries are the same in Cases 1) and 2) and the flow pattern is quite similar in both cases shown in Figures 1.7 (b) and 1.8 (d). However, the distributions of tracer are different in both cases especially between C in case 1) and salinity in Case 2). Such difference is due to the difference of boundary condition of tracer at the inner side boundary, that is, the boundary condition of C in case 1) is constant $C=-1$ as shown in Eq. (17) and that of salinity is the flux type $Q_R=-SR$ as shown in Eq. (25). C is kept to be constant under flowing along the inner side boundary in Case 1) but salinity continues to decrease under flowing along the inner side boundary in Case 2) and temperature also continues to decrease due to surface cooling in Case 2). The boundary conditions of tracer in Case 1) are the same at inner and outer side boundaries and that results in the symmetrical distribution of tracer C to the centre point of the basin. On the other hand, salinity and temperature always decrease at the inner side boundary and those at the outer side boundary are fixed in Case 2). Such difference of boundary conditions at both side boundaries results in the asymmetry of tracers distribution in Case 2).

Case 3)

The basic equations and boundary conditions are the same as Case 2) except the state equation. The nonlinear state equation is followed by Mamaev (1964).

$$(\rho-1.0) \times 10^3 = 28.14 - 0.0135T - 0.00469T^2 + (0.802 - 0.002T)\,(\nu-35). \qquad (26)$$

The results are shown in Figure 1.9. The sharper thermohaline front is formed in the left lower part of the basin. Water temperature and salinity change by 4 °C and 0.8 ppt. between two meshes at the front. The cabbeling effect due to the nonlinearity of state equation makes the sharper front.

Figure 1.8: Horizontal distributions in water temperature (a), salinity (b), density (c) and horizontal velocity in the surface layer in case (2).

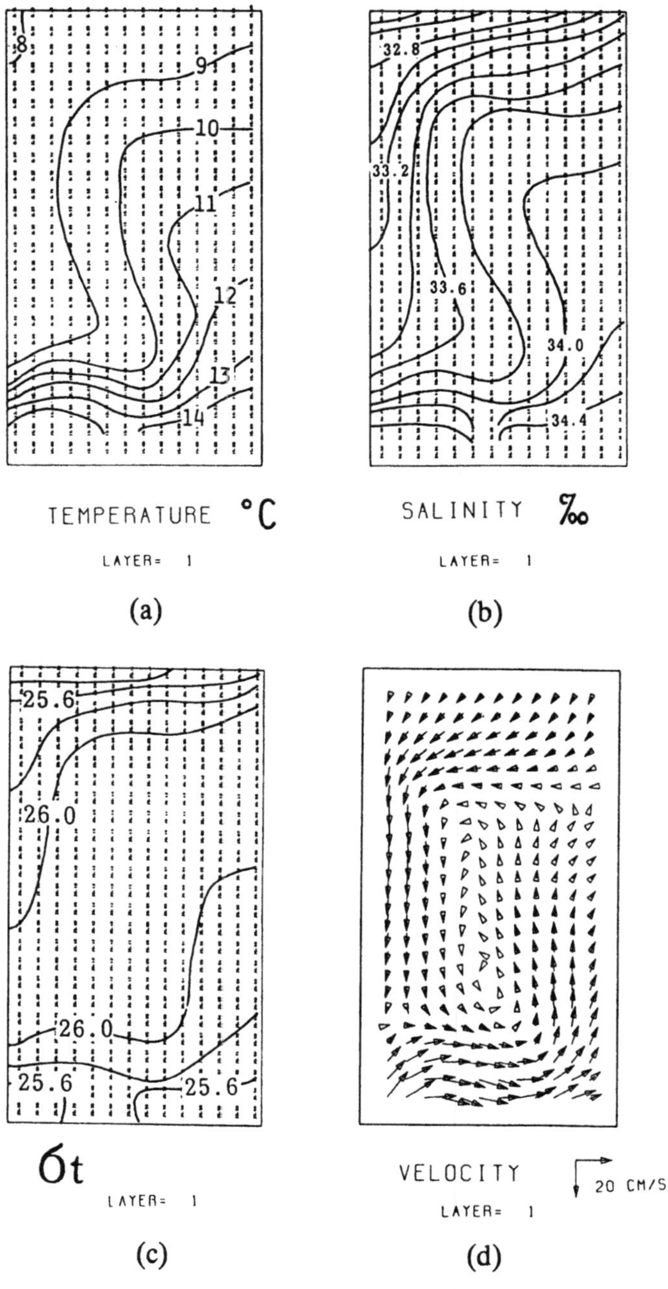

Figure 1.9: Horizontal distributions in water temperature (a), salinity (b), density (c) and horizontal velocity in the surface layer in case (3).

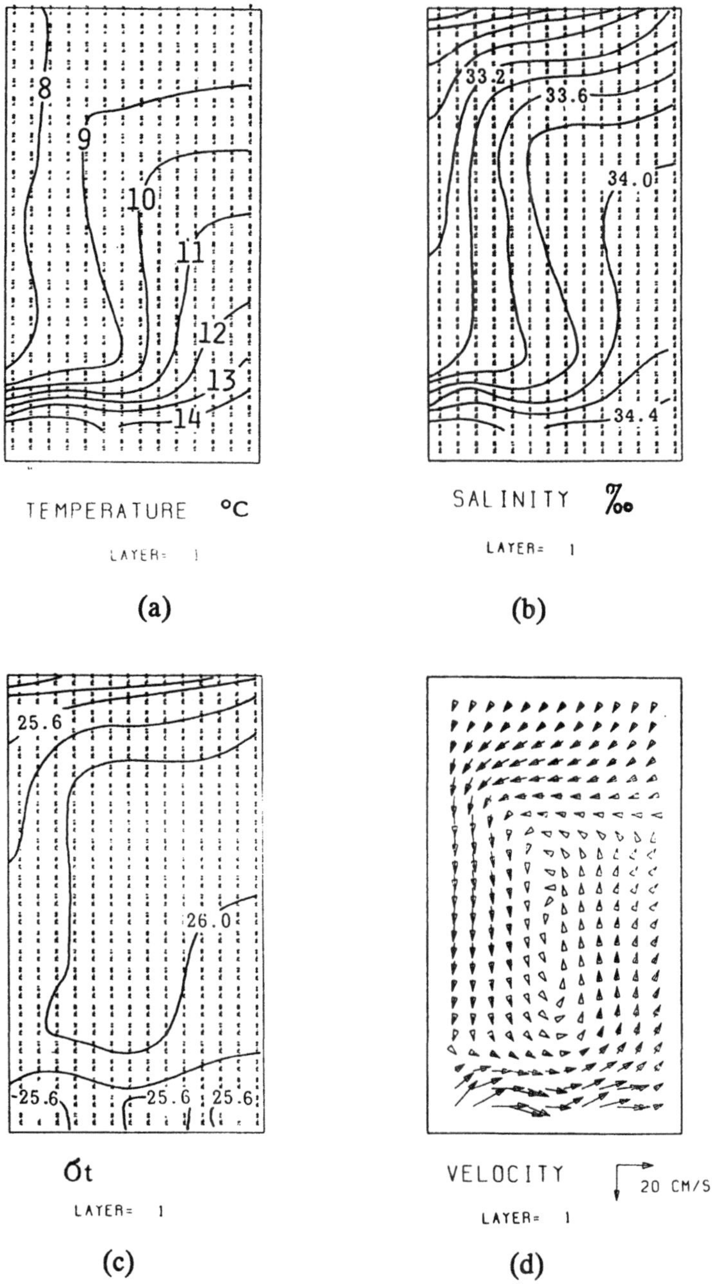

Figure 1.10: Gradients of passive tracer C in case 1 (point-chain line), salinity in case 2 (broken line) and salinity in case 3 (full line) at the thermohaline front.

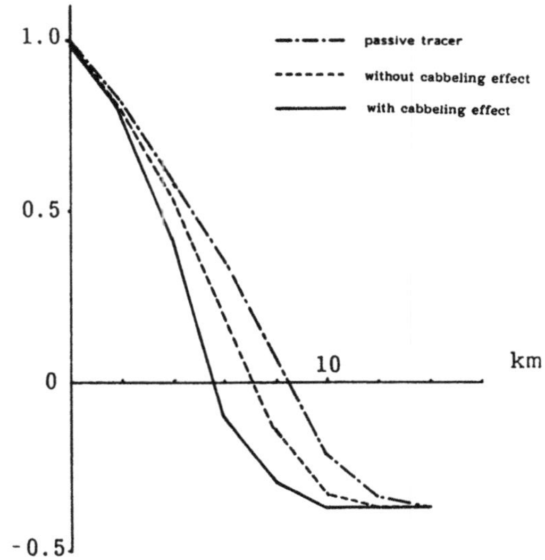

The sharpness of the thermohaline fronts in Cases 1) to 3) are compared. The normalized gradients of passive tracer *C* in Case 1), salinity in Case 2) and salinity in Case 3) in the surface layer at the frontal zone in the W lower part of the basin are shown in Figure 1.10. 1.0 means the value at the open ocean boundary in Figure 1.10. The sharpest front is generated in Case 3) and the cabbeling effect is considered to play an important role for making the sharp thermohaline front.

Case 4)

The basic equations and boundary conditions are the same as Case 3) and the horizontal geometry and bottom topography are altered to be similar to those of Tokyo Bay as Figure 1.11. The results are shown in Figure 1.12. The sharp thermohaline front is formed near the Kan-non-zaki and the calculated horizontal distribution of surface water temperature (Figure 1.12a) is similar to the observed one which is shown in Figure 1.3. The vertical distributions of water temperature along the section normal to the thermohaline front are shown in Figure 1.13. The sharpness of frontal structure weakens from W to E and the pattern along C-C' section is similar to the observed one which is shown in Figure 1.5. The distributions of surface velocity, bottom velocity and vertical velocity from the surface layer to the intermediate layer are shown in Figure 1.14. the surface convergence, large downwelling and outward bottom flow are seen at the thermohaline front region.

Figure 1.11: Horizontal geometry of the numerical model of Tokyo Bay.

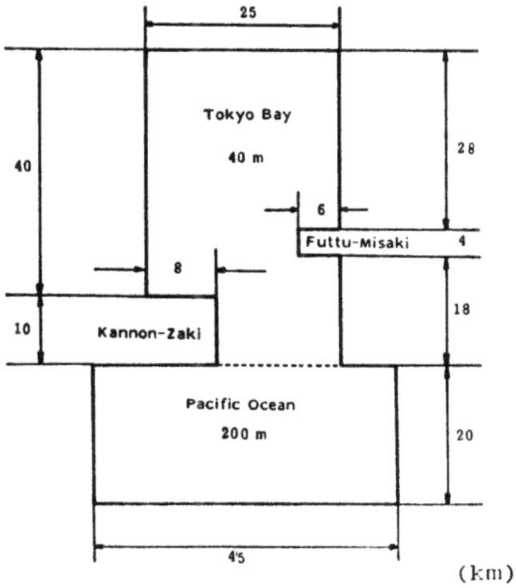

1.4 Discussion

Front is defined as "a convergence on the sea surface larger than diffusion between two different water masses" (Yanagi,1987). We consider the mechanism of surface convergence around the thermohaline front.

In Case 2), the density driven current from the inner side boundary comes across the water mass of high water temperature and high salinity, whose density is nearly the same as that of density-driven current, at the lower end of the W longitudinal boundary. The density-driven current cannot push out the water mass because the density gradient at the outer side boundary pushes its water mass into the basin and the density-driven current turns to left at the collision place. Then the secondary vertical circulation is induced due to the Ekman section in the bottom boundary layer by the horizontal counterclockwise circulation as shown in Figures 1.9 and 1.14. The surface convergence owing to such vertical circulation makes the sharp thermohaline front at the SW part of the basin.

Figure 1.12: Horizontal distributions in water temperature (a), salinity (b) and density (c) in the surface layer in case (4).

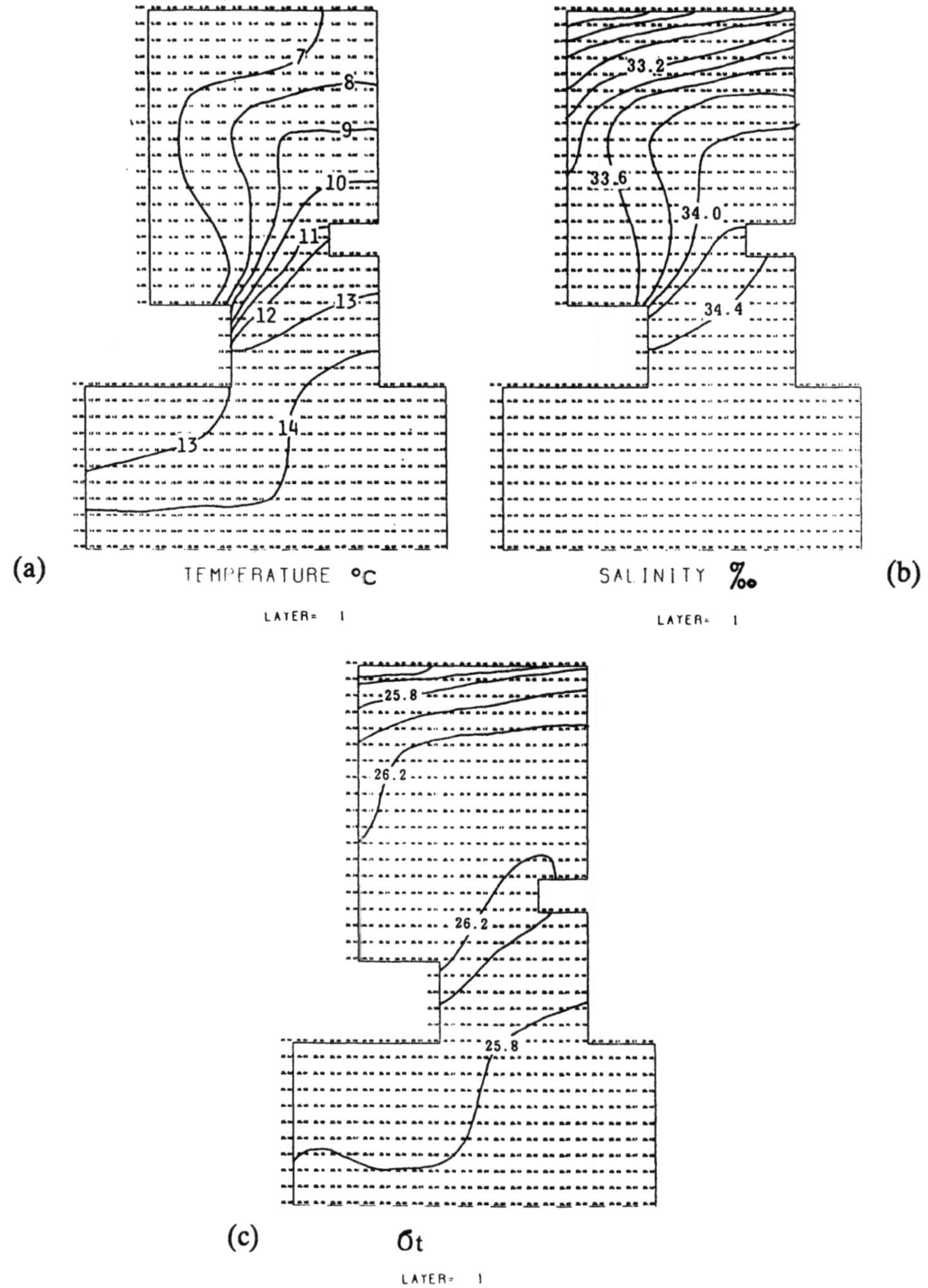

Figure 1.13: Distributions of water temperature (°) along the sections A-A', B-B', C-C'.

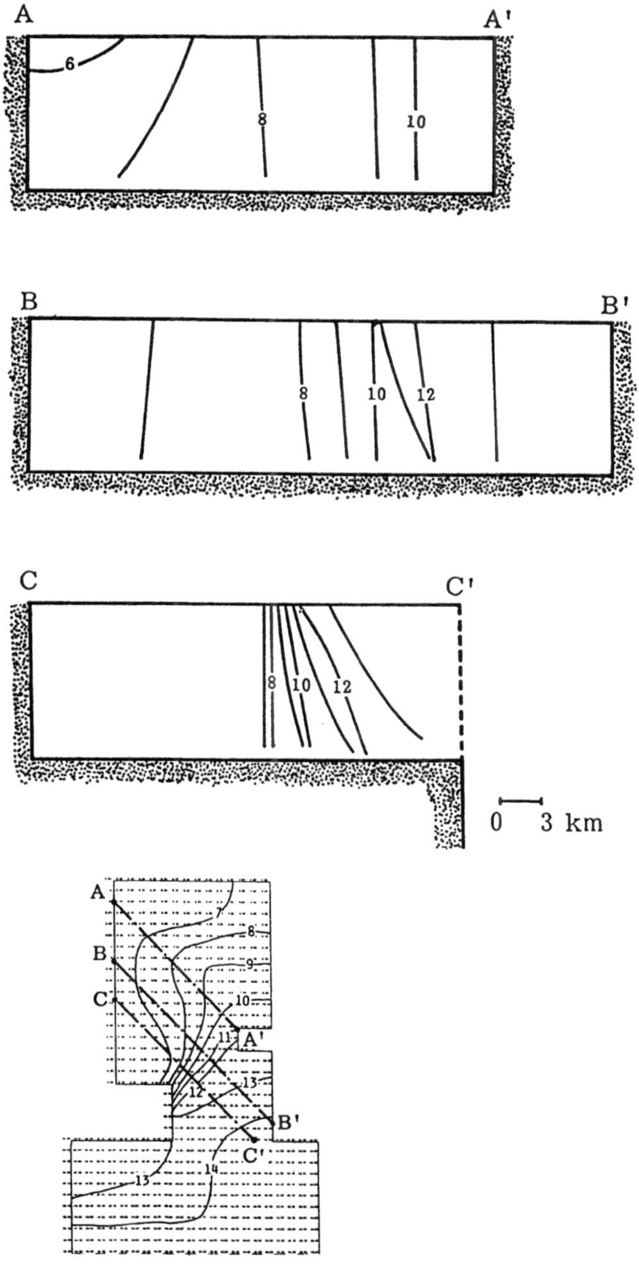

Figure 1.14: Distributions of horizontal velocity in the surface layer (a), that in the bottom layer (b) and downwelling velocity from the surface layer to the intermediate layer (c). The shadow area denotes the region where the downwelling speed is larger than 10^{-2} cm/sec.

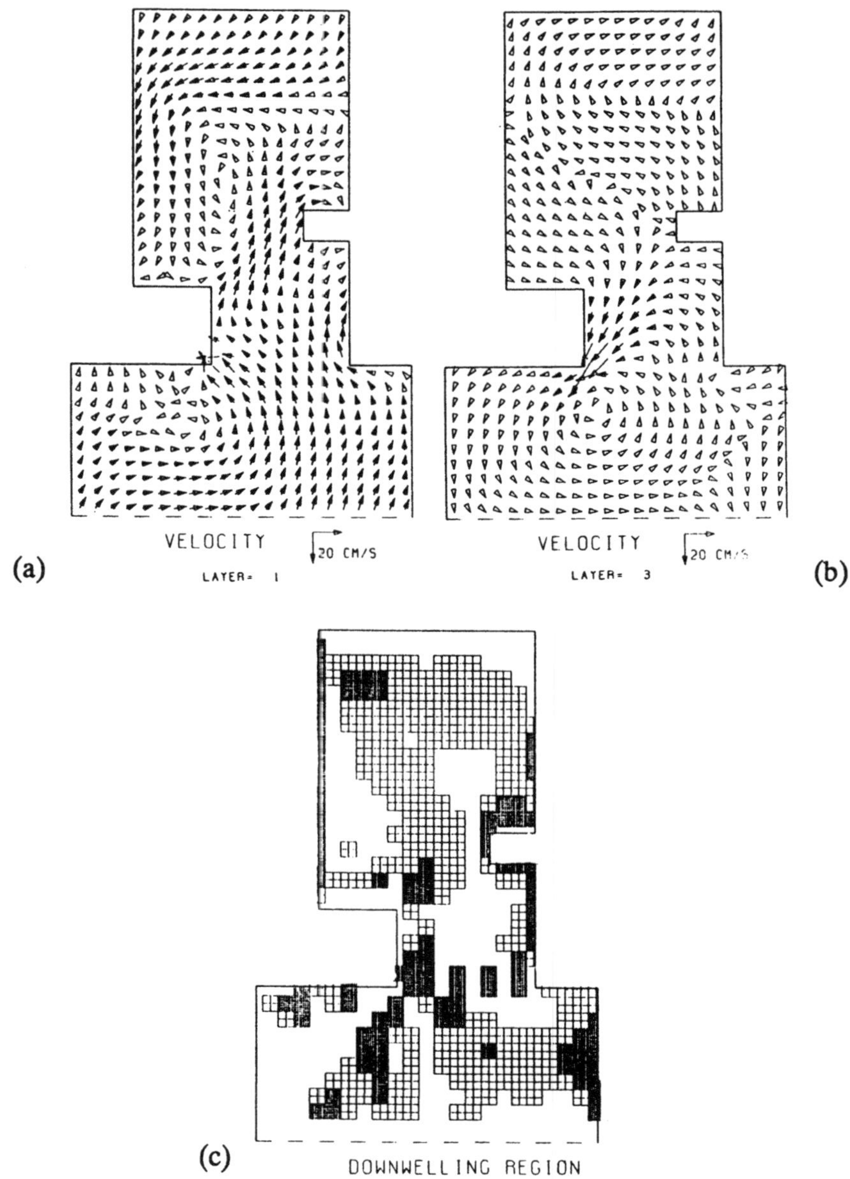

Figure 1.15: Dependence of sharpness (full line) and position (broken line) of the thermohaline front on the colling rate through the sea surface (a), river discharge (b), water temperature at the outer boundary (c) and salinity at the outer boundary (d) in Case (2).

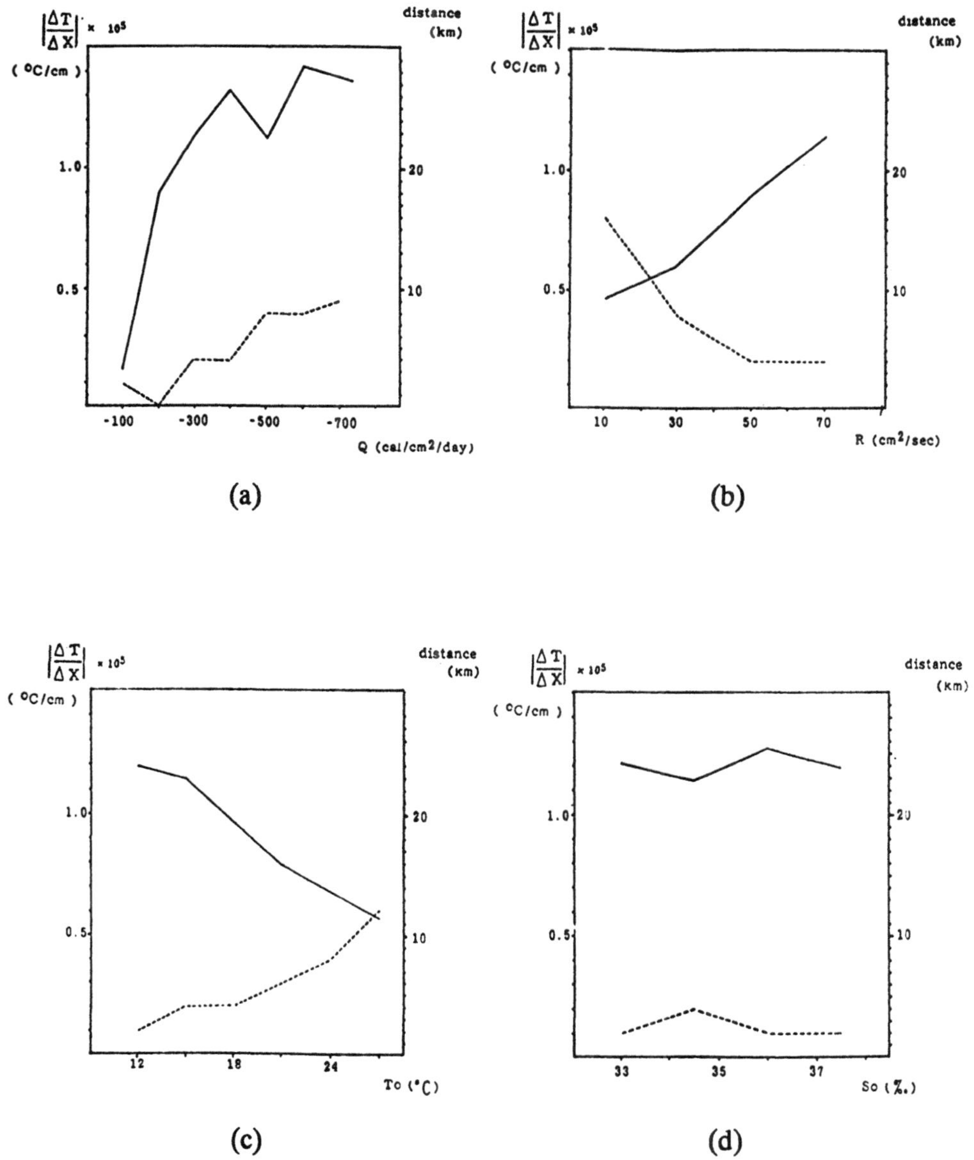

The sensitivities of the sharpness, position of thermohaline front by the change of external parameters are examined in the Case 3) experiment. The sharpness and the position of the thermohaline front are defined as the temperature gradient of frontal region at the left lower part of the basin and the position of thermohaline front as that of maximum temperature gradient at the same region. The results are shown in Figure 1.15. The sharpness and position of thermohaline front increase and retreat, respectively as the rate of surface cooling increases (Figure 1.15 a). On the other hand, the sharpness and position of the thermohaline front increase and advance, respectively as the fresh water discharge increases (Figure 1.15 b). These results coincide with those in the nonrotating case which are already pointed out by Oonishi et al. (1978). The sharpness and position of thermohaline front decrease and retreat, respectively as the water temperature at the outer boundary increases (Figure 1.15 c). These are due to that the supply of large heat from the outer boundary weakens the density-driven current from the inner side boundary. However, the sharpness and position of thermohaline front little change as the salinity at the outer boundary (Figure 1.15 d). The supply of much salt from the outer boundary leads to increase of average salinity in the basin but it little changes the structure of density-driven current from the inner boundary due to salinity efflux.

Acknowledgements

The authors express their sincere thanks to Dr. H. Takeoka of Ehime University and Dr. T. Hibiya, Dr. T. Saino and Dr. T. Ishimaru of University of Tokyo for their helpful discussions. Thanks are also due to Cap. Y. Igarashi, officers and crew of R/V Tansei-Maru for their assist in the cruise.

References

Endoh, M. (1977) "Formation of thermohaline front by cooling of the sea surface and inflow of the fresh water", J.Oceanogr.Soc.Japan 33, 6-15.

Harashima, A., Oonishi, Y. and Kunishi, H. (1978) "Formation of water masses and fronts due to density-induced current system", J.Oceanogr.Soc.Japan 34, 57-66.

Harashima, A. and Oonishi, Y. (1981) "The Coriolis effect against frontogenesis in steady buoyancy-driven circulation", J.Oceanogr.Soc.Japan 37, 49-59.

Lie, H.J. (1985) "Wintertime temperature-salinity characteristics in the South-Eastern Hwanghae (Yellow Sea)". J.Oceanogr.Soc.Japan 41, 291-298.

Mamaev, O.I. (1968) "A simplified relationship between density, temperature and salinity of sea water", Bull.Akad.Sci.USSR.Geophy.Ser. 2, 180-181.

Oonishi, Y. (1978) "The method of numerical study (in inland sea)", in Y. Horibe (ed.), Report of the Co-operative Research (Preservation of Marine Environment), Vol.2, Univ. of Tokyo Press, 107-144 (in Japanese).

Oonishi, Y., Harashima, A. and Kunishi, H. (1978) "Characteristics of a front formed by cooling of the sea surface and inflow of the fresh water", J.Oceanogr.Soc.Japan 31 17023.

Yanagi T. (1980) "A coastal front in the Sea of Iyo", J.Oceanogr.Soc.Japan 35, 253-260.

Yanagi, T. (1987) "Classification of "Siome", streaks and fronts", J.Oceanogr.Soc.Japan 43, 149-158.

Yanagi, T. and Koike, T. (1987) "Seasonal variation in thermohaline and tidal fronts, Seto Inland sea, Japan", Continental Shelf Research 7, 149-160.
Yanagi, T., Isobe, A. and Ishimaru, T. (1988) "Thermohaline front at the mouth of Tokyo Bay", Continental Shelf Research (in press).
Yoshioka, H. (1971) "Oceanic front at the Kii Channel in winter:, Umi to Sora 46, 1-31 (in Japanese).

Tetsuo Yanagi and Atsuhiko Isobe
Department of Ocean Engineering
Ehime University
Matsuyama 790
Japan

2 STABILITY AND FINE STRUCTURE OF WATER MASSES IN THE KUROSHIO FRONTAL ZONE

V.P. Belonozhko, L.P. Muratov and V.A. Sosnin

Abstract

Summary of long term data enabled to identify the fine developed structure stratification zone of inversion type with accurate borders in the region of the subarctic front in the NW Pacific. Statistical characteristics for inversions in different structural zones of this region were defined. All inversions were drew out due to the general orientation of the Kuroshio current. The main features of static and dynamic stabilities of sea water in inversion zone were determined.

Summer-fall, 1987, in the subarctic front (SAF) area of the NW Pacific has provided a large amount of data on fine structure of water masses in this area obtained during a complex marine expedition "Megapoligon". Soviet and foreign-made sounding equipment ("Istok-4", "Rozet") with vertical resolution of 0.5-2.0 m was used. The measurements were taken on hydrological sections 23-24 miles apart, distance between stations being about 15 miles.

The observations have covered the area of 34-41°N and 147-160°E. The results of the research in the subarctic front zone carried out by the Pacific Oceanological Institute, USSR Academy of Sciences, were also analyzed.

All the vertical distributions of hydrological properties under analysis contain the elements of fine structure of inversional type different in scale and intensivity. The data obtained enables us to construct meridional sections of temperature, salinity and probable density. The distribution of hydrological characteristics of the section is rather complicated. A lot of local areas with sharply pronounced parameters are observed, lamina of warm salty waters penetrating into subarctic water structure as well as patches of cold fresh waters spreading into SAF can be observed here. Below the sharply changing layer the temperature is monotonously decreasing down to 900-1000 m in the S stations of the sections. Separate lamina which temperature/salinity characteristics are different from those of the surrounding waters occurs here. Sharp rise of isolines up to 1000 m in the area of S SAF between 34°N and 35°N could be found. This is accompanied by the merging of the basic thermocline and seasonal sharp changing layer hydrological characteristics. Their vertical profiles in this part of the section contain heterogeneities found at 400-700 m depth. The latitudes of 35°N-39°30' N are characterized by interfrontal water zone.
Depth of 250-700 m can be distinguished due to numerous heterogeneities found here in vertical distributions of hydrological characteristics. Northward a rise of isotherms and isoclines could be found, the phenomenon being characterized by the frontal divide between 39°30' and 40°00'N. We found the N subarctic front here. In the temperature field at 40-350 m it is pronounced more sharply. Numerous inversional formations are typical for this layer. They are localized in the horizons of 100-250 m to the N from this frontal divide. Similar picture can be also observed in other hydrological sections made in different seasons of the year. Without citing them, let us indicate the main results of the analysis of spatial distribution of hydrological properties.

V. I. Ilyichev and V. V. Anikiev (eds.), Oceanic and Anthropogenic Controls of Life in the Pacific Ocean, 35–44.

Fine structure elements of inversional type are localized at 850-1000 m in subtripic structure water masses adjacent to the S subarctic front. A certain rise of layer boundary is observed, the latter having a developed fine structure stratification as this layer is moving eastward. More intensive inversional formations were found in an interfrontal zone where they were present at 200-700 m depth. To the N from the SAF they were present in 100-250 m layer. Data analysis measured in different marine expeditions and in different seasons of the year has proved that considerable changes in spatial distribution of heterogeneities with certain hydrological properties is lacking. Only the change of frontal divide position does shift the S and the N boundaries of layers' depth with developed inversional structure. Thus, SAF area marked by the fine structure stratification of inversional type zone. Its depth location is a subject of spacetime changes due to variability of frontal divides' position. The zone is basically characterized as following: in water masses of subtropical structures in the area directly adjacent to the S subarctic front it is present at 830-1000 m; in the waters of interfrontal zone the isolated area can be bounded at 250-700 m depth; in water masses of subarctic structure adjacent to the N SAF temperature inversions are localized at 80-250 m depth. Noteworthy that the depth decrease of the boundaries of the isolated zone occurs eastward. Figure 2.1 shows a summarized zone location with the developed fine structure of an inversional type according to the sections normally oriented towards SAF. Table 2.1 gives the basic characteristics of the isolated zone.

Figure 2.1: Combined scheme for developed fine structure zone position.

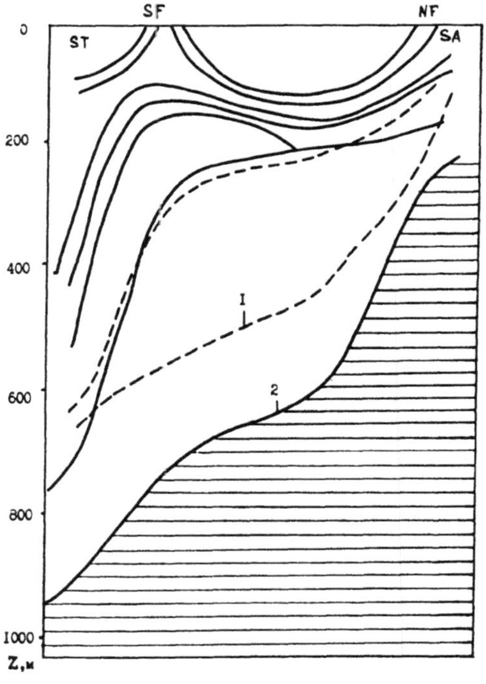

Table 2.1: Fine structure stratification characteristics for 3 structural zones of the subarctic front.

Zones	Observation time	Border depth of inversion zone		Mean characteristics of inverversions			Maximum characteristics of inversions		
		upper border depth	low border depth	H_t (m)	t (°C)	H_s (m)	H_t (m)	t (°C)	H_s (m)
Subtropical	August, 1975	850	940	8	0,2	8	26	0,9	26
	September,1975	830	935	9,3	0,26	10	-	-	-
	November, 1975	810	970	12	0,29	-	-	-	-
	December, 1975	800	1000	14,6	0,34	-	34	1,2	-
	Mean value	822	961	13	0,28	9	30	1,2	26
Interfrontal	August, 1975	250	680	24	0,97	20	116	2,4	116
	September,1975	220	700	27	1,1	25	127	2,5	127
	November, 1976	150	730	46	1,8	-	286	4,3	-
	December, 1976	120	750	49	1,9	-	412	4,7	-
	Mean value	185	715	40	1,7	22	412	4,7	-
Subarctic	August, 1975	80	210	21	0,8	19	39	1,4	39
	September,1975	90	220	25	1,0	21	45	1,8	45
	November, 1976	-	-	-	-	-	-	-	-
	December, 1976	-	-	-	-	-	-	-	-
	Mean value	85	215	23	0,9	13	45	1,8	45

Due to considerable variations of the thermocline structure of the three types mentioned isolated in "Megapolygon" area, it is expedient to analyse fine structure properties separately for each isolated type of water masses. Such a division is based on a well-known positions of the N and the S frontal divides since they serve as natural physical boundaries which divide the water masses of three structures. To the S from the S SAF inversional formations are found at 850-1000 m depth. Mean vertical distribution of hydrological properties here reveals 2,95 inversional formations. Average thickness of a single inversion was equal to 42,5 m. Its maximal value was observed in December, 1976 -62. Figure 2.2 (a) shows a histogram calculated for the entire data array on the distribution of separate inversional formation thicknesses. The distinctly pronounced distribution centre is lacking. Thickness inversion values here are most probably 10-15 m. Figure 2.2 (b) shows distribution histogram of mean temperature profile perturbation values, making it possible to select its most probable value of 0,2-0,4 °C. Salinity field shows the mean thickness of a separate fine structure element to be 8-10 m, while a perturbation value of salinity mean profile is 0,3 °/₀₀. Histogram of vertical size distribution of heterogeneities in the salinity field differs in form from that of for the temperature field (Figure 2.2 (c): maximal values are shifted into the area of smaller scales - 5-10 m. In general the histogram illustrates the entire shift of P (ΔH_s) into the area of smaller scales: vertical scales not higher than 15 m cover 85 % of all the fine structure elements of salinity field under review, and only 64 % belongs to a temperature field.

To evaluate horizontal sizes of fine structure elements of various vertical scales we have calculated mutually coherent functions between separate pairs' realizations of fine structure components of vertical temperature profiles obtained by microphotography. The latter was carried out in the subarctic structures of water masses in two mutually perpendicular sections, latitudinally and longitudinally oriented, the distance between the station sections being 5 miles. We have made calculations for pairs of realizations latitudinally and longitudinally oriented. Mutually coherent function values have been averaged for heterogeneities of vertical scales 5, 10, 15, 20 and 30 m on the basis of 6 values and have constituted, accordingly, 0,17, 0,22, 0,38, 0,68. The value of 90 % nonzero coherency level is 0,18, 0,20, 0,24, 0,26, 0,3. The averaging we have undertaken increases confidence in our values, nevertheless we neglect this point. That is, heterogeneities with vertical sizes more than 15 m can be considered to stretch not less than for 5 miles. Vertical heterogeneity scale is related to a horizontal one as 2×10^{-3}, this value being in good agreement with the known data.

Similar calculations have been carried out for latitudinally oriented pairs of realizations. Mutually coherent functions being averaged for 6 values were as following: $F_5 = 0{,}2$; $F_{10} = 0{,}31$; $F_{15} = 0{,}46$; $F_{20} = 0{,}67$; $F_{30} = 0{,}72$. The distance between realization pairs was also 5 miles. This enables us to suggest that heterogeneity horizontal sizes of 10 m vertical scales can reach 5 miles in a wide direction. In general higher coherency values form realization pairs which are latitudinally oriented indicate the anisotropy of fine structure elements: they are latitudinally stretched.

Interfrontal zone water masses show the inversional formations at 185-715 m depth. Figure 2.3 (a,b,c) illustrate the histograms of thickness value distribution of separate fine structure elements in temperature and salinity fields as well as perturbation values of their mean distribution. Thickness value of a separate inversional temperature formation for interfrontal zone water masses is 40 m, perturbation value on the t(z) profile is about 1,5 °C. Maximal perturbation value of the average temperature distribution was registered in December, 1976, and was equal to 4,7 °C. In salinity field the thickness farthest from the fine structure element might prove to be 20-30 m. In this case a shift of the entire histogram into the smaller scale field is observed.

Figure 2.2 (a,b,c): Fine structure characteristics histograms for the subtropical zone.

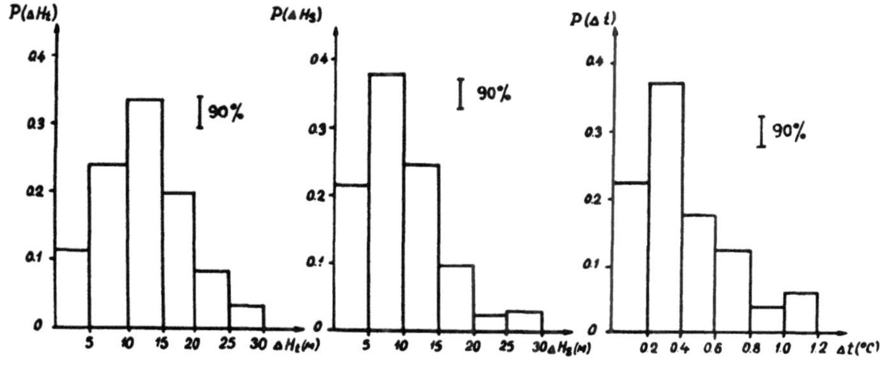

Figure 2.3 (a,b,c): Fine structure characteristics histograms for the interfrontal zone.

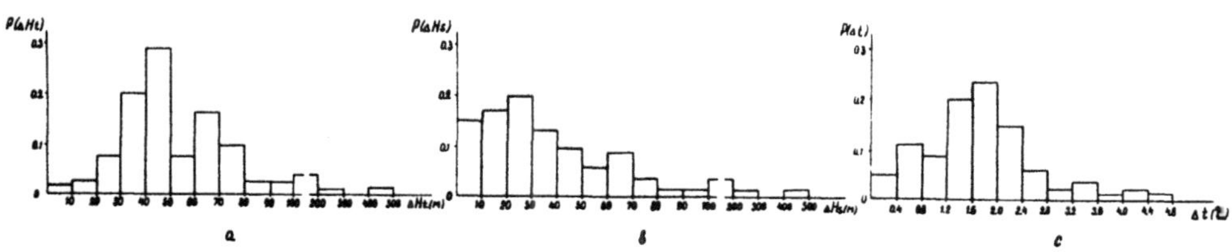

Figure 2.4 (a,b,c): Fine structure characteristics histograms for the subarctical zone.

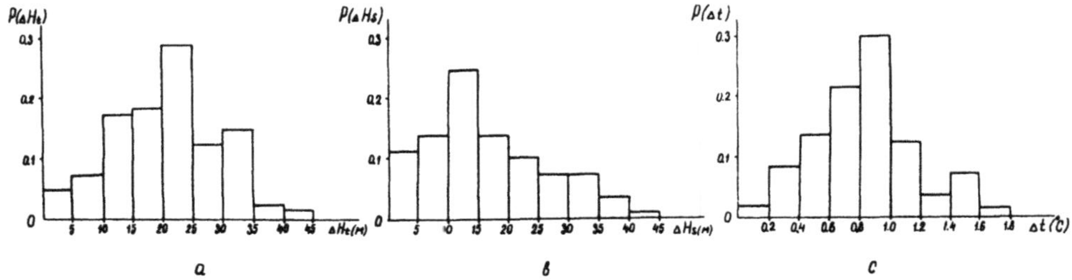

Thickness of 55 % elements under analysis does not exceed 30 m; temperature inversions of these scales can be found for only 17 % data analyzed in this case.

According to microphotography carried out here in two mutually perpendicular sections which are latitudinally and longitudinally oriented we have evaluated horizontal stretching of heterogeneities with various vertical scales. Values of mutually coherent functions have been used in this case calculated for realization pairs, the latter being oriented in latitudinal and longitudinal directions. Sections' length was 30 miles, the distance between sections being 5 miles. Vertical heterogeneities' scales 5, 10, 15, 20 and 30 correspond to values of mutually coherent function calculated for the data which are meridionally oriented and averaged on the following 6 values , accordingly: 0,15; 0,18; 0,24; 0,32; 0,41. Nonzero coherency level with 90 % coverage as recorded, accordingly, 0,18; 0,2; 0,24; 0,26; 0,3. That is, heterogeneity horizontal stretching 20-30 m thickness constitutes meridionally about 5 miles. Vertical heterogeneity size is related to a horizontal one as 2×10^{-3}. Such evaluations carried out of realization pairs, the latter being latitudinally oriented and averaged for the fine structure elements with vertical sizes 5, 10, 15, 20 and 30 prove to be, accordingly, 0,18; 0,23; 0,28; 0,36; 0,51. The values obtained with nonzero coherency level remaining the same, admit the possibility of separate fine structure elements being present which thickness is more than 10 m horizontally and more than 5 miles latitudinally. In this case vertical heterogeneity sizes are related to horizontal ones as 5×10^{-3}.

In water masses of subarctic structure adjacent to the N SAF fine structure of inversional type was observed at 85-215 m depth. An average number of inversions observed here in separate vertical profiles is about 2. Figure 2.4 (a,b,c) show the distribution histograms of typical inversional formations. Thickness value of separate

fine structure element in a temperature field most probably is about 20-25 m, its mean profile perturbations being about 1 °C. Fine structure elements of the salinity field prove to reach only 10-15 m for the heterogeneity thickness value, while the entire histogram is shifted into the smaller scale field: 50 % of heterogeneities in the salinity field have the sizes no more than 15 m; the temperature field is characterized by vertical scales only for 25 % of the analyzed data.
Evaluation of horizontal stretching of heterogeneities with various vertical scales carried out by means of a similar method brought about the following results. Mutually coherent functions average for the six realization pairs oriented in meridional direction for 5, 10,15,20 and 20 heterogeneity thickness proved to be, accordingly 0,21; 0,2; 0,28; 0,31; 0,42. Such evaluations for realization pairs latitudinally oriented are characterized by the values 0,18; 0,26; 0,34; 0,46; 0,61. Nonzero coherency levels with 90 % coverage for the isolated scales proved to be 0,18; 0,2; 0,24; 0,26; 0,3, accordingly. These data analysis enables us to consider the heterogeneities observed in the temperature field to be anisotropic. In this case if heterogeneities of vertical sizes are related to horizontal ones for the realization pairs meridionally oriented this relation is about 2×10^{-3} whereas for longitudinally oriented pairs it approaches 1×10^{-3}.

Hesselberg stability parameters' values have been calculated on the basis of the measurements obtained. $\int E_{\circ}(z)dz$ was calculated on a separate station to obtain integral stability parameters. Its typical distribution along a meridional section is given on Figure 2.5 (curve 1). Variability of meridional stability parameters of integral values is closely related to the location of the frontal divides: southward and northward one can notice a considerable decrease of its values, maximal values being present in a frontal divide field whereas interfrontal zone is also marked by higher values. Figure 2.6 illustrates histograms of density instability distribution of water masses layers whose depth is characterized by three water types. Maximum (except a sharply changing layer) can be distinguished in deep locations of inversional formations in water masses adjacent to S SAF: 850-1000 m. In an interfrontal zone maximum can be also found both in a sharply changing layer and in deep waters where we have isolated inversional formations earlier: 200-700 m. A similar picture can be observed for data obtained in subarctic waters adjacent to N SAF. In this case the histograms reveal maxima at 500-200 m depth. For a number of meridional sections we have made calculations of local Richardson number values. Presence of interlayers of waters whose Richardson number values are less than critical is typical of them. They can be observed at the depths of density inversion locations. Maximal summarized depth of hydrodynamical instable layers of water masses is observed in the area of frontal divides. Figure 2.5 (curve 2) shows its higher values in the interfrontal area as well. The analysis made thus enables us to make the following conclusion:

1: Subarctic front area in the N-W of the Pacific is marked by the presence of a large zone with a well-developed fine structure stratification of an inversional type whose spatial boundaries are distinctly marked. Their location is defined by geographical coordinates of the N and S frontal divides undergoing temporary variability due to seasonal reshaping of thermocline water structure, front meandering, atmospheric influences of synoptical scale etc. Nevertheless, its basic properties are: deepening in the subtropical waters, rise of boundaries and thickness growth in interfrontal zone, thickness decrease and boundaries ries to smaller depths in subarctic water masses are all present in the zone.

Figure 2.5: The characteristics of hydrostatic (1) and dynamic (2) stabilities distribution through the subarctic front.

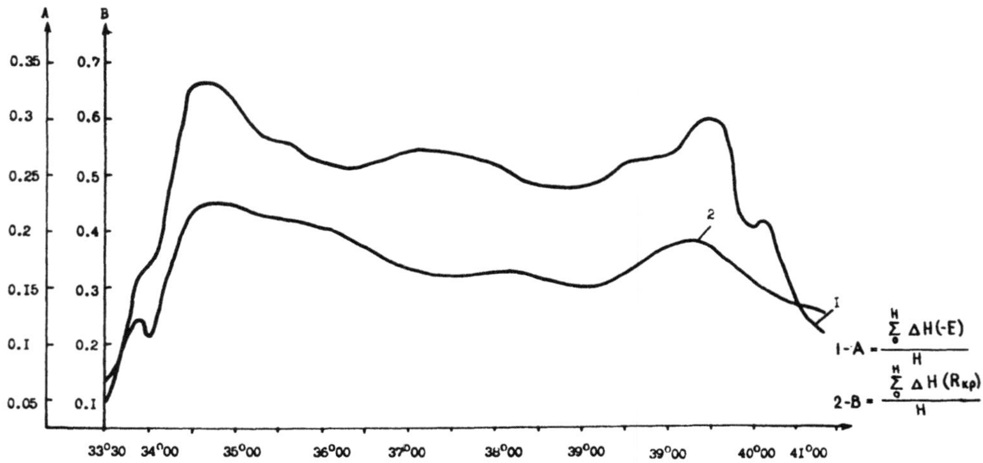

2. Heterogeneity advection on the surfaces close to isopectic ones should be considered to be the main mechanism of inversional structure formation of hydrophysical fields in a subarctic front zone. In this case advection occurs northward and southward laterally to the front. The fine structure observed is a kind of a trace of northward and southward water penetration, the latter forming a layer of lower salinity on intermediate water horizons and forming a warm intermediate layer in a subarctical waters. There are also conditions for differential-diffusional convention development, and a possibility of fine structure generation by internal waves and eddy shedding remains as well.

3. Lateral and longitudinal sizes of fine structure elements possess values typical of three water types mentioned for the N-W Pacific. The same can be referred to a disturbance of an average distribution of the property under review.

Figure 2.6 (a,b,c): Histograms of density instability layers distribution in depth: a - subtropical zone, b - interfrontal zone, c - subarctical zone.

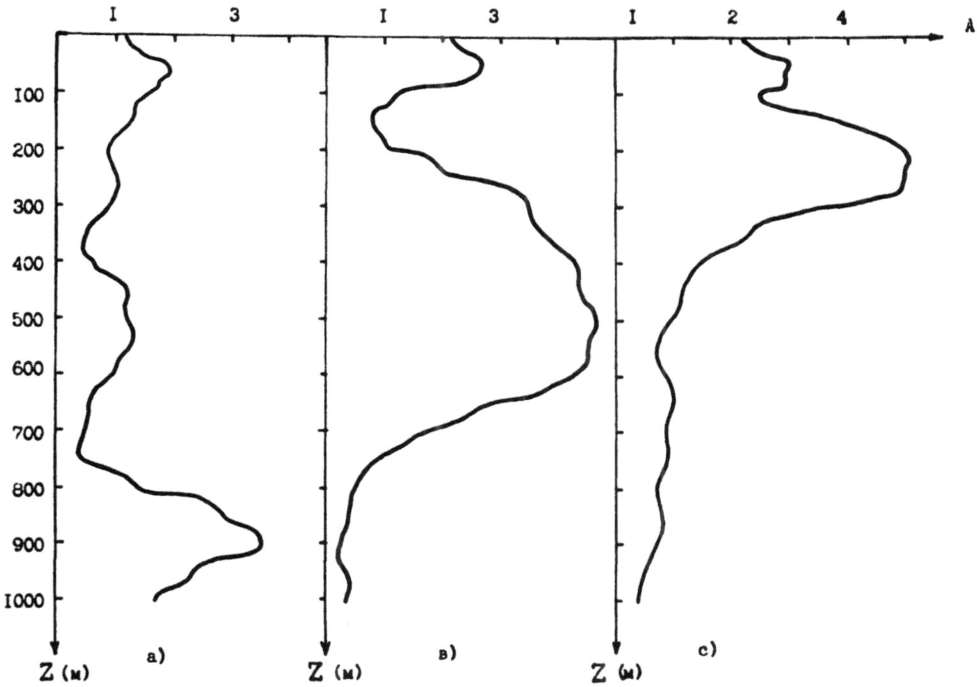

4. Spatial variability analysis of the values which are typical of intensivity of structure forming processes proves most intensified and large-scale inversional formation to develop in an interfrontal zone.

5. Evaluation of horizontal fine structure element sizes proves them to be anisotropic, their lateral outstretching could be noticed, the latter fact might be attributed to general orientation of the Kuroshio flow in the area under study.

6. The regularities of distribution of hydrophysical field characteristics mentioned above enable to predict in the first approximation zone location with a well-developed fine structure proceeding from known positions of the S and N subarctic fronts. The same regularity also enables us to predict inversions, fine structure elements of average thickness, their maximal values as well as disturbance mean summarized values, averaged distribution of hydrological characteristics and their extreme values.

7. The water masses of the isolated zone with a developed fine structure stratification are characterized by a decreased hydrostatic and dynamic stability, the latter fact proving the case of an increased intensivity of small-scale eddy shedding generation and lateral intermixing processes.

3 MINERAL AEROSOL OVER AND DEPOSITION TO THE PACIFIC OCEAN

Mitsuo Uematsu

3.1 Introduction

Mineral particles, which are mainly produced by the weathering of the continental crust, are important for geochemical cycles in the atmosphere and in the oceans. These particles have relatively short residence times in both atmosphere and hydrosphere, and therefore their concentrations are sensitive to changes in weather, climate, perturbations by human's activities, etc. In addition, many trace metals are associated with aluminosilicates (mineral or dust particles), and therefore understanding the behaviour of mineral particles is important because the biogeochemical cycling of at least a dozen of the chemical elements are linked to the transport and fluxes of dust (Ran, 1976; Chester, 1986).

Mineral aerosol can be a source of cloud-condensation nuclei over open oceans (Isono et al., 1971). Dust particles also affect the reflectance of clouds and atmospheric turbidity (Carlson and Benjamin, 1980). Thus the concentrations of mineral aerosol may affect climate over large areas of the globe.

The impact of mineral particles on chemical and biological processes in the water column after air-to-sea transfer occurs as a subject that recently has received increasing attention in the chemical oceanography community (Moore et al., 1984; Duce, 1986; Martin and Gordon, 1988). Mineral aerosol is also major source for the non-biogenic component of deep-sea sediments in the Pacific Ocean (Rex and Goldberg, 1958; Griffin et al., 1968; Windom, 1969, 1975; Prospero 1981).

Very limited data for mineral aerosol concentrations over the ocean is available, but the concentrations and deposition rates of mineral aerosol clearly vary widely over space and time. In the past, most of data from the Pacific was obtained from ships, and thus it was quite difficult to assess the temporal and areal variations of dust concentrations.

In order to characterize the distributions of mineral aerosol in the marine atmosphere and its deposition to the Pacific, it is necessary to make measurements over an extended time and space scale. To this end, as part of the Sea/Air Exchange (SEAREX) program, the SEAREX Asian Dust Study (SADS) was started in early 1981. The SADS or North Pacific Network consisted of seven island stations in the Central N Pacific (Uematsu et al., 1983). In cooperation with the North Pacific Network, six stations on Japanese islands in the Western N Pacific closer to the Asian continent were operated by a Japanese group (Tsunogai et al., 1985). The SEAREX South Pacific Aerosol Network (SEASPAN) was established in early 1983 (Uematsu, 1987). The equipment and procedures used at the three networks were identical. All network stations where mineral aerosol was continuously sampled are shown in Figure 3.1.

V. I. Ilyichev and V. V. Anikiev (eds.), Oceanic and Anthropogenic Controls of Life in the Pacific Ocean, 45–69.

Figure 3.1: Mineral aerosol sampling stations in the Pacific ocean.

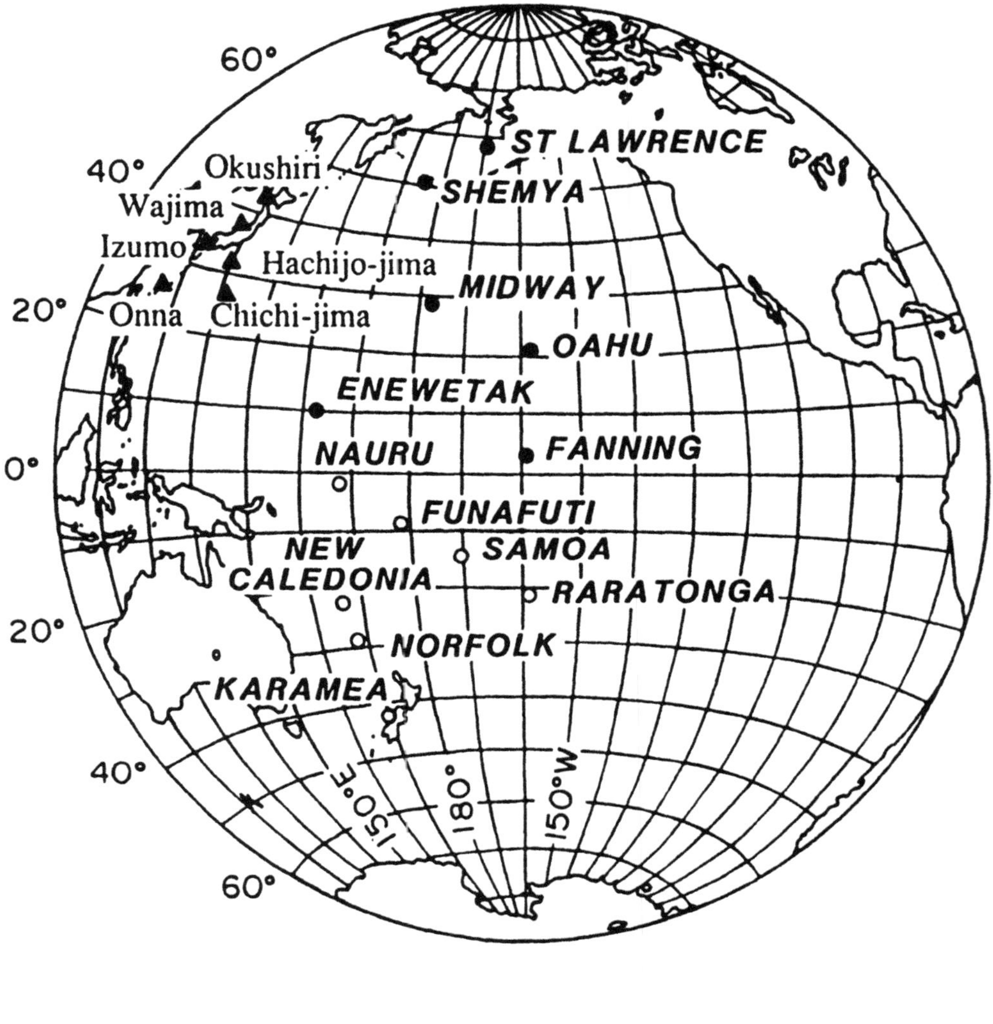

3.2 Characterization of marine aerosol

The composition of marine aerosol, expressed in terms of seven components is shown in Table 3.1 (Okita et al., 1986). The aerosol samples were collected at Hachijo-jima island and at Chichi-jima island which are ≈ 200 km and ≈ 1000 km away from the main island of Japan, respectively. Concentrations of each component varied independently because of the different sources, particle sizes, chemical properties, and transformation during the transport. The sum of the concentrations of seasalt particles, mineral particles, and as the particulate sulphate, nitrate, organics and elemental carbon agreed well with the total mass of the aerosol samples determined gravimetrically. A few percent to twenty percent of the total mass of marine aerosol was attributed to mineral aerosol (0.7 - 4.7 $\mu g\ m^{-3}$) at these stations during the winter. In general, atmospheric sea salt was the predominant component of marine aerosol, and its concentration ranged from 10 to over 40 $\mu g\ m^{-3}$ m, depending largely on the magnitude and the variance of the wind speed (Erickson et al., 1986).

The concentration of mineral aerosol in the remote marine boundary layer is governed by source strengths on the continents, long-range transport pathways, and wet and dry removal processes. There are complex links among these processes, and as a result it is difficult to determine the relative importance of the various processes that influence the atmospheric dust concentrations over the ocean.

3.3 Sources and transport of mineral particles

The main source areas of mineral aerosol are the arid and seasonally-arid regions on the continents where dry, freshly ploughed soils are exposed to strong winds. The possible source areas of mineral particles for the Pacific basin are arid and semi-arid regions in Australia, Central Asia, NW America, and SW America. The predominant surface air flow patterns over these source regions are westerly, however, and as a result the Asian and Australian continents are likely to be the major sources of mineral aerosol over the Pacific Ocean.

3.3.1 North Pacific

In Asia, most of the arid regions are located between 30 °N and 50 °N. The largest desert in Central Asia is the Gobi desert ($\approx$1,300,000 km^2). Approximately 40 % of the Gobi desert is in the Mongolian People's Republic and the rest is in the People's Republic of China. The Takla Makan desert ($\approx$330,000 km^2) is located in W China. In addition, there are several deserts ($\approx$140,000 km^2) in the Alashan and Ordos Plains located in the upper reaches of the Huang Ho (Yellow River).
Dust storm activity in Asia is greatest in the spring (March-May) due to the combined effects of low humidity, high surface wind velocity, and soils freshly ploughed for planting (Koizumi, 1933; Fuch, 1939; Ing, 1972). The mechanism for generating dust storms in the arid region of the Asian continent has been discussed in detail by Merrill et al. (1985). There are over 250 surface meteorological stations on the Asian continent, and by taking the present weather codes from their reports, its is possible to characterize the temporal and geographical distribution of mineral aerosol over Asia.

The dust storm frequency statistics (summed to produce weekly values) for 1981-1986 are presented in Figure 3.2. The frequency of dust storm reports is high in spring and low in summer. There is small peak in fall, but during the winter dust storm activity is low due to frozen soils and to snow cover.

Figure 3.2: Frequency of dust storm reports in Asia (Uematsu et al., 1983). This indicates qualitatively the annual variation of the strength of the mineral aerosol.

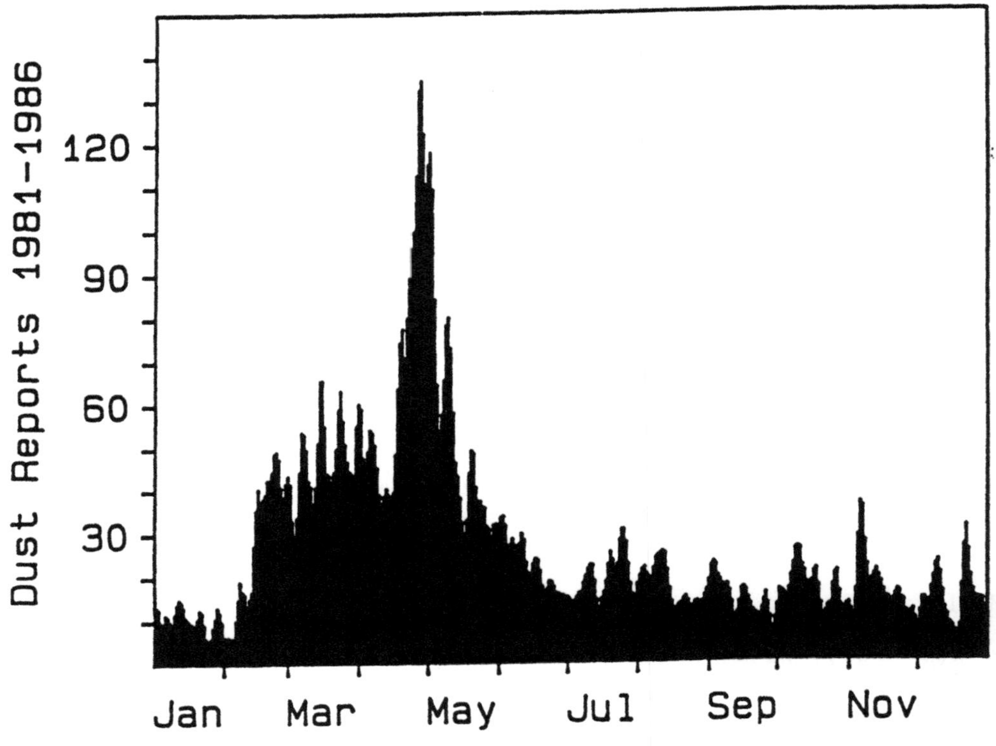

Major dust events (called 'Kosa' in Japan) are observed during spring in Japan. Kosa is yellow or brown airborne dust which is known to be transported to Japan from China. During Kosa outbreaks, colourful phenomena are often observed: e.g., red snow, black rain, yellow fog. During the past 62 years, 85 % of the Kosa events occurred between March and June, based upon visual observations at Nagasaki in S Japan (Arao et al., 1979). The seasonal cycle of Kosa events in Japan correlates with the seasonal variation of dust storm activity over the Asian continent. The large-scale Kosa events observed in Japan always occur shortly after severe dust storm outbreaks in China, which is suggestive of a cause and effect relationship.

Asian dust storm clouds passing over Japan are studied by various techniques, including chemical analyses of the aerosol for various elements (e.g., Mizohata et al., 1976; Kadowaki, 1979; Tanaka et al., 1983; Tsunogai et al., 1985; Suzuki and Tsunogai, 1988), mineralogical analysis (Ishizaka and Ono, 1982), electron microscopic analysis (Okada et al., 1987), atmospheric turbidity measurements (Arao and Ishizaka, 1986), satellite observations (Murayama, 1980; Chung, 1986; Takayama and Takashima, 1986) and laser radar observation (Iwasaka et al., 1983; Kobayashi et al., 1985). According to Iwasaka et al. (1983), laser radar measurements and satellite data indicated that during an event in April of 1979 dust was transported out of Asia in two distinct layers, one at a height of 6 km and the other at 2 km. Air trajectory analysis of the event suggested the dust cloud in the upper layer was transported from the Takla Makan desert and that in the lower layer originated in the Huang Ho basin and the Gobi desert. The dust plume then moved further E, i.e. toward the Central N Pacific. It has been suggested by Merrill et al. (1985) that part of this dust plume reached the site at Enewetak Atoll in the Marshal Islands, approximately 5000 km SE of Asia. During that time, SEAREX experiment were being conducted at Enewetak (Duce et al. 1980).

Table 3.1: Characterization of mineral aerosol on Hachijo-Jima and Chich-jima islands from Okita et al., (1986) ($\mu g\ m^{-3}$).

	Hachijo-Jima		Chichi-jima	
	Jan. 14-20	Dec. 14-20	Jan 14-20	Dec. 14-20
Sea salt	15.8	17.3	9.2	18.5
Soil	4.7	1.0	1.7	0.7
$(NH_4)_2SO_4$	1.5	1.1	1.2	0.8
Other excess SO_4^{2-} (mainly H_2SO_4)	1.3	0.6	1.0	0.7
Organics	2.2	3.1		1.4
Elemental Carbon	1.3	0.8		0.7
NO_3^-	0.5	1.3	0.2	0.4
Total	27.3	25.2	13.3	23.2
Total by weighting	27.9	27.2		26.9

The mineral particles from the Asian continent also have been observed as ice nuclei (Isono et al., 1971) or a series of haze bands over Alaska (Rahn et al., 1977; Rahn et al.,1981). High concentrations of mineral aerosol also have been measured on ships in the western N Pacific by Tsunogai and Kondo (1982) and Hirose et al. (1983). These authors attributed the high concentrations of mineral aerosol to dust storm activity on the Asian continent. Mineral aerosol concentrations measured on the ships are presented in Table 3.2, but is is difficult to assess the seasonal variability of the mineral aerosol concentrations by such short term samplings.

Table 3.2: Mineral aerosol concentrations ($\mu g\ m^{-3}$) measured from ships in the Pacific Ocean.

Location	Concentration range	Collection period	Reference
Eastern Equatorial Pacific, 8°N-17°S	0.04-1.2	Feb.-Apr.	Prospero and Bonatti (1969)
North Pacific, 32-45°N	0.5-1.0	Apr.	Ferguson et al. (1970)
Eastern Mid-Pacific	0.10-0.57	Jan.	Prospero (1979)
North Pacific, 26-28°N	0.73-2.34	May	Prospero (1979)
Pacific	0.45	Feb.-Apr.	Miklishanskiy et al. (1981)
West and central North Pacific, 28-40°N	0.21-8.4	May-Jun.	Tsunogai and Kondo (1982)
Western North Pacific, 30-35°N	1.4-64	Mar.-May	Hirose et al. (1983)
Eastern Equatorial	0.066	Jun.	Maenhaut et al. (1983)
Tropical North Pacific	0.1-0.6	Jun.-Aug.	Raemdonck et al. (1983)
Equatorial Pacific	~0.1	Jun.-Aug.	Raemdonck et al. (1983)

3.3.2 South Pacific

The arid and semi-arid regions account for over 80 % (≈6,400,000 km^2) of the total area in the Australian continent. There are three major deserts in Australia: the Simpson Desert in Central Australia, the Great Victoria Desert in the W and S part of Australia; and the Gibson Desert in the interior and to the W. According to Griffin et al. (1968), kaolinite and illite are the most abundant minerals in the arid interior of Australia. W Australia soil tends to be very rich in kaolinite, whereas the E soil is illite- and kaolinite-rich. However, chlorite is absent from a major fraction of the area of the Australian continent, except for the chlorite-rich zone of New South Wales in SE Australia. Thus it may be possible to identify distinctive source regions for atmospheric dust over the S Pacific based on the mineral composition of aerosol particles.

There was evidence of severe dust storms over Australia in October 1928. During that event, mineral particles were transported to New Zealand within 24 hours via the jet stream in the middle and upper troposphere (Healy, 1970). Windom (1969) also reported that "red snows" in the spring occurred over N and S Islands of New Zealand in 1966 due to the droughts in SE Australia. A combination of air trajectory analysis and mineralogical analysis indicated that dust, collected at Hokitika, the S Island of New Zealand, was of Australian origin (Collyer et al., 1984). Thus, there is considerable evidence that mineral particles are transported out of Australia in large quantities and that much of the material is transported to the E, i.e., to the Tasman Sea and beyond.

3.4 Concentration of mineral aerosol

Recently, the size and chemical composition of aerosol particles from the N China have been studied using a time sequence sampler in 4-hour time steps (Winchester et al. 1981a, 1981b, 1981c; Winchester and Mu-tian 1984; Wang et al., 1982). According to Yang et al. (1981) and Zhou et al. (1981), the highest concentration of aerosol was 6130 $\mu g\ m^{-3}$ during a dust storm over Beijing, which occurred on April 17-20, 1980. The mineral aerosol sampled during the dust storm evidently was transported for about 1500 km, and its origin was in the area of Hami, Xinjiang Provence and the W part of the inner Mongolia (Yang et al., 1981). For comparison, the highest concentration of mineral aerosol measured in the Western N Pacific network was 210 $\mu g\ m^{-3}$ in the mid-May 1981. That dust concentration was measured at Izumo which is located on the coast of the Japan Sea (Tsunogai et al. 1985).

Aerosol samples were collected at the Pacific Network sites in one week intervals for a period of five years using high volume air sampling systems an Whatman 41 filters. The sampling system at each site was located on the climatological windward coast and as close to the ocean as possible. To minimize the possibility of contamination from local sources, the samples was controlled by a wind monitoring system. The sampler was activated only when the wind blew from the ocean. The samples were analyzed for Al by neutron activation analysis (Uematsu et al., 1983). The Al concentrations can be converted to equivalent mineral aerosol concentrations on the basis of the average Al concentration in soils. Mineral aerosol concentrations at the stations in the Pacific Networks are summarized in Table 3.3.

Table 3.3: Concentrations of mineral aerosol ($\mu g\ m^{-3}$) over the Pacific.

Station	Mean	Maximum
	The Western N Pacific network	
Okushiri	32.6	200
Wajima	17.2	80
Izumo	23.5	210
Hachijo-jima	9.7	60
Onna	11.8	150
Chichi-jima	4.4	20
	The Central N Pacific network	
Shemya	0.89	5.0
Midway	0.83	7.8
Oahu	0.65	4.6
Enewetak	0.26	1.9
Fanning	0.05	0.19
	The S Pacific network	
Nauru	0.13	2.0
Funafuti	0.27	7.5
American Samoa	0.017	0.09
Rarotonga	0.11	0.66
New Caledonia	0.11	0.87
Norfolk	0.53	10

[a]Tsunogai et al. (1985) and Suzuki (1987)
[b]Uematsu et al. (1983)

The ratio of the aerosol Al concentration to the aerosol filter ash weight appears relatively constant worldwide, averaging approximately 0.080. This value was calculated from samples form Oahy and Midway (Uematsu et al., 1983). The same ratio is obtained in the aerosol from Miami during Saharan dust events (Prospero et al., 1987) typical of Japanese soil (Okita et al., 1986) and average soil material (Taylor, 1964).

3.4.1 Seasonal Patterns

Figure 3.3 presents the concentrations of mineral aerosol at sites from each zone of the Pacific during the period from 1981 to 1985. As a mid-latitude site at 30 °N, Midway is strongly influenced by the prevailing westerly winds, and it is the station in the North Pacific Network most likely to show the effects of Asian dust storm outbreaks. Seasonal variations of mineral aerosol concentration are found to be generally similar from year to year (Figure 3.3a). Peaks of high mineral aerosol concentrations were observed from February until June. Then the concentrations dropped rapidly and remained low during the summer, but there was evidence of secondary dust transport period in the fall. Atmospheric dust concentrations returned low values through the winter and increased in late February or March. This trend is similar to the pattern of dust storm activity on the Asian continent. It is apparent that the transition from high to low concentration of mineral aerosol was synchronous across most of the N Pacific (Figure 3.4).

According to Bodhain et al. (1981), the measurement of aerosol light scattering during the period from 1974 to 1978 at Mauna Loa showed a large peak in the spring an a small peak in the fall, and this is similar to the pattern of dust concentrations at Midway. The seasonal trend is also apparent from the four year (1979-1982) record of mineral aerosol measurements made at Mauna Loa Observatory (MLO), Hawaii, by Parrington et al. (1983) and Parrington and Zoller (1984). The results from MLO, which is at an altitude of 3400 m, are particularly interesting because the transport of mineral aerosol from Asia to MLO (Shaw, 1980; Darzi and Winchester, 1982; Braaten and Cahill, 1986) occurs above the trade-wind inversion in the free troposphere.

The Fanning station at 4 °N, i.e., in the equatorial zone, was the only station in the N Pacific not exhibiting the seasonal pattern during 1981. Since 1982, peaks of higher than average mineral aerosol concentrations have been observed every spring at Fanning, and the maximum values have varied from year to year. In contrast, the mean dust concentrations during the low dust season (July-January) have remained constant through four years of sampling (Figure 3.3b).

Fanning lies S of the intertropical convergence zone (ITCZ) much of the year, and thus the low concentration of mineral aerosol may be representative of clean Southern Hemisphere air. When the ITCZ moves to S of the island in the spring, air contain relatively a high concentration of mineral aerosol reaches the island. The peak dust concentrations observed in April, 1982 also may have been influenced by the eruption of El Chichon in late March and early April.

Figure 3.3: Temporal variation of the concentrations of mineral aerosol at (a_ Midway (28°N), (b) Fanning (4°N), and (c) American Samoa (14°S) during the period from 1981 to 1985 (Uematsu, 1987).

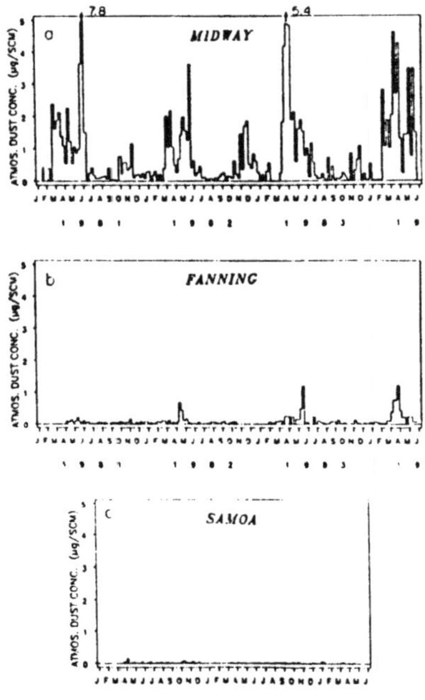

The mineral aerosol concentration at American Samoa is the lowest of all stations in either the North or South Pacific Networks (Figure 3.3c). Furthermore, there are no pronounced seasonal patterns in dust concentrations at Samoa. These results are consistent with the mineral aerosol concentration observed (0.002-0.007 $\mu g\ m^{-3}$) during the SEAREX 1981 Samoa Experiment (Arimoto et al., 1987). The concentrations at American Samoa are among the lowest ever measured near the earth's surface, even including data from the S Pole (Maenhaut et al., 1979). Thus the eolian input from continental sources probably is minimal in this region of the Pacific Ocean.

3.4.2 Areal Variation

In the N Pacific, a latitudinal gradient in the mean annual concentrations of mineral aerosol was observed, with the higher concentrations occurring in the mid-latitudes, (30 °N to 50 °N; Figure 3.5 and Table 3.3). Large arid and semi-arid regions are located in the mid-latitude band of central Asia, and the prevailing winds in the mid-latitude regions also favour transport from W to E.

Figure 3.4: Variation of atmospheric Al concentrations in the transition period from May to July at three site form the central North Pacific network (Uematsu et al., 1983).

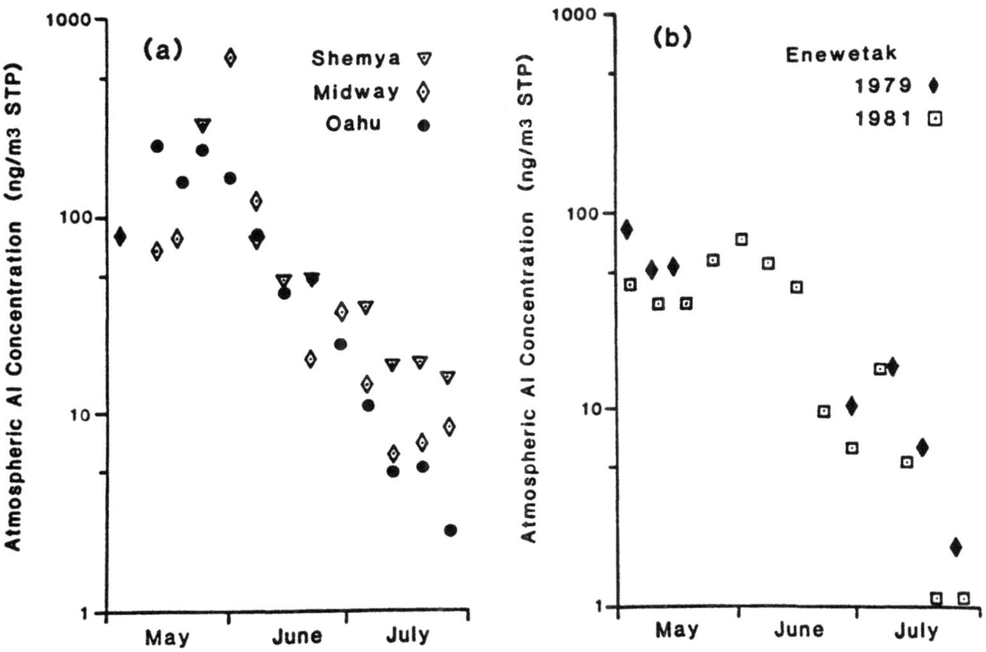

Figure 3.5: Mineral aerosol concentrations, precipitation amounts, and estimated fluxes of mineral particles at Shemya (SH), Midway (MD), Oahu (OA), Enewetak (EN), Fanning (FA), Nauru (NA), American Samoa (SA), Rarotonga (RT), New Caledonia (NC), and Norfolk (NF) from the Pacific networks (Uematsu, 1987).

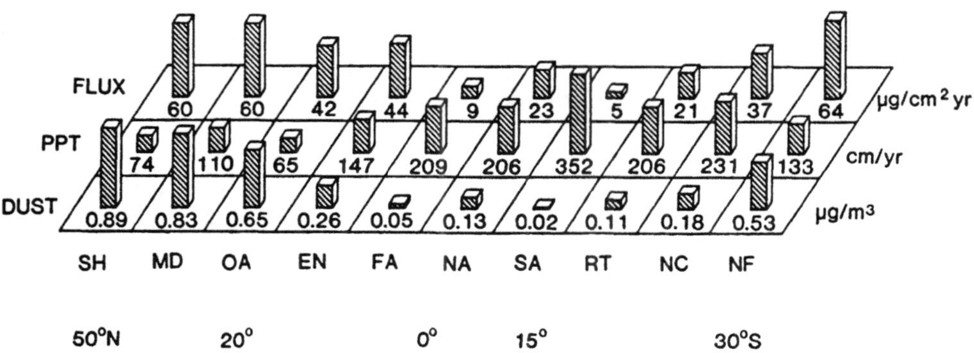

Figure 3.6: The concentrations of mineral aerosol as a function of distance from the estimated source point (40°N, 110°E). Data used was obtained from the stations in the Western N Pacific (Tsunogai et al., 1985) the Central North Pacific Network (Uematsu et al., 1983), and the Nagoya station (Kadowaki, 1979).

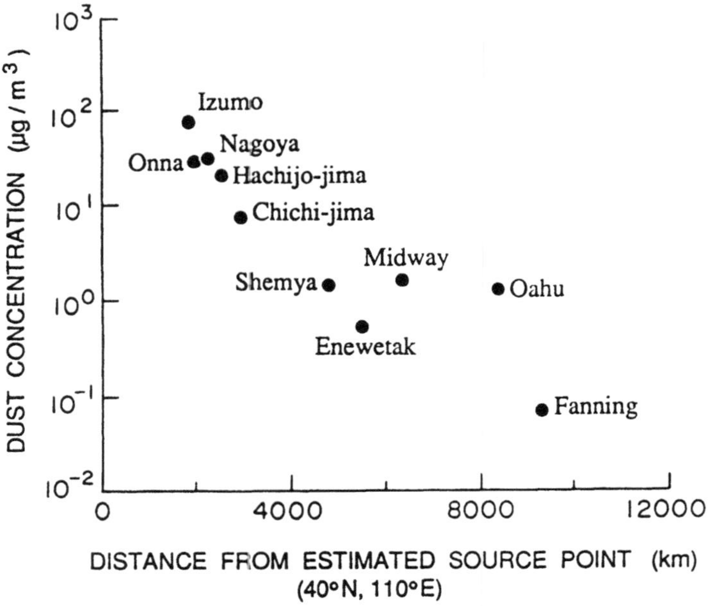

The concentrations of mineral aerosol are higher over the Western N Pacific than over the Central N Pacific. This is largely a result of the distances between the sampling stations and the source for the mineral particles. In Figure 3.6, the concentration of mineral particles from the Western N Pacific Network stations (Tsunogai et al., 1985) and other data (Kadowaki, 1979) are plotted along with the data from the Central N Pacific Network stations (Uematsu et al., 1983). As shown in Figure 3.4, the concentrations of mineral aerosol at the island stations in the N Pacific decrease logarithmically with distance from the estimated source point for the mineral aerosol. Tsunogai et al. (1985) concluded that the concentration of mineral aerosol decreased by half for each 500-6000 km of transport for all latitudes in the Western N Pacific.

Wind transported mineral particles appear to be a significant source of abiogenic marine sediments in the Pacific. Several investigators (Rex and Goldberg, 1958; Griffin et al., 1968; Leinen et al. 1986) have found the highest concentrations of eolian mineral components, such as quarts and illite, in sediments of mid-latitude regions over the N Pacific (Figure 3.7). In the Western N Pacific, the mineralogy of the mineral aerosol is identical to that of the underlying sediments (Blank et al., 1985). This is further evidence that mineral aerosol particles are a major source for sedimentary material. Furthermore, the data suggest that the present transport patterns for mineral aerosol from Asia may be representative of transport to the underlying sediments in the N Pacific over the past few thousand years.

In the S Pacific, the highest mean mineral aerosol concentration was found at Norfolk (Table 3.3). The Norfolk station is located relatively close to Australia in the S Pacific and beneath the prevailing westerly winds. The mean mineral aerosol concentrations at Nauru and Funafuti, i.e., the two SEASPAN stations nearest to the equator, were higher than those at the three SEASPAN stations (American Samoa, Rarotonga, and New Caledonia) located between 14 °S and 25 °S.

New Caledonia and Rarotonga are in the same latitude band (20 °S and 25 °S), but New Caledonia is only 1500 km from Australia. Rarotonga is further from the continental sources; it is approximately 5200 km from Australia and 9200 km from S America. The mean mineral aerosol concentration at Rarotonga, however is about 65 % of the at New Caledonia, suggesting that there is favourable air flow for the long-range transport of mineral aerosol from the continents to this remote marine region.

The flux mineral aerosol to the S Pacific is linked to the distribution of clay minerals in the sea floor sediments. Some minerals (e.g., illite, quartz, chlorite) exhibit high concentrations in a mid-latitude band (30 °S-50 °S), extending eastward from Australia (Griffin et al., 1968; Thiede, 1979; Leinen et al.,1986). This is perhaps due to the transport of Australian material by the prevailing westerly winds that pass over the S portion of Australia. The lowest concentrations of eolian minerals, such as illite, kaolinite, and quartz, in the sediments were found around American Samoa in the S Pacific. In that region of the tropical S Pacific, the in-situ formation of montmorillonite from phillipsite and volcanic glass shards apparently is important: more than seventy percent of the sediment is montmorillonite in the region. The eolian input around America Samoa from continental sources is minimal because of the extraordinarily low concentrations of mineral aerosol there. The kaolinite distribution pattern in sediments of the Western S Pacific (Griffin et al., 1968) shows evidence of two plumes extending from the coast of Australia, one extending N and E, and the other more extensive one to the SE (Figure 3.8). Both Nauru and Funafuti are located in the NE plume. The relatively high mineral aerosol concentrations at Nauru and Funafuti suggest that eolian inputs probably are important in the region. Griffin et al. (1968) suggested the the plume to the SE, which has a pattern similar to that of illite, is most probably derived from the deposition of eolian material from the arid interior of Australia. In contrast, the NE branch reportedly arises from the runoff of coastal rivers draining kaolinite-rich soil.

Figure 3.7: Distribution of quartz in the non-biogenic fraction of deep-sea sediments (Leien et al., 1986). All data are corrected for calcium carbonate and biogenic silica. Contours are in weight percent. Shaded areas have concentrations higher than 15 wt %.

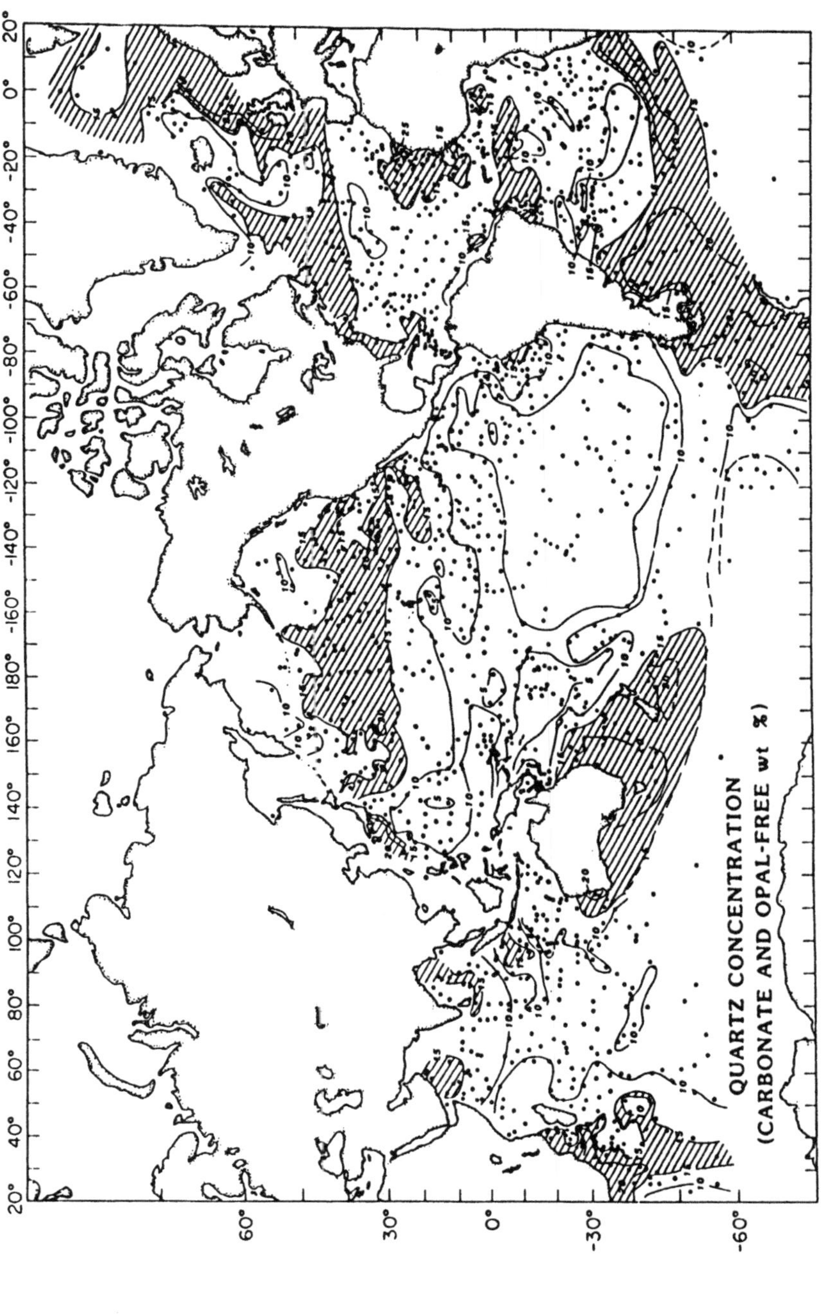

Figure 3.8: Distribution of kaolinite in the < 2μm fraction (%) in Pacific sediments (after Griffin et al., 1968).

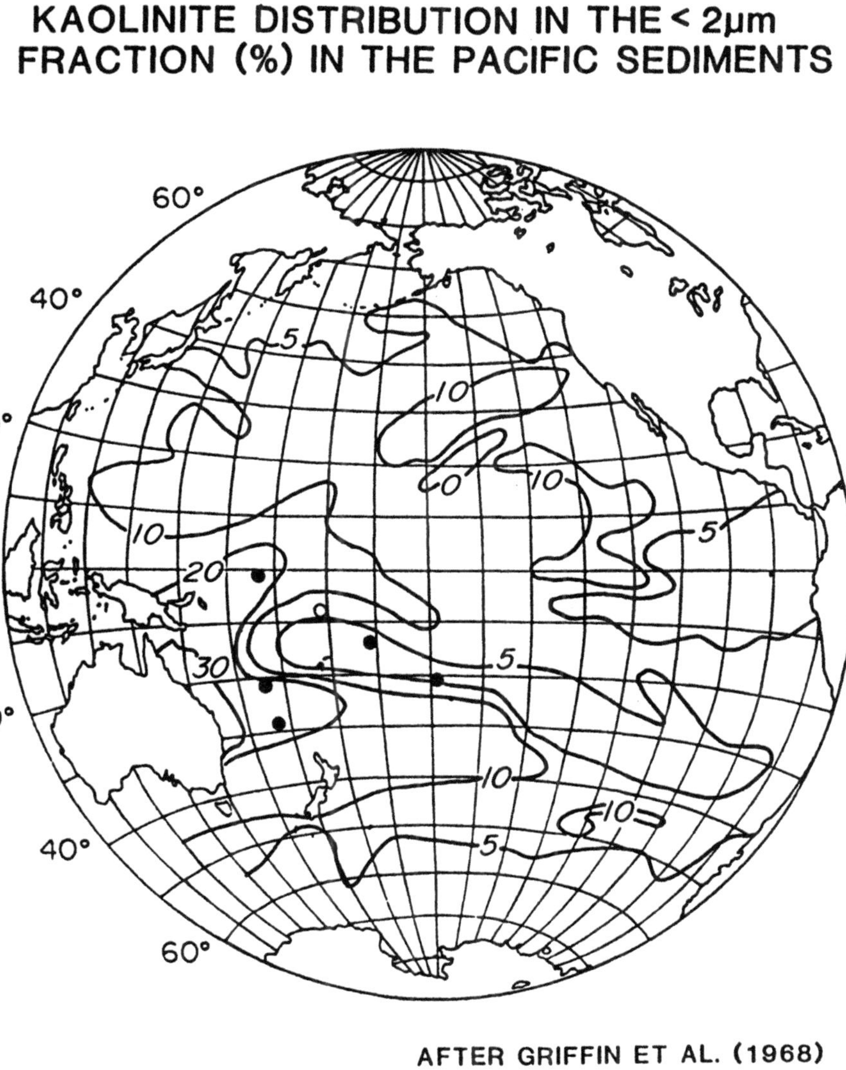

3.5 Deposition of mineral particles to the Pacific Ocean

The deposition rate of mineral particles during a dust storm in the Beijing area was estimated to be 2400 μg cm^{-2} day^{-1} (Liu et al., 1981). At a station in Japan near the coast of the Japan Sea, the deposition rate of mineral particles in "red snow" was found to be 1500 μg cm^{-2} for one overnight event in early February, 1966 (Tsunogai et al., 1972). However, there have been few direct measurements of the present-day deposition of mineral particles from the atmosphere to the N and S Pacific. Based on the rain sampling and dry-deposition sampling at Enewetak and American Samoa, Arimoto et al. (1985,1987) estimated the atmospheric flux of mineral particles to be 15 $\mu g\ cm^{-2}\ yr^{-1}$ and 1.7-23 $\mu g\ cm^{-2}\ yr^{-1}$, respectively.

At four stations in the North Pacific Network, total atmospheric deposition rates of mineral particles were measured concurrently with the aerosol sampling during January 1981 to March 1982 (Uematsu et al., 1985a). Using direct measurement of the total deposition rate of mineral particles, precipitation rate, and the concentration of mineral aerosol at the four stations, the scavenging ratio for mineral dust (i.e., the ratio of its concentration in rain to that in the atmosphere, using the density of air to make the scavenging ration dimensionless) was estimated to be 1000.

Based on this scavenging ratio for mineral aerosol, the measured concentrations of mineral aerosol, and finally the precipitation amount from the network stations, deposition rates for mineral particles were calculated (Figure 3.5). The uncertainties in these calculations are approximately a factor of ≈ 2, even though there are large seasonal and geographical variations of both the precipitation amounts and the mineral aerosol concentrations over the N and S Pacific.

A systematic geographical trend was apparent in the dust fluxes, with higher fluxes occurring at stations located at higher latitudes in both hemispheres. This tendency was similar to the distribution of the mineral aerosol concentration. However, the areal variation of mineral aerosol concentrations was more pronounced than that for the fluxes of mineral particle. The high flux period was coincident with the high-dust season in the N Pacific (Figure 3.9). Strong and persistent westerly winds at high latitudes transport mineral particles from arid regions in Asia to the N Pacific and from arid regions in Australia to the S Pacific. The highest flux periods were observed when rainfall was coincident with mineral aerosol loadings. The annual flux of mineral particles to N Pacific waters appears to be dominated by sporadic dust events of short duration. For example almost half of the annual dust deposition at Midway occurred during a period of only few weeks in the spring (Figure 3.10). Similarly, measurements of the deposition of Saharan dust in rain at Miami (Prospero et al., 1987) showed that 70 % of the annual deposition of mineral particles occurred in only four days.
The highly episodic nature of the dust fluxes suggests that the distribution of mineral particles (Yamamoto, 1984) and related species in sea water vary significantly over short time scales. Furthermore, a strong correlation between flux of mineral particles and the precipitation amount (correlation coefficient, $r = 0.94$) was found at Fanning, suggesting that rain events dominate the air-to-sea transfer of mineral particles.
The estimated dust fluxes in the central N Pacific ranged from 42 to 60 $\mu g\ cm^{-2}\ yr^{-1}$ (Figure 3.5); these values are comparable to the present-day downward flux of mineral particles through the water column as measured by sediment traps (Honjo et al., 1982; Tsunogai et al. 1982; Noriki and Tsunogai, 1986).

Figure 3.9: Comparison of mineral particle flux (a) and precipitation amount (b) at Midway (MD), Oahu (OA), Enewetak (EN), and Fanning (FA) from the Central North Pacific Network during the high-dust season, low dust season and for the entire year of 1981 (Uematsu et al., 1985a).

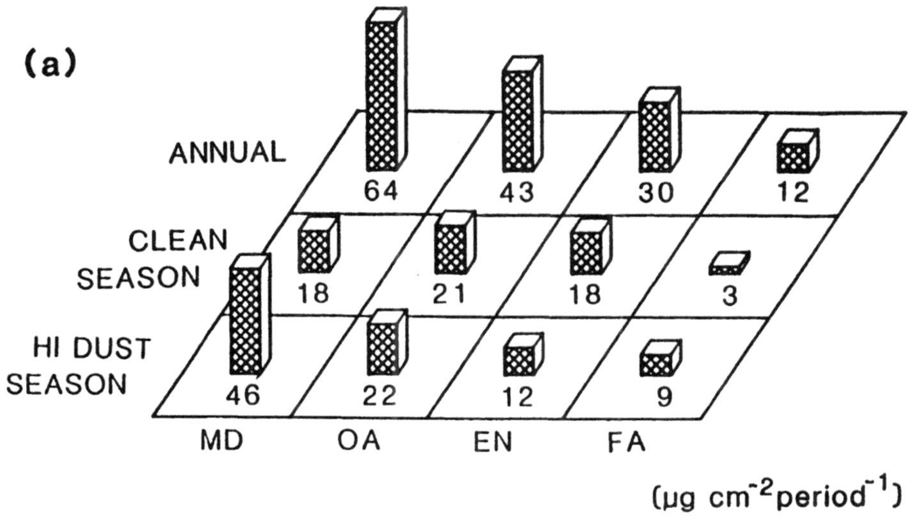

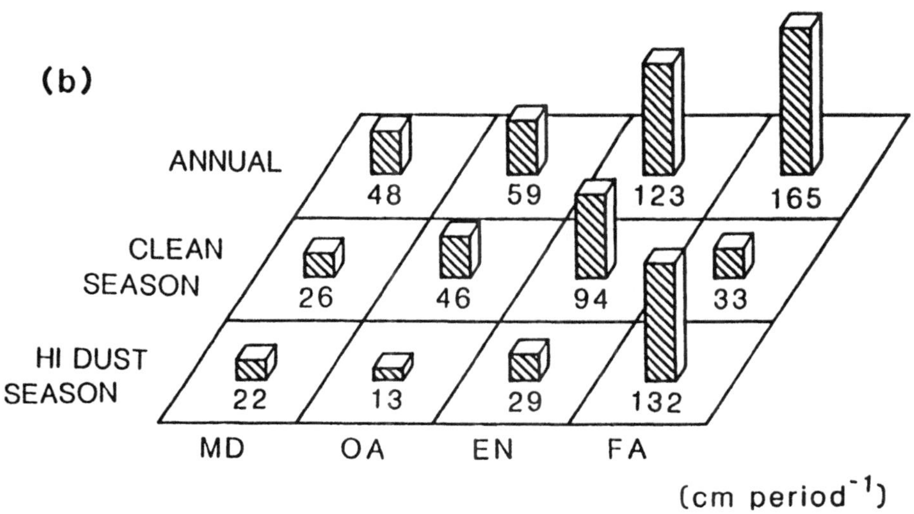

Figure 3.10: Normalized temporal variations of mineral particle flux (solid line), mineral aerosol concentrations (circles), and precipitation amount (dashed line) at Midway in 1981 and part of 1982 (Uematsu et al. 1985a).

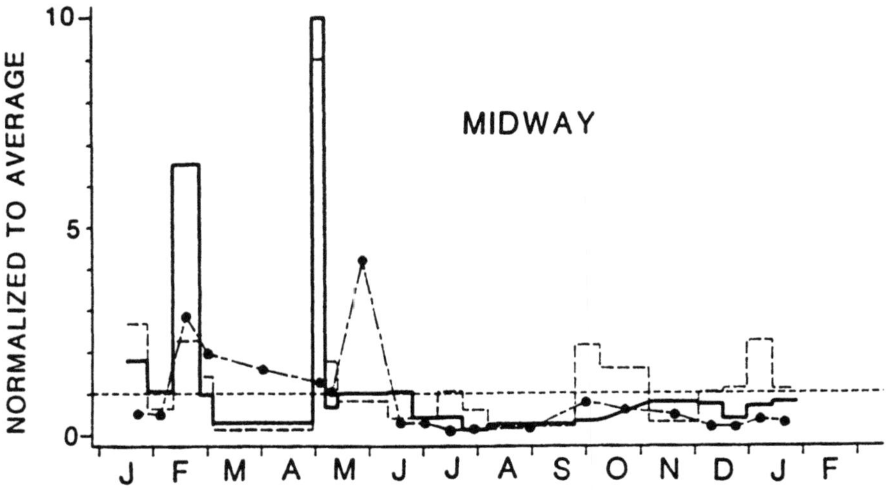

Martin et al. (1985) measured fluxes of particulate aluminum using sediment traps in the Central and Eastern N Pacific during the high- and low-dust seasons. The fluxes of mineral particles (based on aluminum values) with near-surface (150 m) traps during the high-dust season were approximately 5 times larger than those during the low-dust season. These results agreed well with the estimates of the seasonal atmospheric inputs of mineral particles made by Uematsu et al. (1983, 1985a).

Based on the estimated fluxes for the SADS and SEASPAN Networks (both shown in Figure 3.5), and for Western N Pacific Network (Tsunogai et al., 1985) an annual mass flux of $\approx 350 \times 10^{12}$ g yr^{-1} can be calculated for the entire Pacific (Table 3.4). The deposition rate in the Western N Pacific is two orders of magnitude bigger than in the Central Pacific. As a result the mass flux of dust in the Western N Pacific is ten times bigger than that in the remainder of the N Pacific. The atmospheric depositions of dust to coastal regions off the Asian continent is likely to be specially significant due to the magnitude of the flux. The value estimated for the coastal waters is almost equivalent to the flux of 130-440 x 10^{12} g yr-^{1} for the entire Atlantic (Prospero, 1981).

Table 3.4: Estimates fluxes of mineral particle to the Pacific Ocean.

	Deposition rate	
Ocean region	$\mu g\ cm^{-2}\ yr^{-1}$	$x10^{12}\ g\ yr^{-1}$
North Pacific (0-50°N)		
Western North Pacific (25-50°N, W of 150°E)	~5000[a]	~300
Other area in North Pacific	11-62[b]	30
South Pacific (0-50°S)		
West of 180°	23-64[c]	8
East of 180°	5-21[c]	10
Entire Pacific (50°N-50°S)		~350

[a]Tsunogai et al. (1985)
[b]Uematsu et al. (1985a)
[c]Uematsu (1987)

3.6 Influence of mineral particle deposition on surface waters

Stokes settling rates for finely divided atmosphere-derived debris found in bottom sediments indicate that the residence times for particles in the water column should be on the order of years. However, there is evidence that the eolian particles are removed from surface waters more rapidly than can be accounted for by Stokes settling alone. Uematsu et al. (1985b) reported concentrations of mineral particles in the surface waters measured on a cruise from Tokyo to Alaska and back in the summer of 1978. The mean concentration of mineral particles in the surface waters decreased for $\approx 70\ \mu g\ kg^{-1}$ in early July to $9\ \mu g\ kg^{-1}$ in mid-August. In contrast the total suspended particle concentration decreased by only a factor of ≈ 2 (Table 3.5). The high concentrations of mineral particles in the surface water during early July were correlated with high mineral aerosol concentrations measured on the ship. From the decrease in mineral particle concentrations in surface waters, the minimum removal rate of mineral particles to deeper waters during that one month period from July to August, 1978 was $\approx 70\ \mu g\ cm^{-2}$. This is about half the mean annual non-biogenic sedimentation rate averaged over the past several thousand years. According to Tsunogai et al. (1982), the sedimentary flux is approximately $150\ \mu g\ cm^{-2}\ yr^{-1}$.

Table 3.5: Mean total suspended matter (TSM) concentrations and Al concentrations and mineral content of particles from the surface ocean in the Northwestern N Pacific and other regions.

Stations	Date	Concentration in seawater		Mineral in TSP
		TSP (μg kg^{-1})	Al (μg kg^{-1})	(%)
North Pacific[a]				
S1-24[b]	Jul. 5-11	340 (82-1030)[c]	5.5 (1.4-18)	20
S39-57[b]	Aug. 15-19	150 (53-360)	0.72 (0.13-2.2)	6.0
Atlantic Ocean				
North Atlantic[d]		26	0.078	3.8
Sargasso Sea (10 m)[e]		72	0.30	5.3
Indian Ocean[f]		66	0.29	6.5

[a]Uematsu et al. (1985b); [b]same geographical region; [c]concentration ranges show in parentheses; [d]Buat-Menard and Chesselet (1979); [e]Deuser et al. (1983); [f]Chester et al. (1976).

Studies in the Atlantic Ocean have similarly shown a relatively rapid removal and transfer of atmospheric and biogenic material to the deep ocean (Deuser et al., 1981, 1983). In addition, large changes in detrital fluxes at the sediment-water interface have been observed in periods of less than two days in the N Atlantic (Billett et al., 1983).

Mineral particles from atmosphere are important as a source of dissolved Al in sea water. Exposure of atmospheric aerosols to sea water and artificial rain water showed that ≈9 % of the aerosol alumino-silicate Al was soluble (Maring and Duce, 1987). These authors showed that 5-6 % of the Al dissolved very rapidly (<0.6 hr) and an additional 3-4 % dissolved within 60 hr. Equally important, they proposed that the deposition of mineral particles containing 5-10 % dissolvable Al would constitute a major source of dissolved Al in surface waters of the open N Pacific. According to Orians and Bruland (1986), the dissolved Al distribution along a latitudinal transact from 20 °N to 20 °S on the 160 °W meridian was similar to the difference in the mean mineral aerosol concentrations between Oahu (20 °N) and Rarotonga (21 °S). These authors observed a five-fold decrease in dissolved Al surface concentrations going from N to S. This result is another indication that the atmospheric input of mineral particles is an important source for dissolved Al for the central N and S Pacific surface waters.

Other elements associated with mineral particles from atmosphere may be leached in sea water, and some elements, such as Fe, may be important for biological productivity (Menzel and Ryher, 1961; Moore et al., 1984; Duce, 1986; Martin and Gordon, 1988). Equally important is that our understanding of the geochemical cycling of many trace substances in the atmosphere and in the oceans is only fragmentary, and the air-to-sea transfer of mineral particles may be important in ways yet unknown. Therefore, a comprehensive understanding of the importance of eolian mineral particles will require information from biological and chemical oceanographers as well as from atmospheric chemists.

Acknowledgements

I wish to thank Drs. R.A. Duce and J.M. Prospero for their helpful advice, comments and encouragement. I am also grateful to Dr. R. Arimoto for his critical reading of the manuscript. This work was supported by NSF grants OCE 84-05605, OCE 84-05608, ATM 83-11694 and ATM 83-11335 as a part of the SEAREX Program and the North Pacific Tropospheric Transport and Deposition Study.

References

Arao, K. and Ishizaka, Y. (1986) "Volume and mass of yellow sand dust in the air over Japan as estimated from atmospheric turbidity", J.Meteorol.Soc.Japan 64, 79-94.
Arao, K., Makino, Y. and Nagaki, Y. (1979) "Some statistical aspects of yellow sand", Sci.Bull.Fac.Educ.Nagasaki Univ. 30, 65-74.
Arimoto, R., Duce, R.A., Ray, B.J. and Unni, C.K. (1985) "Atmospheric trace elements at Enewetak Atoll: 2. Transport to the ocean by wet and dry deposition" J.Geophys.Res. 90, 2391-2408.
Arimoto, R., Duce, R.A., Ray, B.J., Hewitt, A.D. and Williams, J. (1987) "Trace elements in the atmosphere of American Samoa: Concentrations and deposition to the Tropical South Pacific", J.Geophys.Res. 92, 8465-8479.
Billett, D.S.M., Lampitt, R.S., Rice, A.L. and Mantoura, R.F.C. (1983) "Seasonal sedimentation of phytoplankton to the deep-sea benthos", Nature 302, 520-522.
Blank, M., Leinen, M. and Prospero, J.M. (1985) "Major Asian aeolian inputs indicated by the mineralogy of aerosols and sediments in the Western North Pacific", Nature 314, 84-86.
Bodhaine, B.A., Mendonca, B.G., Harris, J.M. and Miller, J.M. (1981) "Seasonal variations in aerosols and atmospheric transmission at Mauna Loa Observatory", J.Geophys.Res. 86, 7395-7398.
Braaten, D.A. and Cahill, T.A. (1986) "Size and composition of Asian dust transported to Hawaii", Hawaii.Atmos.Environ. 20, 1105-1109.
Buat-Menard, P. and Chesselet, R. (1979) "Precipitation scavenging of aerosol particles over remote marine regions", Nature 321, 508-510.

Carlson, T.N. and Benjamin, S.G. (1980) "Radiative heating rates for Saharan dust", J.Atmos.Sci. 37, 193-213.
Chester, R. (1983) "The marine mineral aerosol", in Buat-Menard, P. (ed.), The role of air-sea exchange in geochemical cycling, D.Reidel, Dordrecht, 443-476.
Chester, R., Cross, D., Griffiths, A. and Stoner, J.H. (1976) "The concentrations of "alumino-silicates" in particulates from some surface waters of the world ocean", Mar.Geol. 22, M59-M67.
Chung, Y.S. (1986) "Air pollution detection by satellites: the transport and deposition of air pollutants over oceans", Atmos. Environ. 20, 617-630.
Collyer, F.X., Barnes, B.G., Churchman, G.J., Clarkson, T.S. and Steiner, J.T. (1984) "A trans-Tasman dust transport event", Weather Clim. 4, 42-46.
Darzi, M. and Winchester, J.W. (1982) "Aerosol characteristics at Mauna Loa Observatory, Hawaii, after East Asian dust storm episodes", J.Geophys.Res. 87, 1251-1258.
Deuser, W.G., Brewer, P.G., Jickells, T.D. and Commeau, R.F. (1983) "Biological control of the removal of abiogenic particles from the surface ocean", Science 219, 388-391.
Deuser, W.G., Ross, E.H. and Anderson, R.F. (1981) "Seasonality in the supply of sediment to the deep Sargasso Sea and implications for the rapid transfer of matter to the deep ocean", Deep-Sea Res. 28A, 495-505.
Duce, R.A. (1983) "The impact of atmospheric nitrogen, phosphorus, and iron species on marine biological productivity, In Buat-Menard, P. (ed.), The role of air-sea exchange in geochemical cycling, D. Reidel, Dordrecht, 497-529.
Duce, R.A., Unni, C.K., Ray, B.J., Prospero, J.M. and Merrill, J.T. (1980) "Long-range atmospheric transport of soil dust from Asia to the tropical North Pacific: Temporal variability", Science 209, 1522-1524.
Erickson, D.J., Merrill, J.T. and Duce, R.A. (1986) "Seasonal estimates of global atmospheric sea-salt distributions. J.Geophys.Res. 91, 1067-1072.
Ferguson, W.S., Griffin, J.J. and Goldberg, E.D. (1970) "Atmospheric dusts from the North Pacific short note on a long-range eolian transport", J.Geophys.Res. 75, 1137-1139.
Fuchi, H. (1939) "On dust-storms in China and Manchoukuo", J.Meteorol.Soc.Japan Ser. 2 17, 473-486.
Griffin, J.J., Windom, H. and Goldberg, E.D. (1968) "Clay mineral distribution in the World Ocean", Deep-Sea Res. 15, 433-459.
Healy, T.R. (1970) "Dust from Australia- A reappraisal", Earth Sci.J. 4, 106-116.
Hirose, K., Dokiya, Y. and Sugimura, Y. (1983) "Effect of the continental dust over the North Pacific Ocean: Time variation of chemical components in maritime aerosol particles in spring season", J.Meteorol.Soc.Japan 61, 670-677.
Honjo, S., Manganini, S.J. and Cole, J.J. (1982) "Sedimentation of biogenic matter in the Deep Ocean", Deep-Sea Res. 29, 609-625.
Ing, G.K.T. (1972) "A dust storm over Central China, April 1969", Weather 27, 136-145.
Ishizaka, Y. and Ono, A. (1982) "Mass size distribution of the principal minerals of yellow sand dust in the air over Japan, J.Hungarian Meteorol.Serv., 86, 249-253.
Isono, K., Komabayashi, M., Takeda, T., Tanaka, T., Iwai, K. and Fujiwara, M. (1971) "Concentration and nature of ice nuclei in rim of the North Pacific Ocean, Tellus. 35B, 189-196.

Iwasaka, Y., Minoura, H., and Nagaya, K. (1983) "The transport and spatial scale of Asian dust-storm clouds: a case study of the dust-storm event of April 1979", Tellus. 35B, 189-196.
Kadowaki, S. (1979) "Silicon and aluminum in urban aerosols for characterization of atmospheric soil particles in the Nagoya area, Environ.Sci.Technol. 13, 1130-1133.
Kobayashi, A., Hayashida, S., Okada, K., and Iwasaka, Y. (1985) "Measurements of the polarization properties of kosa particles by a laser radar in spring 1983", J.Meteorol.Soc.Japan 63, 144-149.
Koizumi, H. (1933) "Study of yellow sand", Kokumin Eisei 9, 983-1026.
Leinen, M., Cwienk, D., Heath, G.R., Biscaye, P.E., Kolla, V., Thiede, J., and Dauphin, J.P. (1986) "Distribution of biogenic silica and quartz in recent deep-sea sediments", Geology 14, 199-203.
Liu, T.S., Gu, X.F., An, Z.S., and Fan, Y.X. (1981) "The dust fall in Beijing, China on April 18, 1980", Geol.Soc.Am.Spec.Pap. 186.
Maenhaut, W., Zoller, W.H., Duce, R.A., and Hoffman, G.L. (1979) "Concentration and size distribution of particulate trace elements in the South Polar atmosphere", J.Geophys.Res. 84, 2421-2431.
Maenhaut, W., Raendonck, H., Selen, A., Grieken, R.V., and Winchester, J.W. (1983) "Characterization of the atmospheric aerosol over the eastern equatorial Pacific 88, 5353-5364.
Maring, H.B., and Duce, R.A. (1987) "The impact of atmospheric aerosols on trace metal chemistry in open ocean surface sea water, 1. Aluminum", Earth Planet. Sci.Lett. 84, 381-392.
Martin, J.H., and Gordon, R.M. (1988) "North-East Pacific iron distributions in relation to phytoplankton productivity", Deep-Sea Res. 35, 177-196.
Martin, J.H., Knauer, G.A., and Broenkow, W.W. (1985) "VERTEX: The lateral transport of manganese in the North-East Pacific", Deep-Sea Res. 32, 1405-1427.
Menzel, D.W., and Ryther, J.H. (1961) "Nutrients limiting the production of phytoplankton in the Sargasso Sea, with special reference to iron", Deep-Sea Res. 7, 276-281.
Merrill, J.T., Bleck, R., and Avila, L. (1985) "Modelling atmospheric transport to the Marshall islands", J.Geophys.Res. 90, 12927-12936.
Miklishanskiy, A.Z., Yakovlev, Y.V., Savel'yev, B.V., and Atrashkevich, V.V. (1981) "The entry of trace elements into the atmosphere from the oceans", Geokhimiya 6, 927-169.
Mizohata, A., Matsunami, T., and Mamuro, T. (1976) "Elemental composition and size distribution of continental aerosol particles", Ann.Rep.Rad.Ctr. Osaka 17, 1-6.
Moore, R.M., Milley, J.E., and Chatt, A. (1984) "The potential for biological mobilization of trace elements from aeolian dust in the ocean and its importance in the case of iron", Oceanol.Acta 7, 221-228.
Murayama, N. (1980) "The duststorms over the sea as revealed by GMS satellite", Umi to Sora 55, 149-168.
Noriki, S., and Tsunogai, S. (1986) "Particulate fluxes and major components of settling particles from sediment trap experiments in the Pacific Ocean", Deep-Sea Res. 33, 903-912.
Okada, K., Kobayashi, A., Iwasaka, Y., Naruse, H., Tanaka, T., and Nemoto, O. (1987) "Features of individual Asian dust-storm particles collected at Nagoya, Japan", J.Meteorol.Soc.Japan 65, 515-521.

Okita, T., Okuda, M., Murano, K., Itoh, T., Kanazawa, I., Hirota, M., Hara, H., Hashimoto, Y., Tsunogai, S., Ohta, S., and Ikebe, Y. (1986) "The characterization and distribution of aerosol and gaseous species in the winter monsoon over the Western Pacific Ocean", J.Atmos.Chem. 4, 343-358.

Orians, K.J., and Bruland, K.W. (1986) "The biochemistry of aluminum in the Pacific Ocean", Earth Planet.Sci.Lett. 78, 397-410.

Parrington, J.R., and Zoller, W.H. (1984) "Diurnal and longer-term temporal changes in the composition of atmospheric particles at Mauna Loa, Hawaii", J.Geophys.Res. 89, 2522-2534.

Parrington, J.R., and Zoller, W.H., and Aras, N.K. (1983) "Asian dust: Seasonal transport to the Hawaiian Islands", Science 220, 195-197.

Prospero, J.M. (1979) "Mineral and sea salt aerosol concentration in various ocean regions", J.Geophys.Res.84, 725-731.

Prospero, J.M. (1981) "Eolian transport to the world ocean", In Emiliani, C. (ed.), The Sea, John Wiley, New York, 801-874.

Prospero, J.M., and Bonatti, E. (1969) "Continental dust in the atmosphere of the eastern equatorial Pacific", J.Geophys.Res. 74, 3362-3371.

Prospero, J.M., Nees, R.T., and Uematsu, M. (1987) "Deposition rate of particulate and dissolved aluminum derived from Saharan dust in precipitation at Miami, Florida", J.Geophys.Res. 92, 14723-14731.

Raemdonck, H., Maenhaut, W., and Andreae, M.O. (1986) "Chemistry of marine aerosol over the tropical and equatorial Pacific", J.Geophys.Res. 91, 8623-8636.

Rahn, K.A. (1976) "The chemical composition of the atmospheric aerosol", Tech.Rep.Univ.Rhode Island, Kingston.

Rahn, K.A., Borys, R.D., and Shaw, G.E. (1981) "Asian desert dust over Alaska: Anatomy of an Arctic haze episode", Geol.Soc.Am.Spec.Pap. 186, 37-69.

Rex, R.W., and Goldberg, E.D. (1958) "Quartz contents of pelagic sediments of the Pacific Ocean", Tellus. 10, 153-159.

Shaw, G.E. (1980) "Transport of Asian desert aerosol to the Hawaiian Islands", J.Appl.Meteorol. 19, 1254-1259.

Suzuki, T. (1987) "Study on transport of chemical substances from lands to oceans through the atmosphere. PhD thesis, Hokkaido University, Hakodate.

Suzuki, T., and Tsunogai, S. (1988) "Daily variation of aerosols of marine and continental origin in the surface air over a small island, Okushiri, in the Japan Sea", Tellus. 40B, 42-49.

Takayama, Y., and Takashima, T. (1986) "Aerosol optical thickness of yellow sand over the yellow sea derived from NOAA satellite date", Atmos.Environ. 20, 631-638.

Tanaka, S., Tamura, S., Hashimoto, Y, and Otoshi, T. (1983) "Long range transportation of soil dust from Asian continent to Japan and its influence to the atmosphere in Japan", J.Japan Soc.Air Pollut. 18, 263-270.

Taylor, S.R. (1964) "Abundance of chemical elements in the continental crust: A new table", Geochim. Cosmochim. Acta 28, 1273-1285.

Thiede, J. (1979) "Wind regimes over the late Quaternary southwest Pacific Ocean:, Geology 7, 259-262.

Tsunogai, S., and Kondo, T. (1982) "Sporadic transport and deposition of continental aerosols to the Pacific Ocean", J.Geophys.Res. 87, 8870-8874.

Tsunogai, S., Saito, O., Yamada, K., and Nakaya, S. (1972) "Chemical composition of oceanic aerosol", J.Geophys.Res. 77, 5283-5292.

Tsunogai, S., Uematsu, M., Noriki, S., Tanaka, N., and Yamada, M. (1982) "Sediment trap experiment in the northern North Pacific: Undulation of settling particles", Geochem.J. 16, 129-147.
Tsunogai, S., Suzuki, T., Kurata, T., and Uematsu, M. (1985) "Seasonal and areal variation of continental aerosol in the surface air over the western North Pacific region", J.Oceanogr.Soc.Japan 41, 427-434.
Uematsu, M. (1987) "Study of continental material transported through the atmosphere to the ocean", J.Oceanogr.Soc.Japan 43, 395-401.
Uematsu, M, Duce, R.A., Prospero, J.M., Chen, L.-Q., Merrill, J.T., and McDonald, R.L. (1983) "Transport of mineral aerosol from Asia over the North Pacific Ocean", J.Geophys.Res. 88, 5343-5352.
Uematsu, M, Duce, R.A., and Prospero, J.M. (1985a) "Deposition of atmospheric mineral particles in the North Pacific Ocean", J.Atmos.Chem. 3, 123-138.
Uematsu, M., Duce, R.A., Nakaya, S., and Tsunogai, S. (1985b) "Short-term temporal variability of eolian particles in surface waters of the North-Western North Pacific", J.Geophys.Res. 90, 1167-1172.
Wang, M.-X., Winchester, J.W., Cahill, T.A., and Ren, L.-X. (1982) "Chemical elemental composition of wind-blown dust, April 19, 1980, Beijing", Kexue Tongbao 27, 1193-1198.
Winchester, J.W., and Mu-tian, B. (1984) "Fine and coarse aerosol composition in an urban setting: A case study in Beijing, China", Atmos.Environ. 18, 1399-1409.
Winchester, J.W., Lu, W.-X., Ren, L.-X., Wang, M.-X, and Maenhaut, W. (1981a) "Fine and coarse aerosol composition from a rural area in North China", Atmos. Environ. 15, 933-937.
Winchester, J.W., Darzi, M., Leslie, A.C.D., Wang, M.-X., Ren, L.-X., and Lu, W.-X. (1981b), In Macias, E.S., Hopke, P.K. (eds), Atmospheric Aerosol: Source/Air Quality Relationships, Am.Chem.Soc. Symposium Ser. 167, 287-301.
Winchester, J.W., Wang, M.-X., Ren, L.-X., Lu, W.-X., Hansson, H.-C., Lannefors, H., Darzi, M., and Leslie, A.C.D. (1981c) "Nonurban aerosol composition near Beijing, China", Nucl.Instrum.Meth. 181, 391-398.
Windom, H.L. (1969) "Atmospheric dust records in permanent snowfields: Implications to marine sedimentation", Geol.Soc.Am.Bull. 80, 761-782.
Windom, H.L. (1975) "Eolian contributions to marine sediments", J.Sediment..Petrol. 45, 520-529.
Yamamoto, S. (1984) "Concentration and behaviour of detrital mineral grains suspended in the water columns of open sea in the Western Pacific", J.Oceanogr.Soc.Japan 40, 80-89.
Yang, S.-J., Qian, Q.-F., Zhou, M.-G., Qu, S.-H., Song, X.-M., and Li, Y.-Y. (1981) "Some properties of the aerosols during the passage of a dust storm over Beijing, China, April 17-20, 1980", Geol.Soc.Am.Spec.Pap. 186, 159-167.
Zhou, M.-G., Qu, S.-H., Song, X.-M., Li, Y.-Y., Yang, S.-J., and Qian, Q.-F. (1981) "Some properties of the aerosols during a dust storm over Beijing area:, Kexue Tongbao 26, 901-905.

Mitsuo Uematsu
Department of Marine Sciences and Technology
School of Engineering
Hokkaido Tokai University
Sapporo, Hokkaido, 005, Japan

4 STATISTICAL STRUCTURE OF OCEANIC FIELDS IN THE SUBARCTIC FRONTAL ZONE

V.V. Anikiev, T.H. Zadonskaya, V.I. Ilyichev, E.N. Shumilin and V.V. Yarosh

Abstract

Results of complex investigations of thermohaline structure, hydrochemical (Alk, pH, O_2, Ca, Mg and $PO4^{3-}$) and hydrobiological (biomass of mesoplancton) characteristics of water mass in Kuroshio subarctic frontal zone are presented. For the data processing standard statistical methods including calculation of moments and histograms, factor and correlation analysis were used. Most variables correspond to the law of normal distribution. About 70% of the total data dispersion gives three-dimensional spatial variability of frontal zone characteristics. Compact description of the three-dimensional hydrochemical frontal zone structure has been obtained through a regression model, estimations for correlation bonds among individual parameters and spatial coordinates have been made.

4.1 Introduction

Biological productivity and, consequently, fishery situation in any area depend on hydrophysical, hydrochemical and hydrobiological factors. When making various term prognosis for fishery, one usually uses empiric evidences about relationships of the marine environment parameters, derived from a great number of observations in the area. Carrying out similar investigations in Kuroshio frontal zone is urgent both due to highly exchangeable features of water masses in a relatively limited area and to dominating influence of those area processes on fishery situation throughout the whole NW Pacific. Hydrochemical and hydrobiological water mass parameters were studied to reveal how their spatial variability depends on hydrological structure of the frontal zone.

In this paper we attempted to process all the data obtained by statistical methods.

4.2 Methods and Hydrological Situation

Complex investigations including simultaneous observations over thermohaline structure of water masses, as well as their hydrochemical (Al, pH, O_2, Ca, Mg and PO_4^{3-}) and hydrobiological (biomass of mesoplancton - BM) characteristics have been carried out during a cruise. Observations were conducted on the latitudinal transects during two surveys (Figure 4.1):

(1) June 1987, distance between sections being 60 miles;
(2) August 1987, distance between sections being 30 miles.

Hydrochemical parameters were determined by standard methods with 1 to 5% error for each sample, collected with 5-litre bathometer at O m, 10 m, 25 m, 50 m, 75 m, 100 m, 150 m horizons. Mesoplancton biomass was collected after preliminary trawling with Judy net at 0-50 m, 50-100 m, 100-150 m depths. Afterwards, the results were extrapolated to the middle of the water column. Hydrological situation in both areas was characterized by two fronts, the N and the S, with the eddy formations of opposite signs.

V. I. Ilyichev and V. V. Anikiev (eds.), Oceanic and Anthropogenic Controls of Life in the Pacific Ocean, 71–82.

Figure 4.1: Arrangement of stations (• - 1st survery, × - 2nd survey).

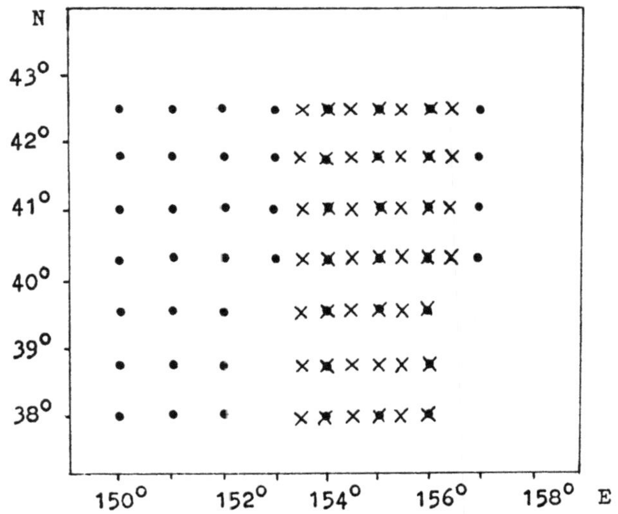

In order to reveal the pattern of the thermohaline structure influence on variability of hydrochemical parameters and plankton biomass, Brunt-Vaisala frequency (N) has been calculated. It characterized vertical variability of buoyancy forces.

4.3 Statistical Methods in Data Processing

To reveal the factors determining variability of hydrochemical parameters and plankton biomass in the area and, hence, the pattern of their dependence on thermohaline structure of water masses in the frontal zone, the data set obtained has been processed using all mathematical statistics methods.

At the first stage statistical characteristics (moments, histograms) were calculated to establish a type of description law. At the second stage changes of those characteristics with depth were studied. This may reveal the influence of vertical thermohaline structure. Correlation matrix enabling the authors to establish groups of interrelated values, was derived at the third stage. All this enabled to propose a hypothesis, that along with the factor analysis relative contributions of separate processes into parameter variability are observed. And in the long run, by deriving schemes of spatial distribution of separate factors it seems possible to notice domination of some processes in some areas of the interfrontal zone in formation of hydrochemical fields.

4.4 Result and Discussion

After initial statistical processing of the data obtained, the parameters under consideration could be divided according to the pattern of distribution of probable characteristics into three groups (Figure 4.2, Table 4.1):

(1) The probable density is close to even distribution with a sufficient negative excess (t°C, PO_4^{3-}).
(2) Though the distrubution parameter is close to a normal one, some deviations could be found (S‰, pH, O2, Mg, Ca). Maximal deviation is typical for S‰ and pH.
(3) Sharp assymetric distribution with great positive excess (Al, BM and N) is observed.

Table 4.1: Function moments of distribution of variables during two surveys

Variable	Average	Mean square Deviation	Asymmetry	Excess
t°C	9.96/12.7[1)]	4.80/5.27	-0.11/0.23	-0.84/-0.86
S ‰	33.9/34.1	0.54/0.37	-0.51/-0.41	0.71/0.01
pH	8.18/8.14	0.09/0.07	-0.21/0.07	-0.27/-0.89
Alk, mg-equiv/l	1.91/1.95	0.042/0.018	-1.25/-0.66	2.04/0.27
O_2, mg/l	5,71/5.10	0.72/0.50	0.24/-0.20	-0.07/0.25
Mg, mg-equiv/l	110/111	2.7/1.8	0.43/0.04	0.12/0.56
Ca, mg-equiv/l	19.5/20.2	0.49/0.35	0.65/-0.18	1.40/-0.96
PO_{43-}, 10^{-6} g/l	20.4/16.4	10.0/11.6	-0.52/-0.19	-0.55/-1.32
BM, mg/m^3	493/368	631/417	3.18/2.46	13.5/6.95
N, 10^{-3} 1/s	10.3/12.1	4.5/6.1	1.58/1.46	2.85/2.11

[1)] Slanted figures belong to the 1-st and 2-nd surveys, respectivly.

The first group of deviation from the norm is associated with a considerable variability of constituting parameters, as a result of which a smoothing distribution throughout the whole range of variations occurs. The third group of values concentrates mainly within small intervals, but there are also few considerable deviations in various directions, which determine the type of distribution.

Most probable factor of constant effect is a vertical thermohaline structure of water masses. Therefore, it is logical to consider the pattern of vertical variability of the parameters under study.

As could be seen from Figure 4.3, monotonous and marked changes with depth in the first group of vertical profiles (t°C, PO_4^{3-}) are observed. This determines their distribution over the whole range of variations.

Figure 4.2: Histograms of marine water parameters (• - 1st survey, ° - 2nd survey).

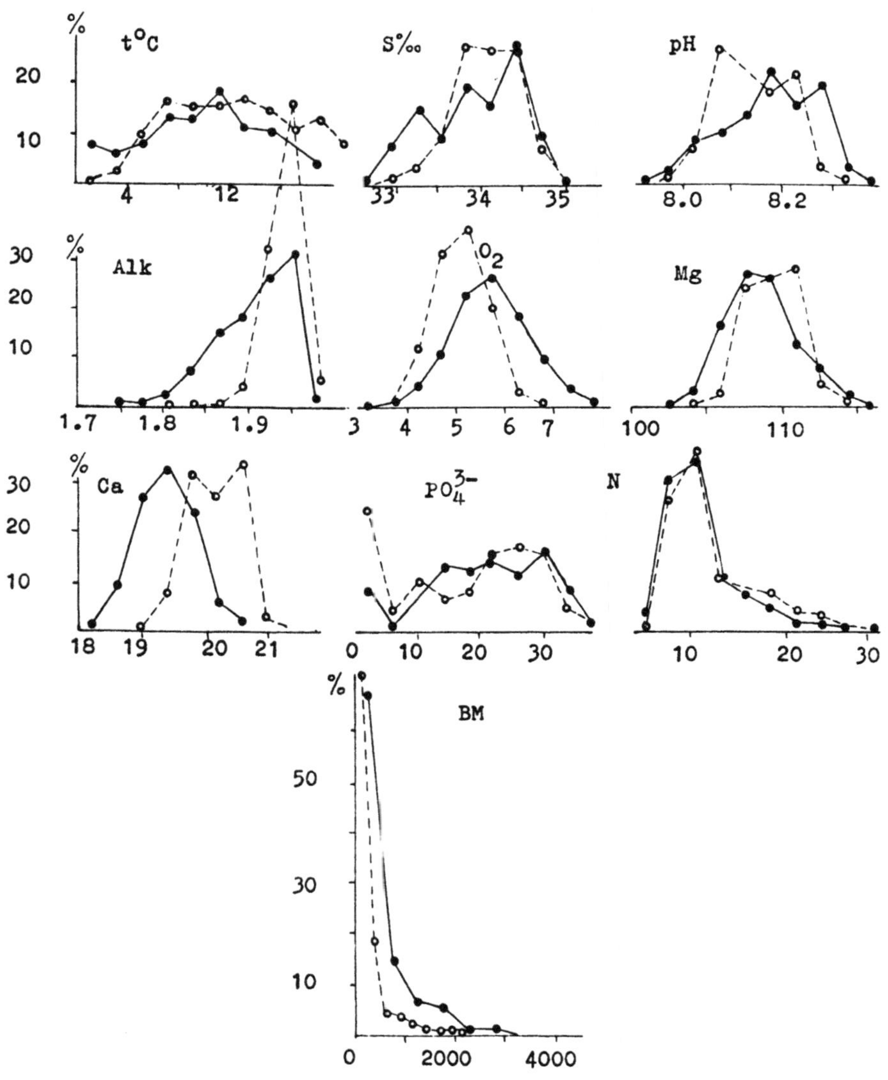

Figure 4.3: Vertical profiles of marine water parameters.

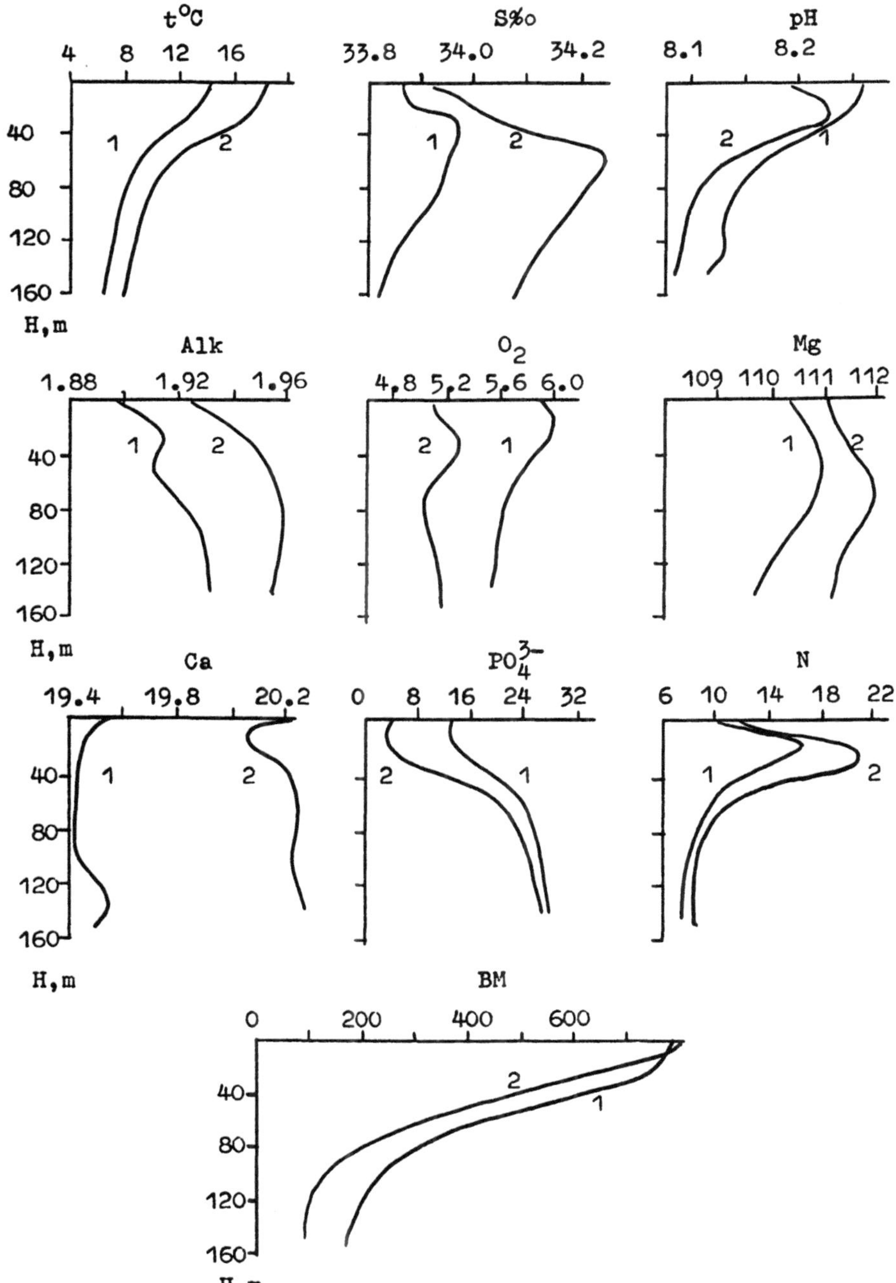

Intermediate maxima and minima are noticed in the second group. This may be one of the reasons for disagreement from normal distribution.
Typical vertical profiles for two indices from the third group (Al and BY) have not been found, whereas a strongly pronounced maximum for N at the depth of density jump could be observed.
From the above presented data it follows that the revealed agreements or disagreements with the normal law of parameter distribution are mainly due to regulation in their vertical structure. Its absence from the third group may be caused by various reasons not related to hydrology and, in particular, to the following parameters:

(1) CO_2 exchange between atmosphere and ocean in the upper layers as well as phytoplankton mass increase in the subsurface horizons may regulate Al variability;
(2) Zooplankton biomass depends first of all on primary production and only afterwards on abiotic factors;
(3) Due to sharp alteration of layers with different buoyancy as well as forces gradient in the frontal zone and asymmetric increase of N value, the deviation of this layers occured also in the density jump layer.

Additional information on processes controlling variability of individual variables may be obtained through their intercorrelation. Spatial and temporal limits of areas under investigation may affect the structure of correlation matrix. This is confirmed by graphs of the two surveys (Figure 4.4). It is difficult to suggest now without doubts whether changes in hydrochemical and hydrobiological structure of water masses have occurred during one month or the change in correlation bonds was influenced by the scale of the events only. However, absence of correlation between salinity and temperature in the second survey indicates fast thermal restructuring during the past period, because of heat diffusion higher by order of 2 than that of salt. An average 3°C increase in water temperature confirms warm water influx.

High correlation between temperature and pH in both areas confirms the earlier suggestion on the possibility of significant influence of CO_2 exchange between the ocean and atmosphere on the sea water carbonate system. With temperature increase, the volubility of CO_2 decreases and pH value transforms into the alkali area. Existence of stable correlation in both areas for t°C - PO_4^{3-} and S‰ - O_2 reflects a well known fact that deep cold waters are enriched in biogens and oxygen. They favor biomass development with elevation.

Significant correlation of Ca and Mg with temperature and salinity in the first area and absence of those in the second one agree to the hypothesis on changes in hydrochemical structure, caused by influx of subtropical waters. Moreover, in this case plankton considerably influences the hydrochemical water conditions, resulting in formation of correlation between the following parameters: BM - pH, PO_4^{3-} -Al, N - pH. Dependence of pH on the Brunt-Vaisala frequency also demonstrates the role of plankton in formation of equilibrium within carbonate system, since buoyancy forces are able to create definite vertical structure of plankton biomass in the ocean.

Figure 4.4: Graphs of correlation bonds for 1st (a) and 2nd (b) surveys.

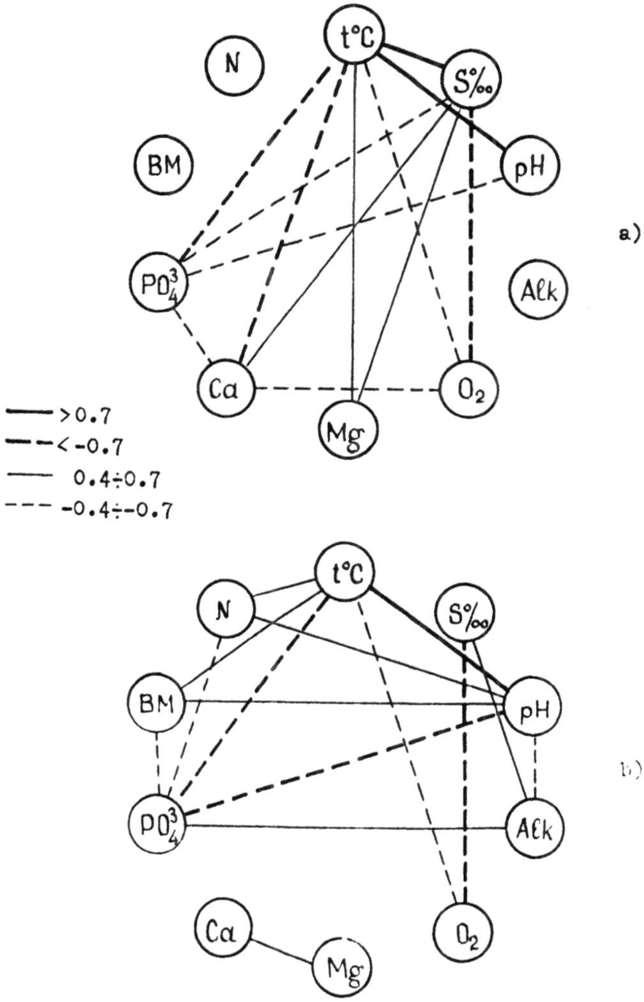

To clarify the possible reasons for hydrochemical parameters variability more precisely, all the data obtained was subjected to a traditional factor analysis, based on the principal components with further VARIMAX processing.
In order to understand the role of spatial variability, station coordinates (longitude λ and latitude λ) as well as sampling depth were additionally introduced into the number of variables. Prior to factor analysis all the variables were standardized by substracting the mean values and normalization by root mean square.

According to the data obtained for two stations, 3 factors controlling 70% of variability (Table 4.2), could be distinquished.

Table 4.2: Factor loading for individual variables.

Factor Variable	F_1	F_2	F_3	Communality
t°C	-0.64/-0.85	-0.72/-0.42	0.09/-0.03	0.93/0.90
S ‰	-0.10/0.07	-0.86/-0.89	0.17/0.11	0.77/0.81
pH	-0.78/-0.88	-0.23/-0.03	0.27/0.12	0.73/0.79
Alk	0.18/0.60	-0.08/-0.53	0.84/-0.07	0.75/0.64
O_2	-0.21/0.05	0.80/-0.87	-0.25/0.04	0.75/0.75
Mg	-0.13/-0.01	-0.44/-0.33	0.43/0.77	0.39/0.72
Ca	0.04/0.07	-0.67/-0.15	-0.14/0.76	0.47/0.60
PO_4^{3-}	0.62/0.92	0.65/0.04	0.26/-0.06	0.88/0.85
BM	-0.52/-0.57	0.37/0.12	0.11/-0.03	0.42/0.34
N	-0.63/-0.63	0.05/-0.16	-0.12/-0.18	0.41/0.46
H	0.90/0.91	0.01/-0.11	0.11/0.03	0.82/0.84
ϕ	-0.03/0.15	0.83/0.79	0.03/-0.15	0.70/0.67
λ	-0.08/-0.03	0.09/0.34	0.89/0.69	0.81/0.58
Factor contribution, %	21.8/33.1	34.0/22.8	14.2/12.9	

The first factor - F_1, determined by vertical stratification, includes such variables, as t°C, pH, P04 and N. For the second survey Al is added. Maybe due to thise, the contribution of this factor to variability of the parameters makes 33.1 and 21.8% for the second and the first surveys, respectively.

The second factor - F2, is determined by latitude, that is its' witnesses of t°C, S‰, °2 and Mg change along meridian, as well as Ca and PO_4^{3-} variances for the first survey and Al -for the second. Thus, gradients of the frontal zone determined the data body in a latitudinal direction by 34% and 22.8% during the surveys.

Figure 4.5 shows spatial distribution of F_2 with negative and positive values. Zero isoline corresponds to the boundary between subarctic and subtropical waters. Frontal zone meandering is clearly observed.

Figure 4.5: Spatial distribution of factors F_2 (a) and F_3 (b).

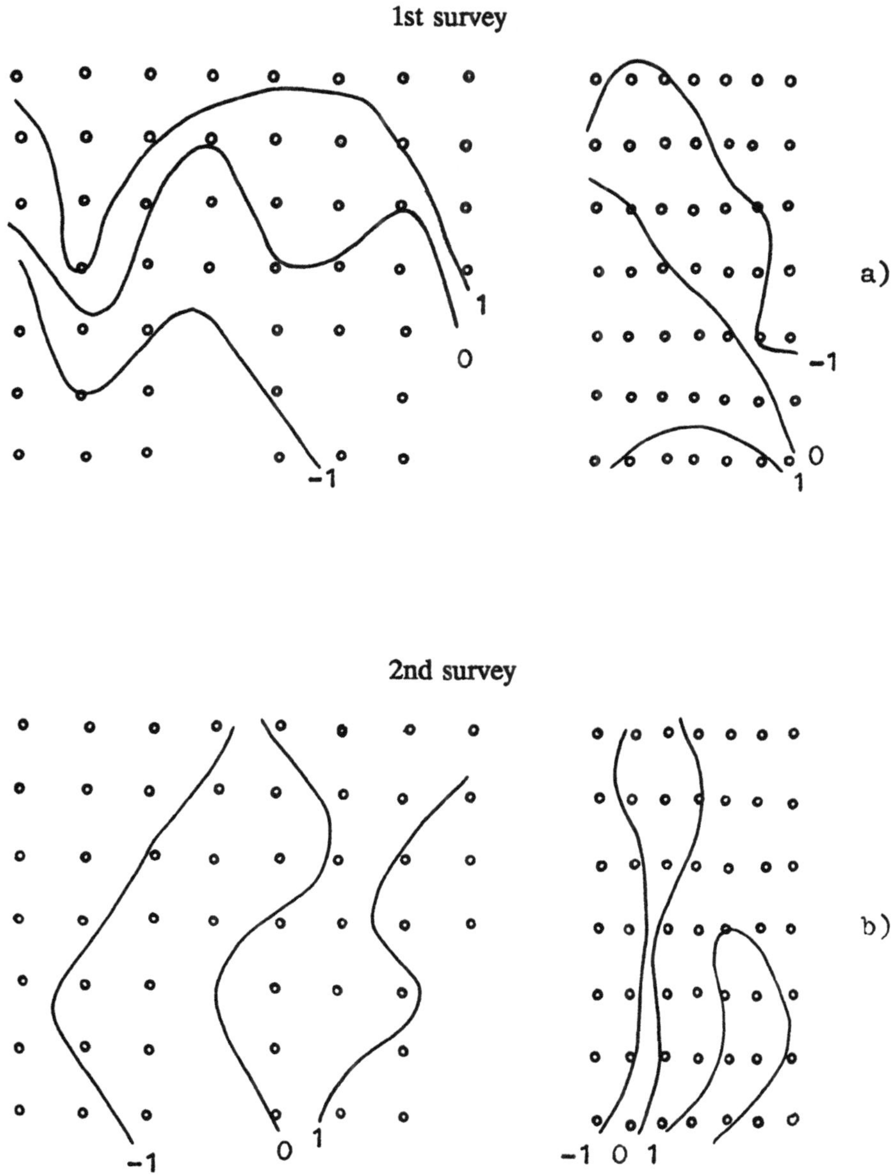

The third factor, Fe, showing meridional variability, includes Alk and Mg for the first survey, and Mg and Ca, for the second one. Emergence of this factor may be caused by variation of carbonate system under the influence of eddy structures and deviation of frontal zone from its latitudinal position (Figure 4.5). It is also possible, that these variations partially are due to temporary fluctuations of water mass hydrochemical structure during the fortnight survey. However, in any case an interaction within water masses with different conditions of the carbonate system in their latitudinal movement is observed. This factor provides from 13 to 14% of the total variability.

The results of the factor analysis enable to derive stationary regression model of the oceanic fields in a frontal zone, in which only spatial coordinates for sampling positions (H - horizont, ` - latitude and λ - longitudes) are used as independent variables. Such model provides data reproducibility not less that 80%. Regression coefficient and values of Fisher statistics F for standardized equations are given in table 3. Threshold statistical value for a given data array is 2.62 at 0.05 confidence level. Thus all the regression equations are significant with sufficiently high probability. Generally, the structures and coefficients of equations for those two surveys are close to each other by relative matrices of factor loadings (table 2). With denormalizing equations with mean and standard deviations, presented in Table 1, the repression model obtained is a compact description of the frontal zone structure and could be applied for practical purposes.

Several sets of regression coefficients, which may testify to a degree and sign of dependence of some parameters on spatial coordinates in a definite direction could be obtained.

As seen from Table 4.3 the subtropical water influx into the area of the second survey has resulted in a change of sign of such parameters, as S%o, Mg and Ca content, and simultaneous correlation increase for the rest of the indices. This was caused by depth decrease of the second survey area which most likely was the main reason for correlation drop in latitudinal direction, except for S%o, Al and N.

Since maximal gradients are confined to the subarctic front, the warm water intrusion, mentioned above, may cause either change of a correlation sign for most parameters, or a decrease of regression coefficients (pH, Mg). At the same time correlation for N increases.

From what was said above we could make the following conclusions on nature of hydrochemical and hydrobiological changeability in the upper layer (O - 140 m) of the Kuroshio frontal zone.

Most variables correspond to the law of normal distribution. But such parameters, as mesoplancton biomass and Brunt-Vaisala frequency have sharply asymmetric distribution with the great positive excess, which may testify to definite processes directed upon them. Obviously, those parameters could be used as indicators for restructuring of the frontal zone.

Table 4.3: Coefficients of regression equation and significance of Fisher statistics, F.

Variables	Coefficients for			
	H	ϕ	λ	F
t°C	-0.596/-0.763	-0.598/-0.450	0.148/-0.133	322.5/520.2
S ‰	-0.057/0.150	-0.678/-0.698	0.152/-0.137	110.9/144.9
pH	-0.621/-0.754	-0.233/-0.117	0.289/0.062	140.0/182.0
Alk	0.257/0.542	-0.087/-0.200	0.609/-0.250	98.0/84.6
O_2	-0.224/-0.055	0.662/0.616	-0.198/0.298	127.9/115.3
Mg	-0.106/0.034	-0.385/-0.242	0.300/0.253	37.3/18.3
Ca	-0.030/0.079	-0.446/-0.128	-0.032/0.202	33.1/8.8
PO_4^{3-}	0.579/0.809	0.486/0.167	0.202/-0.109	220.0/291.4
BM	-0.388/-0.496	0.160/-0.400	0.118/-0.021	33.1/42.5
N	-0.445/-0.481	-0.019/-0.125	-0.047/-0.193	32.4/51.2

Three-dimensional spatial changeability of the frontal zone variables elucidates 70% of the total data dispersion, but each trend has its own typical parameters:

(1) Vertical variability (t°C, pH, PO_4^{3-}, N and Alk);
(2) Latitudinal variability (t°C, S‰, O_2, Mg, Ca and PO_4^{3});
(3) Meridional fluctuation (markers only for Alk, Mg and Ca).

It should be noticed that spatial distribution of the three factors reflect the location of boundaries between water masses of the subarctic and subtropical geneses.

Changes in structures of hydrochemical and hydrobiological fields due to subtropical water influx may take place during a month. A changed structure of the correlation matrix for the data obtained in surveys 1 and 2, may prove this statement.

Compact description of the three dimensional hydrochemical frontal zone structures has been obtained through regression model and has been estimated for the correlation bonds among individual parameters.

When analyzing the data a number of regularities obtained earlier have been confirmed.

References

1) Oceanology. Chemistry of the Ocean., v.1, 1979, Eds.: Bordovsky O.K. and Ivanenkov V.N., Nauka, Moscow (in Russian).

2) Fyodorov,K.N. Fine structure of hydrophisical fields in the ocean. In: Oceanology. Hydrophysics of the Ocean, v.1, 1978, Eds.: Kamenkovich V.M. and Monin A.S., Nauka, Moscow, pp. 113-147 (in Rusian).

3) Oceanology. Biology of the Ocean, v.2, 1977, Ed.: Vinogradov M.E., Nauka, Moscow, (in Russian).

4) Oceanology. Biology of the Ocean, v.1, 1977, Ed.: Vinogradov M.E., Nauka, Moscow, (in Russian).

Anikiev, V.V. and Zadonskaya, T.H.
USSR Academy of Sciences
Far Eastern Branch
Pacific Oceanological Institute
ul. Radio, 7
690032 Vladivostok
USSR

5 SOME RESULTS OF STUDY OF CURRENTS ON THE SOUTHERN SHELF OF VIETNAM

Nguen Tien Dung and Le Phuoc Trinh

5.1 Introduction

The region under investigation is part of the Sunda shelf of SE Asia. It is conventionally limited by 7 ° to 12 °N and 105 ° to 110 °30' E. The N boundary is contiguous with the coastal zone of Central Vietnam. The dominant climatic factor of the region as well as the whole of S China Sea is the monsoon: the NE monsoon blows from December to March (in winter) and the SW monsoon blows from June to September (in summer). Numerical calculations of the wind current fields by means of the two-dimensional non-linar model of the low-water theory have been carried out for obtaining a general pattern of currents in this region. Certain results of geostrophic calculations and instrumental data of observations over the currents have been used as well.

5.2 Task formulation and its algorithm

Taking into account such peculiarities of the region as low-latitude position, shallow water, influence of steady seasonal monsoon, etc., a non-stationary set of equations was singled out from the low water theory. Non-linear members, friction on the bottom, Coriolis parameter variability have been also taken into consideration. After depth-averaging in a barotropic variant, the set of equations appears as

$$\frac{\partial u}{\partial t}+u\frac{\partial u}{\partial x}+v\frac{\partial u}{\partial y}-fv=-g\,\frac{\partial \xi}{\partial x}-\frac{r|U|u}{\rho_0 H}+\frac{\tau_x}{\rho_0 H} \tag{1}$$

$$\frac{\partial v}{\partial t}+u\frac{\partial v}{\partial x}+v\frac{\partial v}{\partial y}+fv=-g\,\frac{\partial \xi}{\partial x}-\frac{r|U|u}{\rho_0 H}+\frac{\tau_y}{\rho_0 H} \tag{2}$$

$$\frac{\partial \xi}{\partial t}+\frac{\partial (HU)}{\partial x}+\frac{\partial (HV)}{\partial y}=0 \tag{3}$$

where

V. I. Ilyichev and V. V. Anikiev (eds.), Oceanic and Anthropogenic Controls of Life in the Pacific Ocean, 83–92.

$$U = \frac{1}{H}\int_{\xi}^{H} u dz \quad ;$$

$$V = \frac{1}{H}\int_{\xi}^{H} v dz \quad ;$$

are components along OX and OY of the averaged current velocity. $|U| = (U^2 + V^2)^{1/2}$ is current velocity modulus. X-axis is oriented eastwards, Y-axis is N-oriented, $H = M + \xi$ is depth. M is the distance from the bottom to a basal plane (non-disturbed sea surface). ξ - is free surface level fluctuation; f - is Coriolis parmeter; g- is gravity acceleration; ρ_0 - is mean sea water density ($\rho_0 = 1$); r- is bottom friction coefficient (r = 0.0032); τ_x, τ_y are tangential wind stress components.

Boundary conditions

At the solid boundary

$$U|_{\eta} = V|_{\eta} = 0 \tag{4}$$

At the liquid boundary

$$\frac{\partial U}{\partial \eta} = 0 \; ; \; \frac{\partial V}{\partial \eta} = 0 \tag{5}$$

where η is normal to the lateral boundary.
Initial condition: the absence of motion at the initial moment at t = 0.

$$U = V = \xi = 0 \tag{6}$$

The set (1)-(3) with boundary conditions (4)-(6) was approximated using the implicit finite-difference scheme.

Not dwelling on the extent of resolution and the method of solution which are set forther (Baklanovskaya et al., 1979; Pham Ngoc Qui, Nguyen Tien Dung, 1982) we will present only a modified set and corresponding recurrent formulas, used in calculations.

The modified set is as follows:

$$-r_{j-\frac{1}{2}} U_{i,j-\frac{1}{2}}^{t+\frac{1}{2}} + \xi_{i,j}^{t+\frac{1}{2}} + r_{j+\frac{1}{2}} U_{i,j+\frac{1}{2}}^{t+\frac{1}{2}} = A_{i,j}^{t} \tag{7}$$

$$-r_j \xi_{i,j}^{t+\frac{1}{2}} + r'_{j+\frac{1}{2}} U_{i,j+\frac{1}{2}}^{t+\frac{1}{2}} + r_{j+1} \xi_{i,j+1}^{t+\frac{1}{2}} = B_{i,j+\frac{1}{2}}^{t-\frac{1}{2}}$$

$$-r_{j-\frac{1}{2}} V_{i-\frac{1}{2},j}^{t+1} + \xi_{i,j}^{t+1} + r_{j+\frac{1}{2}} V_{i+\frac{1}{2},j}^{t+1} = A_{i+\frac{1}{2},j}^{t+\frac{1}{2}} \tag{8}$$

$$-r_i \xi_{i,j}^{t+\frac{1}{2}} + r'_{i+\frac{1}{2}} V_{i+\frac{1}{2},j}^{t+1} + r_{i+1} \xi_{i+1,j}^{t+1} = B_{i+\frac{1}{2},j}^{t}$$

Recurrent equations:

$$\xi_{i,j}^{t+\frac{1}{2}} = -P_j U_{i,j+\frac{1}{2}}^{t+\frac{1}{2}} + Q_j$$

$$U_{i,j-\frac{1}{2}}^{t+\frac{1}{2}} = -R_{j-1} \xi_{i,j}^{t+\frac{1}{2}} + S_{j-1} \tag{9}$$

$$\xi_{i,j}^{t+1} = -\overline{P}_i V_{i+\frac{1}{2},j}^{t+1} + \overline{Q}_i$$

$$V_{i-\frac{1}{2},j}^{t+1} = -\overline{R}_{i-1} \xi_{i,j}^{t-1} + \overline{S}_{i-1} \tag{10}$$

where

$$P_j = \frac{r_{j+\frac{1}{2}}}{1+r_{j-\frac{1}{2}}R_{j-1}} \quad , \quad Q_j = \frac{A^t_{i,j}+r_{j-\frac{1}{2}}S_{j-1}}{1+r_{j-\frac{1}{2}}R_{j-1}}$$

$$R_j = \frac{r_{j+\frac{1}{2}}}{r_{j+\frac{1}{2}}+r_jP_j} \quad , \quad S_j = \frac{B^{t-\frac{1}{2}}_{i,j+\frac{1}{2}}+r_jQ_j}{R_{j+\frac{1}{2}}+r_jP_j}$$

P_i, Q_i, R_i, S_i are defined in the same way.
r, r', A, B are coefficients and free members of difference equations. The set (9), (10) is solved by the one-dimensional regularization method with alternating direction (Samarsky, Nikolayev, 1978). The process is considered steady when the condition

$$\frac{|U^{t+1}_{i,j}-U^t_{i,j}|}{|U^{t+1}_{i,j}|} \leq \varepsilon = 10^{-4}$$

is fulfilled.

5.3 Solution of the task for the given region. Initial data

Certain difficulties in utilizing the above-mentioned algorithm for the region under investigation exist. As was mentioned at the beginning, three sides of the region are surrounded by sea, i.e. there is a long band of liquid boundary. Many works (e.g. Eremeyeva, 1974; Kozlov, 1971; Numerical methods..., 1968) mentioned significant role of adopted boundary values on the liquid boundary, their validity and the extent of influence on the final result. In this case, all calculations have been done for the whole S China Sea, calculation parameters being chosen in a way to apply the obtained result for the region under consideration. Data on sea depths was borrowed from the Nha Trang Department of Marine Geology and Geography Archives. The 1961-1970 wind field data was received from the Ho-Chi-Minh Hydrometeorological Observatory and the Department of Marine Physics (IMR, Nha Trang). The information was processed and reduced to calculation modes according to Sukhovey (1971). Tangential wind stress components were estimated by the formula:

where

$$\tau_x = \gamma \rho_a |w| w_x \quad ; \quad \tau_y = \gamma \rho_a |w| w_y \tag{11}$$

γ = 0.0025 is resistance coefficient;
ρ_a = 1.15 * 10^3 g/cm^3 is air density;
$|w|$ is wind velocity modulus with W_x, W_y components.

5.4 Analysis of results

Figures 5.1a, 5.1b present the steady current field on the S Vietnamese shelf for two seasons, characterizing the SW and NE monsoons. Average wind velocities are from 6 to 8 m/s.

Figure 5.1: The steady current field on the South Vietnamese shelf for two seasons a) in summer, b) in winter.

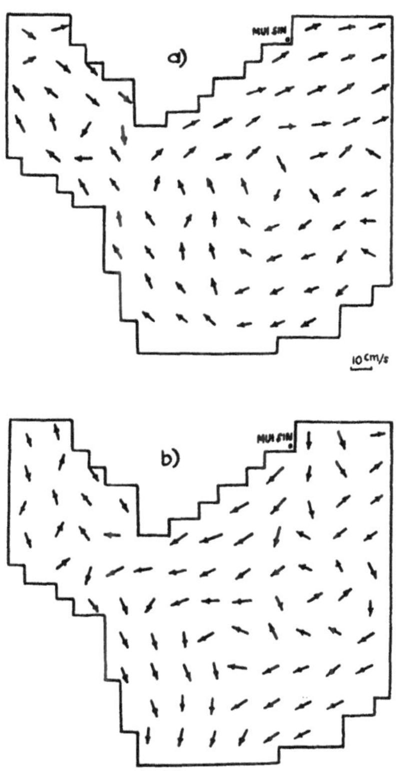

a) In summer (Figure 5.1a) one can observe a wide band of current oriented NE along the S coast of Vietnam. Its maximum velocity is up to 12 m/s. The flow originates in the E Gulf of Thailand at its N entrance where Gulf waters partly blend with those coming from the Equator. On the N boundary in the Muisin area, the flow changes its direction for more eastward one, breaks away from the shore, goes to the high sea and ceases to be a "longshore" current. The current direction in the stream nearly coincides with that of wind. The approximate estimate of the stream size, determined from its intensity and steady direction yielded 120-150 km across and 700-800 km in length. On the SE shelf margin an anticyclonic eddy is clearly seen, its centre being at 6°N, 109°E. Compared with the geostrophic eddy centre (Wo Wan Lanh et al., 1985) it shifts rather far westward into the heart of the shelf and becomes absolutely shallow. In the geostrophic current field this center is deep offshore, its periphery extending far westward where it blends with the longshore flow at the central shelf zone.

b) In winter the picture is nearly the same, but the direction is opposite (Figure 5.1b), i.e. the shelf current diagram is completely transformed. In winter the longshore current runs to SW. The maximum current velocity is 17 cm/s. Judging by vector distribution, the winter flow is a seasonal extension of the constant S current along the coast of Central Vietnam. In the Muisin region where this current branches, the winter longshore flow emerges. Seaward of the S tip of the peninsula of Indochina, on branch flows equatorwards and the other on deflects to the Gulf of Thailand. Here the flow dissipates and vanishes. Though seasonal longshore currents are of similar sizes, the winter one is more intensive. The summer anticyclonic eddy is replaced by the winter cyclonic one. The latter is centered at approximately 7°N, 110°E and shifted a little to NE with respect to the summer eddy. Unlike the geostrophic current winter eddy (Wo, Wan Lanh et al., 1985) which is fully contained on the shelf but cannot be isolated as an individual eddy, this one is rather distinct. The mean eddy velocity modulus is 5 to 8 cm/s.

c) It is possible to compare the evaluated result with that obtained in the course of observations over the current using automomous buoy stations (ABS). These buoy stations were set up in 5 points (Figure 5.2), three of them in the coastal zone and two in the offshore shelf zone. Figs. 3a, 3b show current frequency roses built from measurements at stations V and II. These series of observations have been chosen as typical currents of the SW and NE monsoons, respectively. The black arrows show the mean flow velocity vectors for each layer. For summer monsoon (Figure 5.3a) at all levels water transfer in the NNW to ENE sector occurs. The frequency of the main transfer varies within 22% (ENE) to 13% (NNW) at 5 m level, and with 22% (ENE) to 19% (NNW) at 25 m level.

The outward form of roses calls for attention. It is thick across and at the same time elongated towards the angle of the main velocity vector turn. This implies that in addition to a constant longitudinal current, the summer season has a well-developed transverse water motion.

Judging from the roses (Figure 5.3b), a southwestward flow forms on the shelf when the NE monsoon blows. At the levels of 11 and 19 m, the frequency of SW to WNW currents is 69.44 to 86.0%, respectively. Approaching the bottom, the main current turns to the left relative to the upper layers. At 29 m, the frequency of velocities is SSW to WSW directions is 12.9%, while at 35 m it is 84.8% for SSE to SW directions.

Figure 5.2: The map of mean current at 5 m in summer.

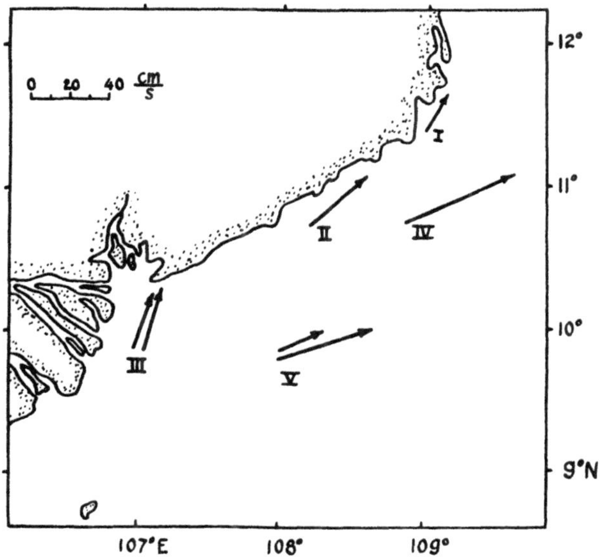

The averaged current velocity vectors at all ABS provided the basis for plotting maps of particular levels. Figure 5.2 presents the map of mean currents at 5 m in summer. Field observations have confirmed the existence of the main longshore NE flow on the shelf, this flow being stronger and is down the wind in the open part of the shelf. Near the shore, being less intensive, it tends to run parallel to the coast line.

Of particular interest is the vertical structure of the transverse velocity component. When the longshore flow is steady and the lateral component is coherent with the water mass which runs either from the shelf zone or enters it depending on season. Vertical compensation of water masses results in commonly known upwelling and downwelling and they, in turn, are the major shelf water motion mechanisms. Fig.4a shows vertical variability of the lateral velocity component on ABS IV during the SW monsoon. Depending on the current, the water column is divided in two layers in the lateral direction. In the upper layer (from surface to 15 m) a positive velocity component is evident. It means that water leaves the shelf zone and runs offshore. In the lower layer (from bottom to 15 m) the velocity component is negative, that is, the offshore water penetrates into the coastal zone. The situation changes when the NE monsoon blows. It is evident from Figure 5.4b that there are water inflow in the upper layer and water outflow in the lower one. The reverse lateral flows (with respect to the upper ones) provide for compensation of water outflow in summer and inflow in winter. The lateral current structure is of two-layered directed character both in winter and in summer.

Figure 5.3: Current frequency roses: a) in summer (buoy stations No.2) b) in winter (buoy station No. 5).

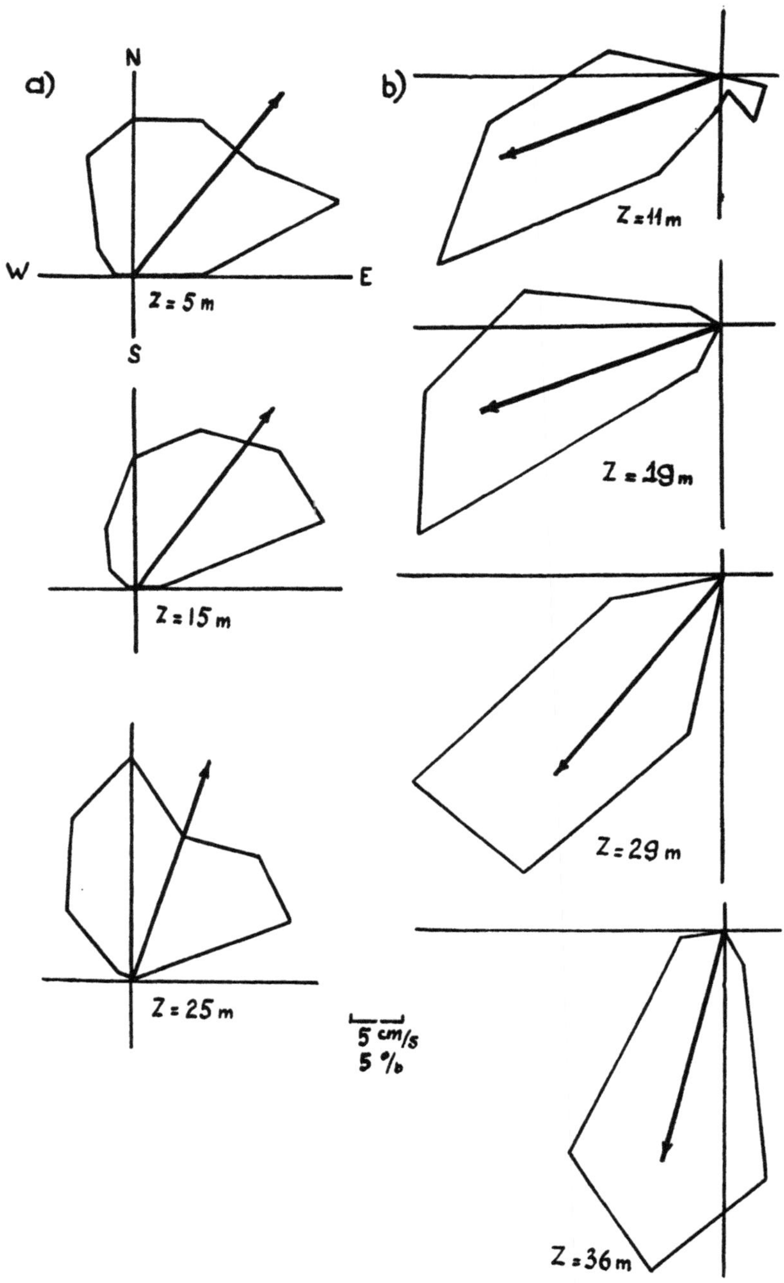

Figure 5.4: Vertical variability of the velocity component of current a) in summer (buoy station No. 4), b) in winter (buoy station No. 2).

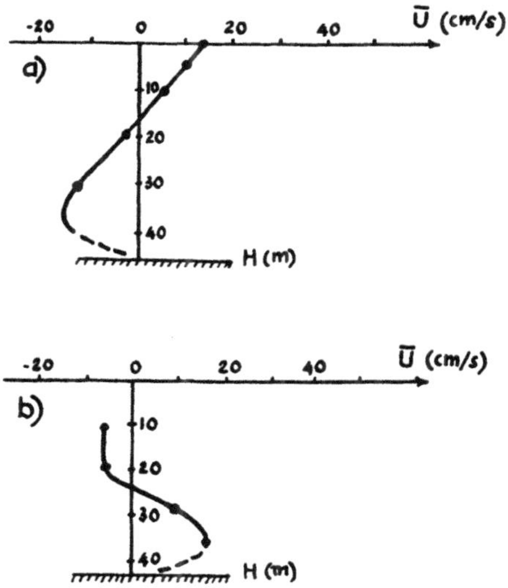

5.5 Summary and conclusions

1. Shelf currents consist of two main components:
   a) seasonal longshore SW flow in winter and NE flow in summer;
   b) local average-scale eddy of anticyclonic type in summer and cyclonic in winter

2. Monsoons are a major control of current fields which results in a pronounced seasonal rearrangement of water mass structure. It is evidenced by a complete transformation of the water mass structure. This transformation is characterized by changing to the opposite of movement directions when switching from season to season.

3. In the lateral direction the vertical current structure in a flow is two-layer-oriented. The presence of the countercurrent in the lower layer results in natural upwelling and downwelling events on the shelf depending on a season.

References

1. Baklanovskaya, V.F.; Paltsev, B.V.; Chechel, I.I.: Boundary value problems for St. Venantś equations on the plane. J.of Comp.Math. and Math. Physics 19, 708-725, 1979 (in Russian)

2. Eremeyeva, V.F.: Calculation current and level in the shallow sea. In: The problems for sea dynamics. Gidrometeoizdat, Moscow 1974 (in Russian).

3. Kozlov, V.F.: Results of approach calculation integral circulation in the Japan sea. J. Meteorology and Hydrology No. 4, 1971 (in Russian)

4. Lanh Vo Van; Tran Van Sam; Nguyen Tien Dung; et al: The geostrophic current in the South China sea. Report of Program Atlas of Vietnam. Hanoi 1985 (in Vietnamese)

5. Numerical methods of solution atmosphere and ocean problems. Gidrometeoizdat, Moscow 1968 (in Russian).

6. Qui Pham Ngoc; Nguyen Tien Dung: Calculation of current in the shallow sea. Papers of Institute of Math. Appl., No. 82201, Ho-Chi-Minh 1982 (in Vietnamese)

7. Samarsky, A.A.; Nicolayev, E.C.: Methods of mesh equations solutions. Nauka, Moscow 1978 (in Russian)

8. Sukhovey, V.F.: Recovery hydrological fields on the base of data of observations. Marine research No. 3, Sevastopol 1971 (in Russian)

Nguen Tien Dung and Le Phuoc Trinh
Institute of Marine Research
National Center of Science Research
SRV

6 ON VERTICAL MIXING IN SHELF AND FRONTAL ZONES OF THE OCEAN

V.V. Navrotsky, A.Yu. Lazaryuk and S.V. Simonenko

Shelf end frontal oceanic zones are characterized by high intensity of biological processes. Hydrophysical specificity is due to large horizontal gradients of all kinds, complicated vertical structure of hydrophysical properties, great importance of internal waves, turbulence, 3-D circulation and vertical mixing. Being interrelated, all these phenomena must be studied in complex.

This paper deals with the internal gravitational waves (IW). We consider them to be the most universal part of the mechanisms responsible for fine structure creation and large vertical transport of properties and admixtures. According to the results previously obtained, short IW are most effective in this respect, their generation being a separate hydrodynamical problem.

The analysis of the experimental data obtained in 1984-1986 in the Japan Sea (Ilyichev, Navrotsky,1987a; Ilyichev, Navrotsky,1987b) indicates a possible formation of an internal hydraulic "jump" seaward before the shelf boundary due to the running of a tidal wave against the continental slope (see Fig. 1,2).

Calculations made for the two-layered liquid using Kong's model (Long,1970) have shown that the calculated jump amplitude is in good agreement with the observed lowering of a thermocline. However we could not obtain such an important feature as layers formation within the two-layered model.

Multi-layered model and correspondent algorithm of Sue (Sue,1976) were used for the simplest case of three-layered liquid. H_i, h_i indicate the layers' thickness before the internal hydraulic jump (IHJ) and after it (H_i is for the lower homogeneous layer, H_2 is for the middle one, and H_3 is for the upper layer). Using same notations

$$\zeta_i = \frac{h_i}{H_i}, \quad \alpha(\zeta) = \frac{2\zeta}{(1+\zeta)(2+\zeta)}, \quad f(\zeta) = \frac{-\zeta^3}{(1+\zeta)^2(2+\zeta)} \tag{1}$$

integrating momentum and continuity equations in a hydrostatic approximation and with a rigid surface condition, we can figure out the necessary relations in a system moving with the stationary jump as follows:

$$\alpha(\zeta_i) = \frac{\rho_{i+1}}{\rho_i}\frac{u_{i+1}^2}{u_i^2}\alpha(\zeta_{i+1}) - \frac{g}{u_i^2}\left(1-\frac{\rho_{i+1}}{\rho_i}\right)\sum_{j=i+1}^{3} H_j\zeta_j, \quad i=1,2 \tag{2}$$

V. I. Ilyichev and V. V. Anikiev (eds.), Oceanic and Anthropogenic Controls of Life in the Pacific Ocean, 93–108.

$$\sum_{j=1}^{3} (h_j - H_j) = \sum_{j=1}^{3} H_j \zeta_j = 0 \tag{3}$$

$$\Delta E = F_d - F_u = \frac{1}{2}\sum_{i+1}^{3} \rho_i H_i u_i^3 f(\zeta_i) = -\frac{1}{2}\rho_1 H F_D \leq 0 \tag{4}$$

Here ρ_i, u_i are density and velocity in the i-th layer, F_u, F_d are energy fluxes up and down of a jump.
The equations (2)-(4) were solved for values

$$\Delta \rho = \frac{\rho_1 - \rho_3}{\rho_1} = 10^{-3} \quad ,$$

$\rho_2/\rho_1 = 0.9997$, $\rho_3/\rho_2 = 0.9993$, $H_3 = 20$ m, $H_2 = 10$ m and for thw two cases: 1) $H_1 = 157$- m ($H = \Sigma H_i = 1600$ m), 2) $H_1 = 170$ m ($H = 200$ m).
The dependencies of $\zeta_2\zeta_3$ on $U = V + C$ (V_z) = const is fluid velocity up of a jump, c is the jump velocity, are given at Fig. 3 for the first and second vertical modes. Obviously the second-mode jump with enlarging middle layer can arise by moderate values of fluid velocity (v = 0.22 m/s). Such a jump can move toward the shore because of tidal current periodicity.

Short IW observed in shelf zones can be generated as a result of hydraulic jump breaking or simultaneously with it in a kind of bore with a wave-like tail. Just above the shelf boundary we observed solitone-like waves of large amplitude (up to 10-12 m) which were absent on the shelf where short wave groups with the amplitude of 3-4 m were typical. Hence the generation of the short IW by an IHJ breaking is supposed to occur through the short-living solitone. Solitone breaking mechanism and short IW generation were described earler (Djordjevic, Redecopp, 1978).
A similar IHJ formation mechanism (with following IW generation) can be represented in a kind of bore on a sloped thermocline but the presence of long wave is necessary, the latter possessing the properties of an internal wave. We suppose that the main IW generators in the frontal zones are inertial and tidal motions superimposed on large horizontal gradient of density, since in the buoyant forces

$$F = \zeta g \frac{\partial \rho}{\partial z} + \xi g \frac{\partial \rho}{\partial x} \tag{5}$$

the second term proportional to horizontal displacement and horizontal gradient of density $\partial\rho/\partial x$ becomes of great importance. In a general case waves which frequency is mf + mω_t (f is inertial frequency, ω_t is tidal frequency and m,n are integers) can be generated due to nonlinear interactions.

Dynamical processes near the shelf boundary result in the fact that large spatial heterogeneities can occur there. According to the types of N(z) we have divided the water area under investigation into 4 basic zones: sea, slope, shelf-1 and shelf-2. Keeping in mind the waves which can be described by linear equations under the WEB zero approximation, we have calculated dispersion characteristics in the named zones.

Fig.4 shows N(z) curves and eigen functions for the first three modes with ω = 4 c/h, while Fig.5 illustrates dispersion curves ω (k) in the zones. Of importance here is the fact that in both slope and shelf zones resonance nodes are well pronounced, these are the regions of ω (K) curves' convergencies where a resonance energy exchange between the neighbour modes occurs with nonlinear interaction taken into account. The results of Navrotsky (1976) show that in a physical space such an interaction takes place on stable layers of fine structure which occurrence can be a result of IW propagation.

Fig. 6 clearly illustrates the difference between the zones as far as ω (K) is concerned and that of the group velocity Cg(K) for the first three modes. The sharpest differences are seen for the first mode. In shelf zones 3, 4 the curves Cg(k) for almost every frequency are found lower than in the sea and on the slope. In a slope zone the group velocity increases as compared with that of the sea for w <6 c/h, decreases for ω > 6 c/h. In the shelf zone adjacent to the slope the pronounced deviations from sea dependencies occur in a high frequency zone (ω > c/h).

Fig.7 illustrates a spatial structure of group velocity Cg(x) for the first three modes. On the slope all the Cg values are maximum while on the shelf they are decreasing: 5-6 fold for the first mode at a 10 mile distance from the shelf boundary. Maximum group velocities on the slope and their decrease on the shelf result in the fact that internal wave energy is carried out as high velocity from the slope area, where perturbations of maximum amplitude are generated, while in the shelf area the energy is accumulated. Thus, both the wave guide geometry and the mean density field structure result in IW energy concentration in the shelf area. However it is noteworthy that density field structure leading to this result is formed mainly by the internal waves themselves.

Fig.8 illustrates sections of mutual correlations R (Δz, Δt) of profiles T(z) which enables us to make a conclusion about temperature perturbations propagating to the depth with the average velocity 1-2 m/h. Fig. 9 and 10 show similar sections, the parameter being not time but space coordinate Δx normally towards the shelf boundary. Section 4 relates to the case of well developed I, while section 6 shows a case when IW intensity is considerably decreased. One can easily notice that propagation of vertical temperature perturbations correlates with internal waves development.

Figure 6.1

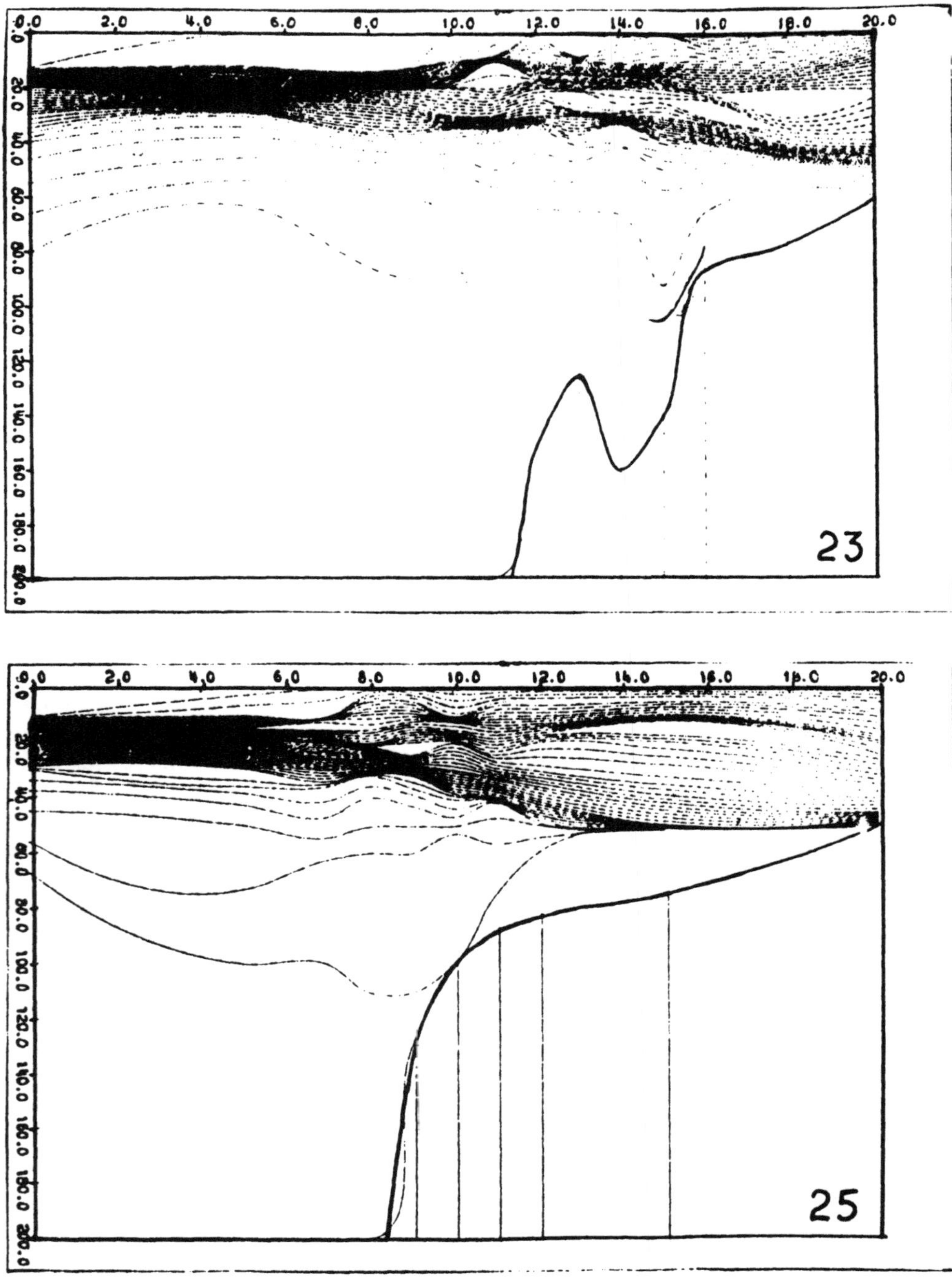

Figure 6.2

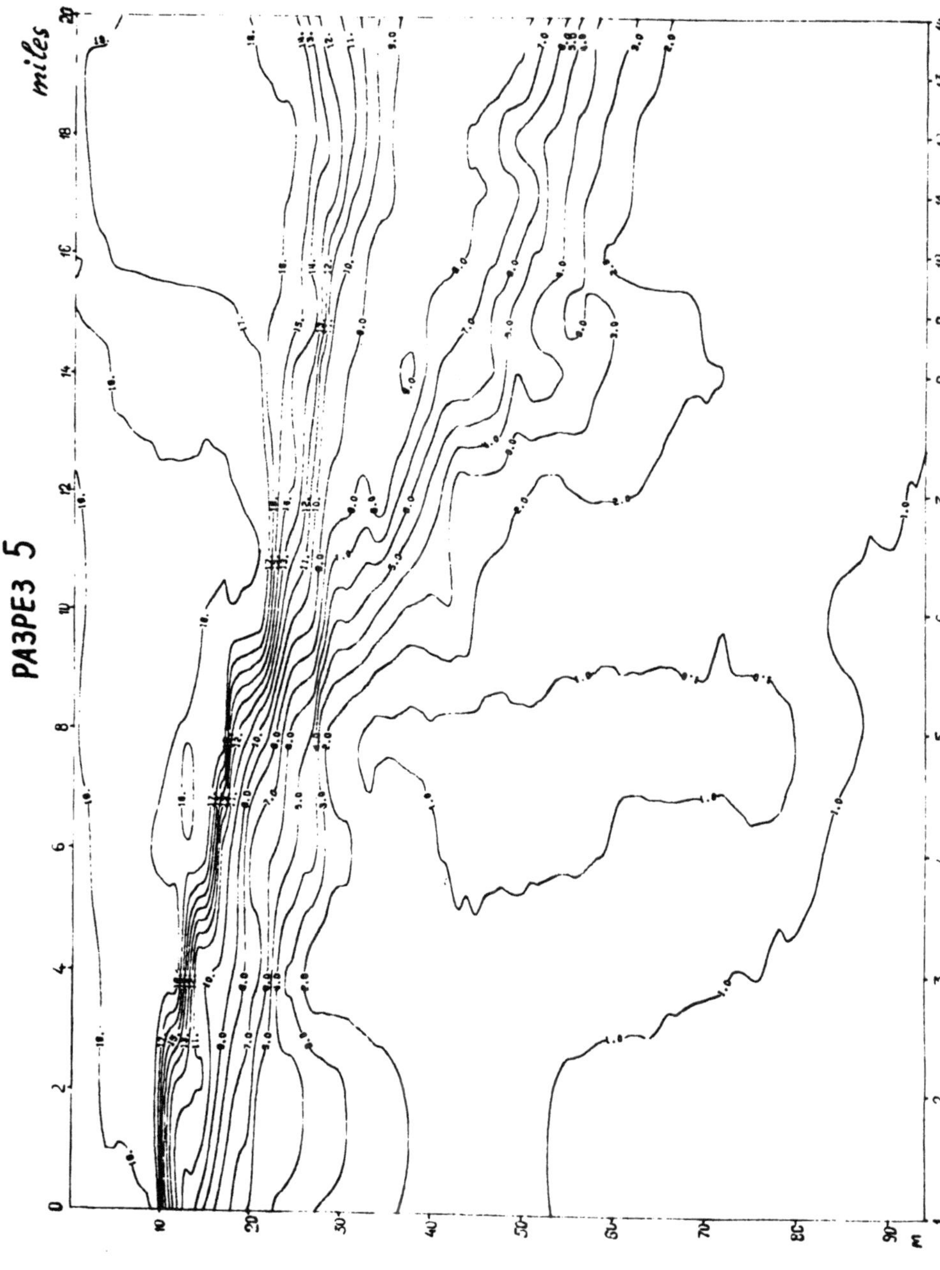

Figure 6.3

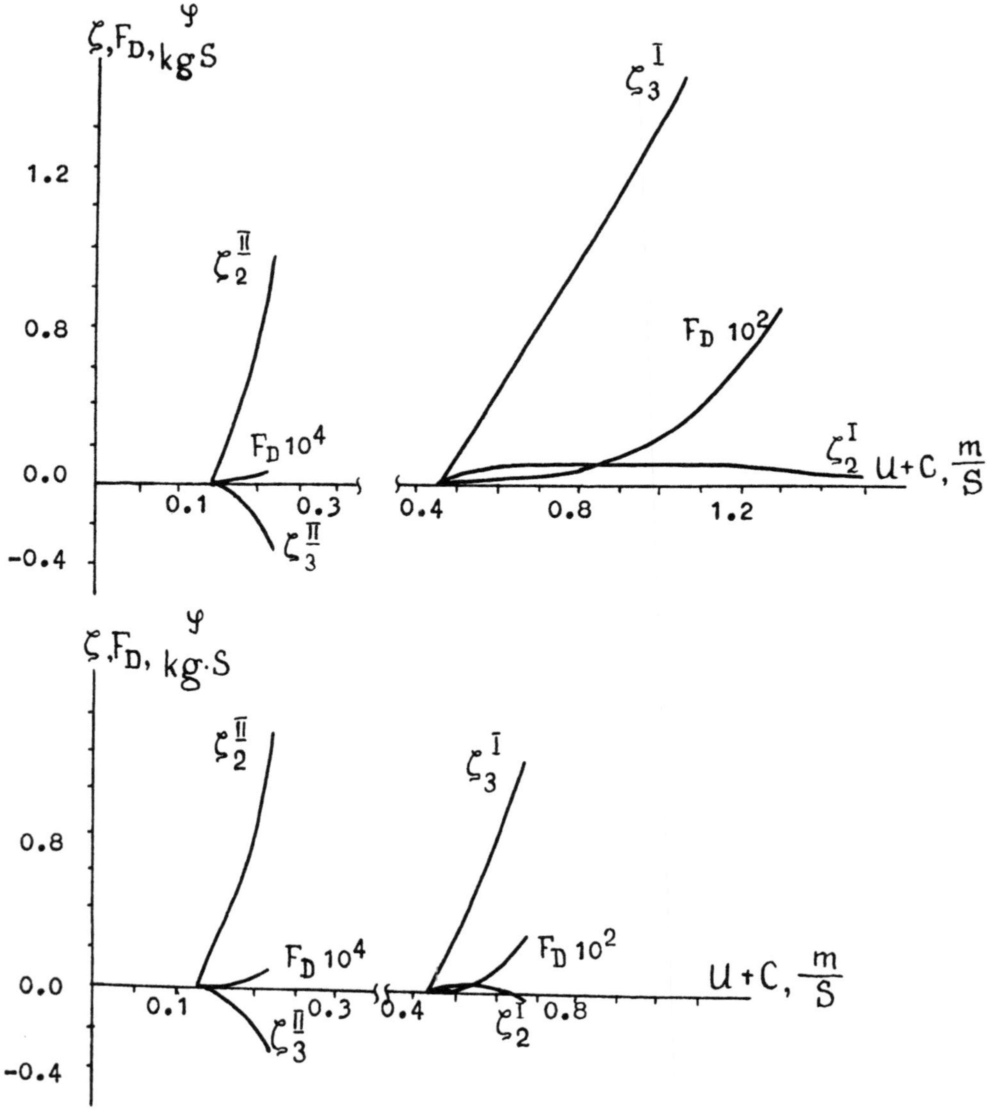
φ
ζ, F_D, kg S
1.2
0.8
0.0
-0.4
ζ₂ II
F_D 10⁴
ζ₃ II
0.1
0.3
0.4
0.8
1.2
ζ₃ I
F_D 10²
ζ₂ I
U+C, m/S
φ
ζ, F_D, kg·S
0.8
0.0
-0.4
ζ₂ II
ζ₃ I
F_D 10⁴
F_D 10²
U+C, m/S
0.1
0.3
0.4
ζ₂ I
0.8
ζ₃ II

Figure 6.4

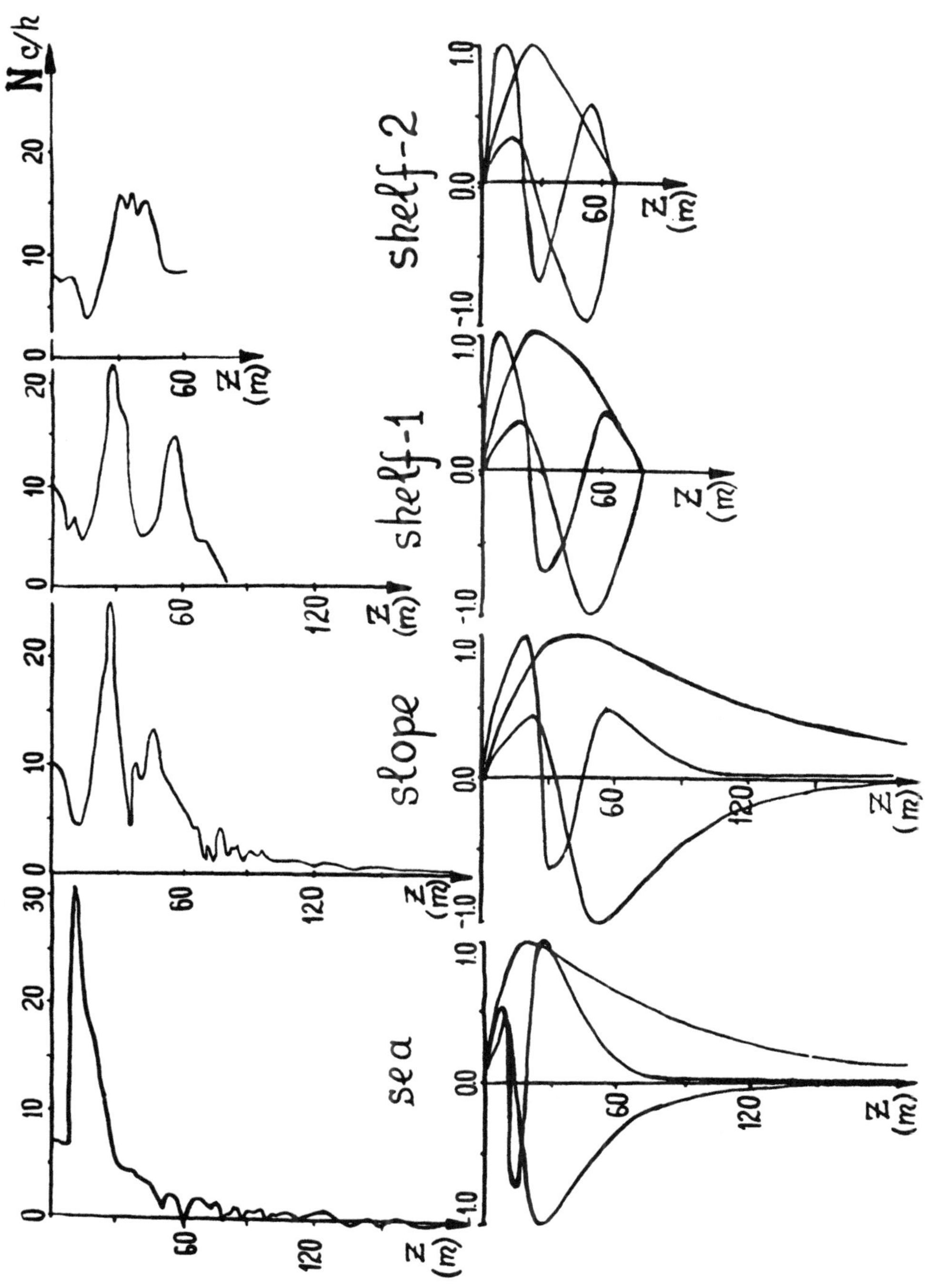

Figure 6.5

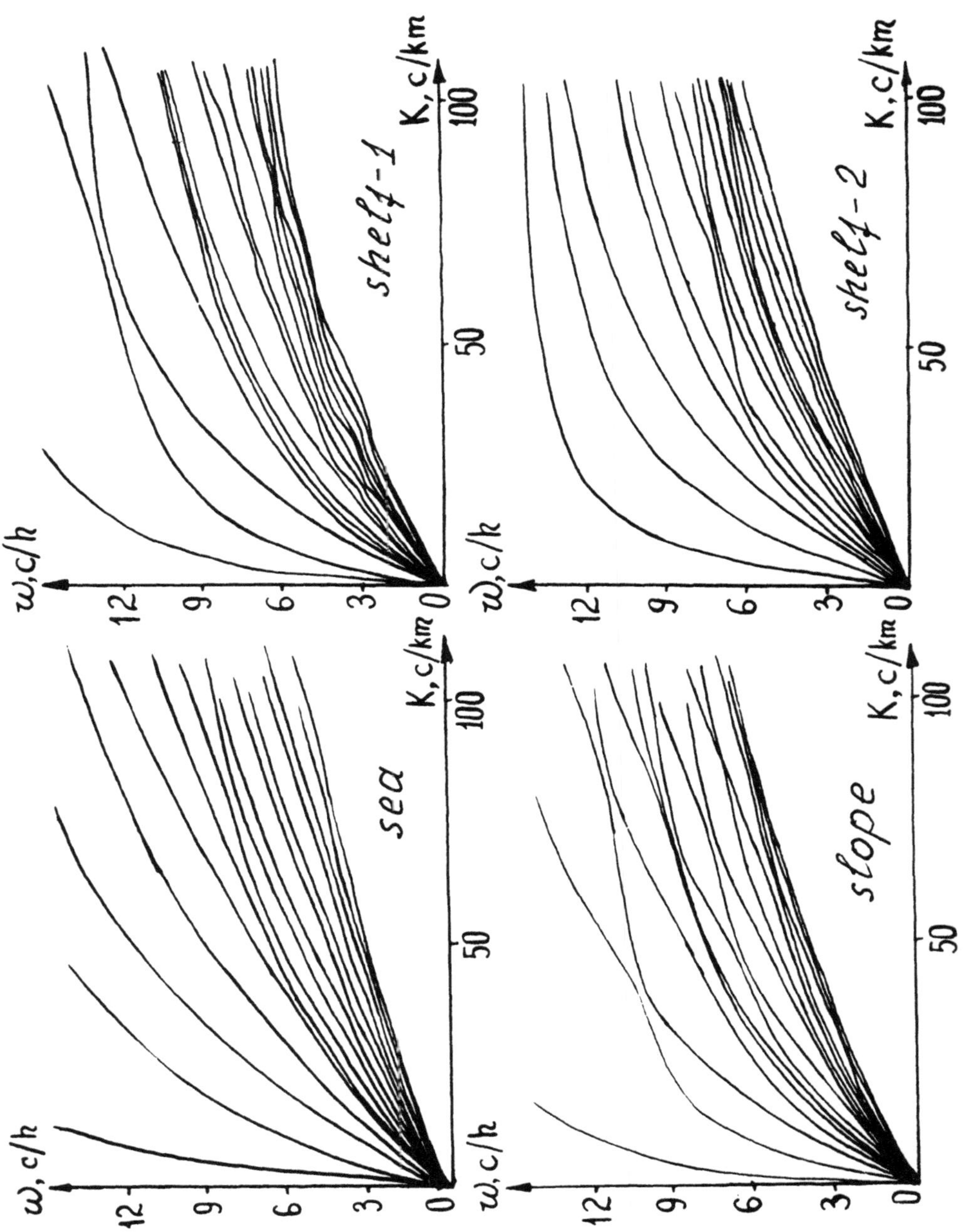

Figure 6.6

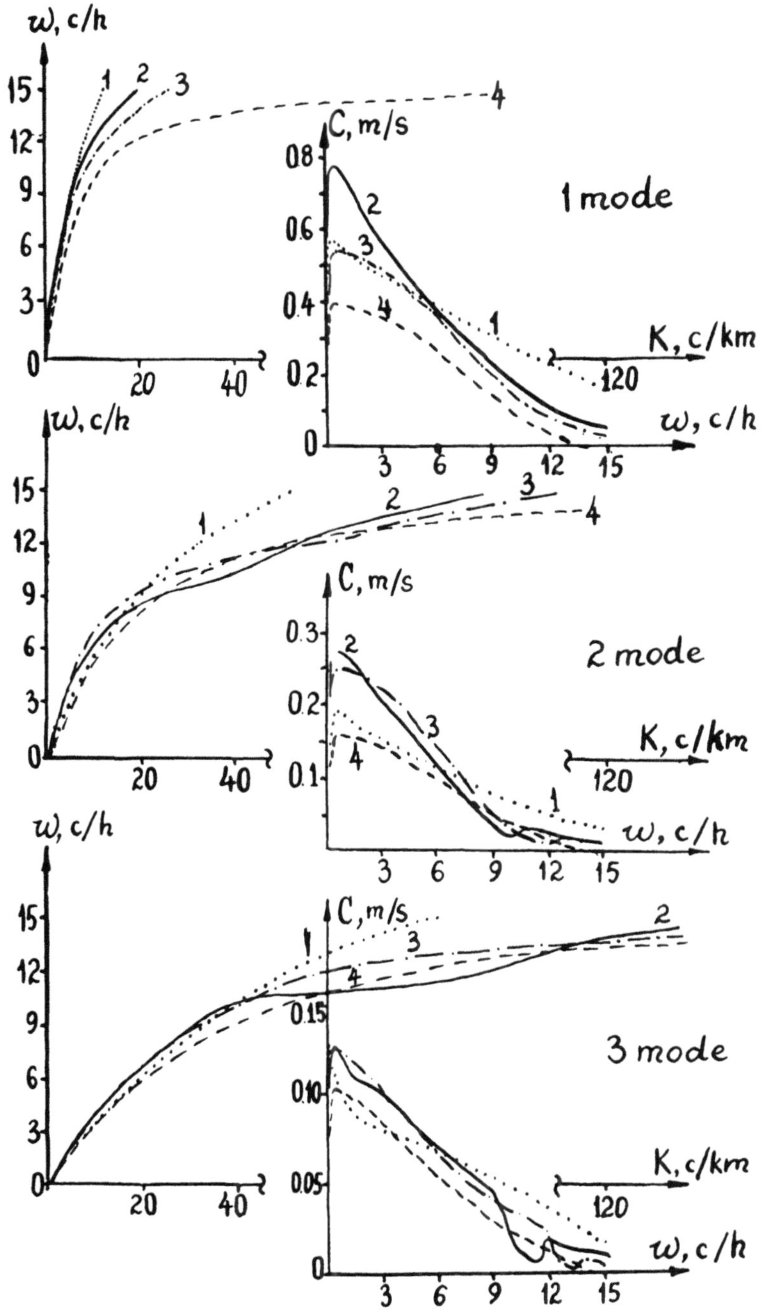

Figure 6.7

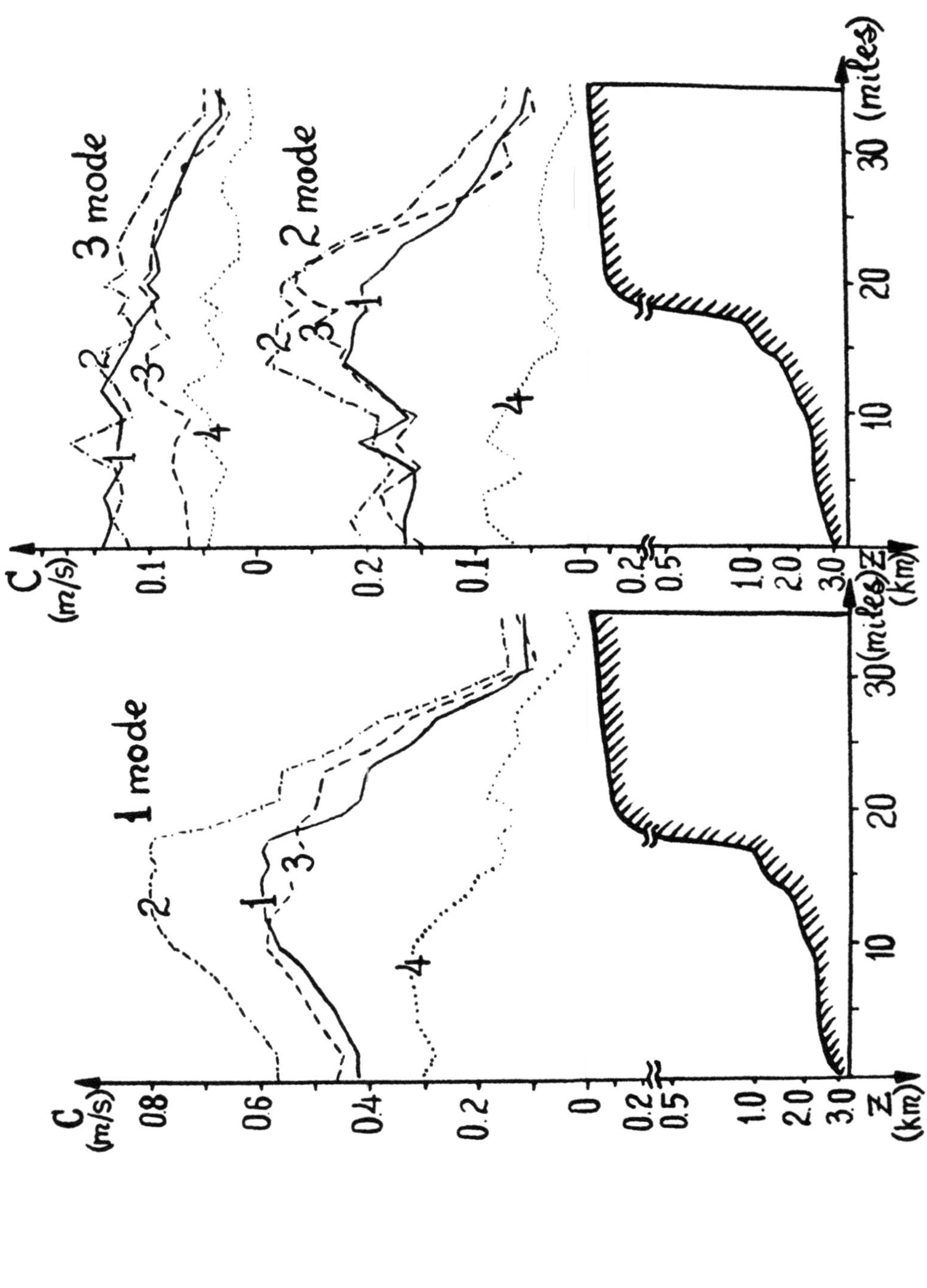

Figure 6.8

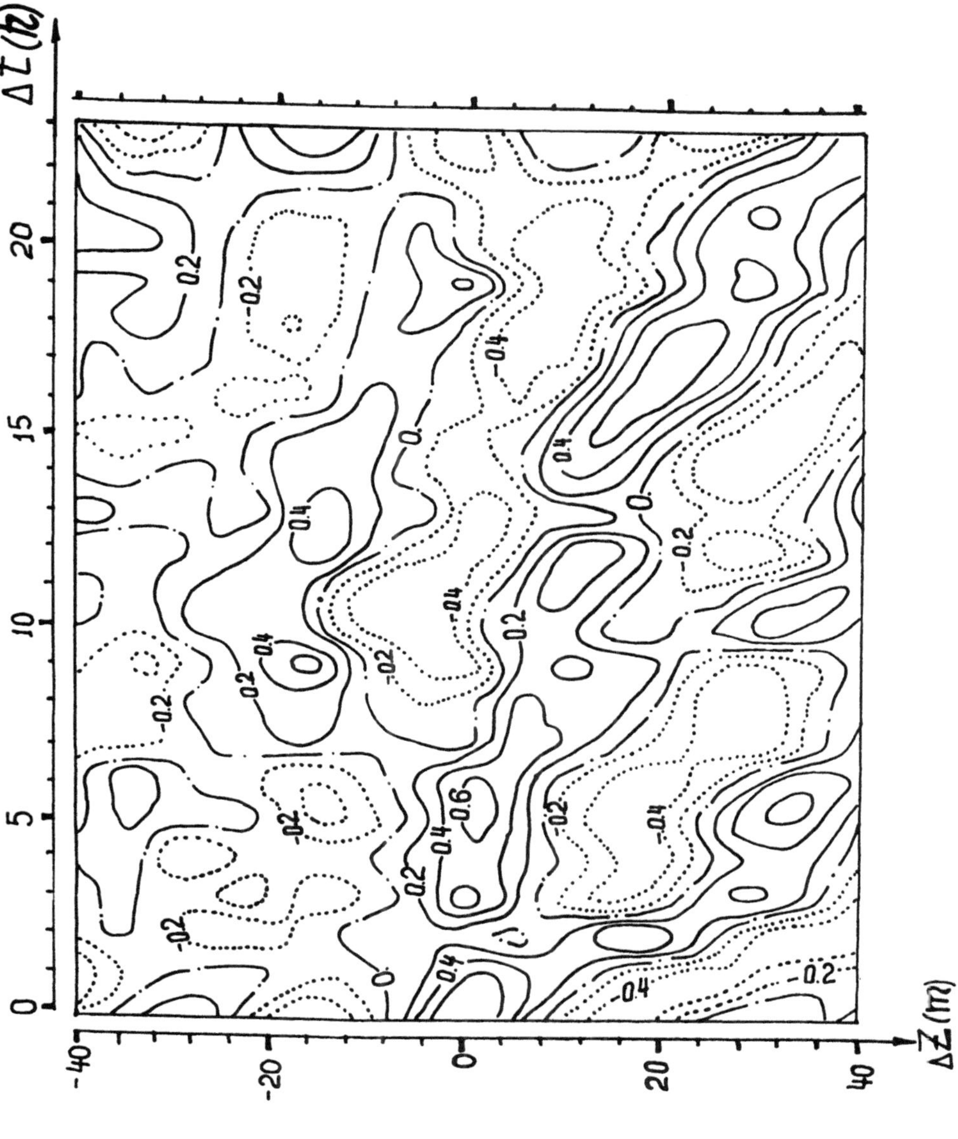

Figure 6.9

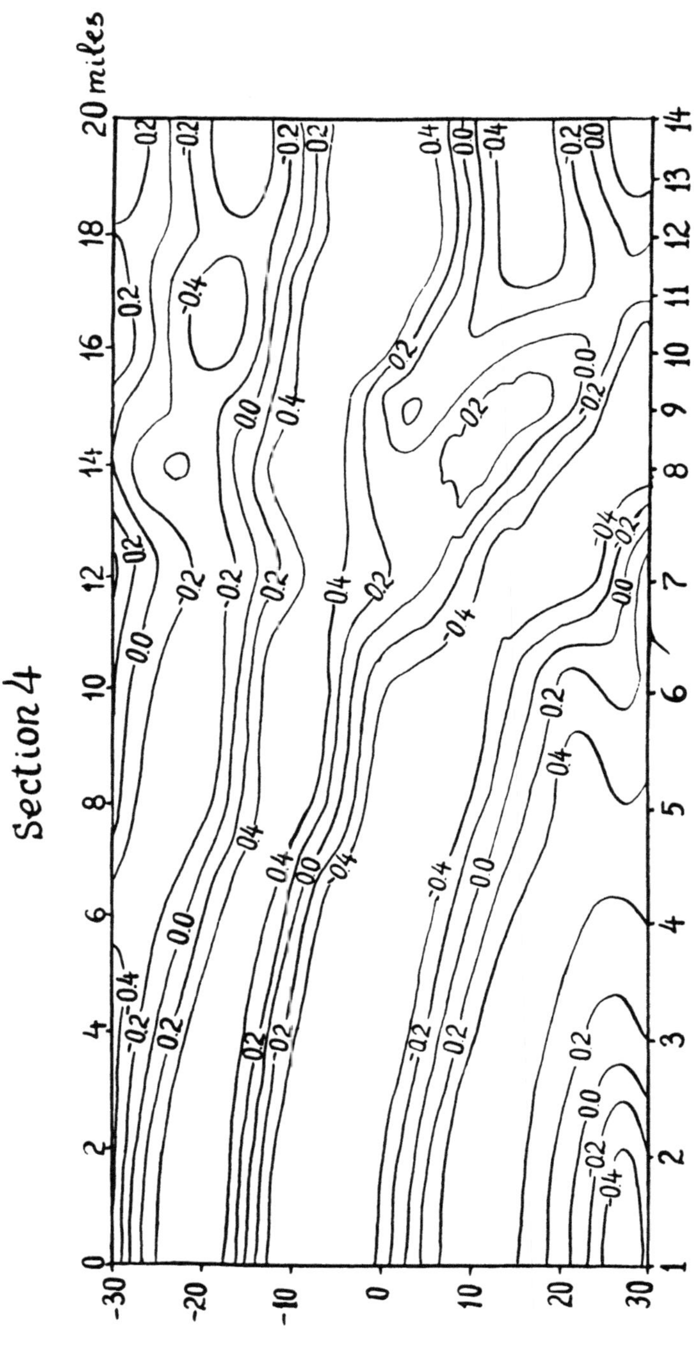

Figure 6.10

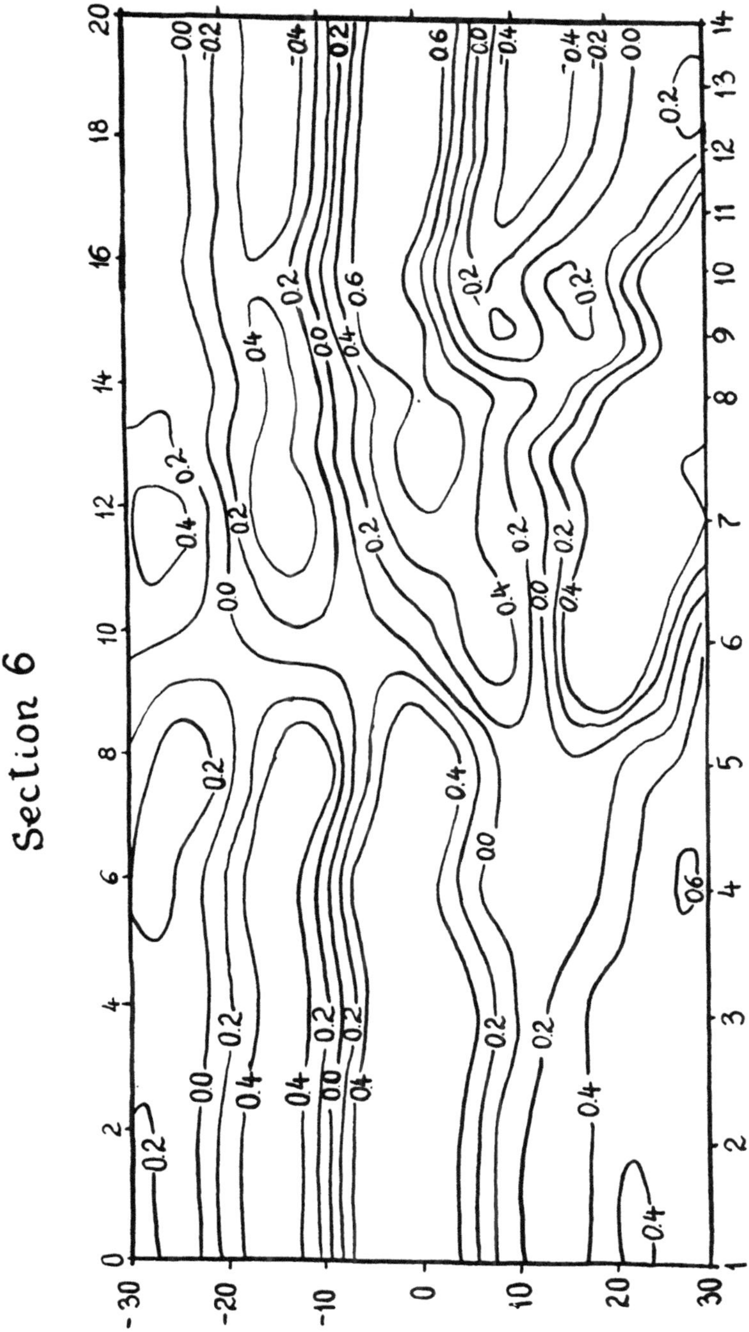

Let us briefly discuss the possible basic mechanisms of IW influence on vertical intermixing and TS formation.

1. IW breaking, turbulation, intermixing, FS formation as a result of mixed layers alternation with the stable ones. This is a traditional mechanism which was discussed in many studies (Musk, Garrett, 1973) and which was observed by Woods (1968) in the upper thermocline in the Mediterranean Sea. The specificity of this mechanism is due to the fact that to break IW one needs to concentrate their energy in thin layers and, if we exclude the upper boundary of the seasonal thermocline, this necessitates a preliminary presence of a fine structure.

2. Thermocline dividing into layers is possible due to mean deformations during nonlinear waves propagation. Corresponding theory of fine structure (FS) formation was suggested by Voronovich et a.(1976), nevertheless quantitatively one can obtain a better agreement with the data observed while using nonlinear Magaard waves (Magaard, 1965) like:

$$\zeta = \frac{\rho_0}{\partial \bar{\rho} / \partial z} \sin(kx - \omega t) \sin\{\frac{\tau_1}{H} (z - \xi)\} \tag{6}$$

Such nonlinear deformations without the account of molecular effects disappear after waves propagation and do not change general heat content of a stratified layer.

3. Because of nonmonotonic vertical structure of the internal waves with the nonmonotonic changes of their vertical amplitudes, they can modulate the effects of small-scale three-dimensional turbulence, if present. This process can be described using Kapitsa's effective wave coefficient (Navrotsky,1976). At values of the background coefficient of the order 1-10 cm^2/s FS is formed during the time of 10-20 wave periods.

4. Horizontal turbulence at scales of hundreds of meters can interact with horizontal gradients forced by larger-scale waves and this results both in a general increase of vertical transfer and in a nonmonotonic vertical distributions due to modulation by IW vertical structure (Ilyichev, Navrotsky,1987b).

5. Stokes transport in short IW superimposed on larger-scale horizontal gradients leads to the FS formation for about twenty four hours.

Indeed, for sine-shaped waves like:

$$\zeta = A(z) \sin\phi = a \sin mz \sin(kx - \omega t) \tag{7}$$

a horizontal component of Stokes transport

$$\overline{u_s} = \xi \overline{\frac{\partial \tilde{u}}{\partial x}} + \zeta \overline{\frac{\partial \tilde{u}}{\partial z}} = \frac{1}{2} a^2 m^2 c \cos 2mz \tag{8}$$

can reach the values of 5-10 cm/s for the internal waves observed in a shelf zone. By horizontal gradients in a shelf and frontal zones about 1 grad/km an additional temperature difference up to 1° on a vertical scale of one mode can arise in a day which results in a fine structure stratification which spectrum depends on the wave-number spectrum of the IW

6. Interaction of the short IW Stokes transport with the horizontal gradient field formed by long internal waves represent an evident generalization of the previous mechanism. Since mean horizontal gradients are small in an open sea with IW lacking, such parametric interaction of various scale IW might be the most universal mechanism of FE formation and of vertical transport increase in an open sea. Quantitative evaluations in (Ilyichev, Navrotsky, 1987b) give values of effective coefficients up to 50 cm²/s in the shelf area and up to 10 cm²/s in the ocean. It is possible to simulate this process by means of an equation for mean temperatures, local heat fluxes being calculated under linear harmonic decisions (zero approach of WKB method), while the heat flux divergence is calculated using the horizontal changes of an amplitude and a grave number (the next WKB approximation):

$$\frac{\partial \overline{T}}{\partial t} + \frac{\partial \overline{\tilde{u}_1 \tilde{T}}}{\partial x} = 0 \tag{9}$$

$$Q_x = \overline{\tilde{u}\tilde{T}} = -\frac{1}{2} A'Ac \frac{\partial \overline{T}}{\partial z} \tag{10}$$

$$Q_z = \overline{\tilde{w}\tilde{T}} = \frac{1}{2} A'Ac \frac{\partial \overline{T}}{\partial x} \tag{11}$$

$$A(x) \sim N^{-1/2}(x),\ k(x) \sim N^{-1}(x) \tag{12}$$

according to (12), A' $A\omega k^{-1}$ combination in (10) and (11) appears not to depend on x, and equation (9) is like the following:

$$\frac{\partial \overline{T}}{\partial t} + \frac{1}{2} a^2 m^2 c / \cos 2mz / \frac{\partial \overline{T}}{\partial x} = 0 \qquad (13)$$

Using the observed IW parameters, the solutions of this equation cover a vertical transfer with effective coefficients 10-50 cm^2/s as well as formation of fine vertical structure during several hours in the shelf zones and 20-30 hours in an ocean. Thus an increase of vertical transfer in a stratified ocean and formation of vertical fine structure can be produced by various complex processes, and axons of these internal gravitational waves may be predominant. It is obvious that vertical transfer of heat, oxygen, organic matter and other admixtures is necessary for high biological productivity in the ocean.

References

Djordjevich, V.D., Redecopp, L.G. (1978) The fission and disintegration of internal solitary waves moving over two-dimensional topography. J.pays.Ocean. 8,6,1016--1024 Ilyichev V.I., Navrotsky, V.V.(1987a) Vertical structure of hydrophysical characteristics and internal waves near shelf boundary. Doklady Ac.Nauk USSR, Vol.294, No.1, p.216-220.(in Russian)

Ilyichev V.I., Navrotsky, V.V.(1987b) Wave-induced transfer in stratified flours. Doklady Ac.Nauk USSR, Vol.294, No.2, pup.462-467.(in Russian)

Long, R.R. (1970)Blocking effects in flow over obstacles. Tellus 22, 471-480

Munk, W., Garrett, C. (1973) Internal wave breaking and micro structure. Boundary Layer Met. 4, 37-45

Magaard, L.(1965) Zur theorie zweidimensionaler nichtlinearer Wellen in stetig geschichteten Median. Kiel Meersforsch. 21, 22-31

Navrotsky, V.V.(1976) Internal waves and vertical temperature structure in the ocean. In: Specificities of structure and dynamics of Pacific waters. pp. 101-117, Vladivostok.

Sue, C.H.(1976) Hydraulic jumps in an incompressible fluid. J.Fluid Mach. 73,1,33-47

Voronovich, A.G.; Leonov, A.I.; Miropolsky, U.Z.(1976) To the theory of formation of fine structure of hydrophysical fields in the ocean. Okeanologiya 14, 5, 750-759

Woods, J.D.(1968) Wave-induced shear instability in the summer thermocline. J.Fluid Mach. 32,4,791-800.

V.V. Navrotsky and A.Yu. Lazaryuk
Pacific Oceanological Institute
USSR Academy of Sciences, Far East Branch
Vladivostok
USSR

7 A NUMERICAL MODEL OF THE FORMATION OF COLD FRONT AND ITS DERIVED CIRCULATION ON THE SHELF

Tangdong Qu and Dunxin Hu

Abstract

Based on the conceptual model given by Hu (1986a,b) on frontcaused upwelling in the Yellow Sea, a two dimensional idealized numerical model is proposed in this paper to simulate and describe the physical process of the formation of cold front and its derived circulation, in which only surface heating, thermally driving and tidal mixing are taken into account. As a result, calculated sectional distribution of water temperature, horizontal and vertical circulation patterns are presented, which are well consistent with observations and Hue's conceptual model. It is concluded that cold front and hence derived upwelling must exist during heating seasons, due to horizontal difference of mixing strength on the shelf like in the Yellow Sea.

7.1 Introduction

Since about 14 years ago, when Simpson and Hunter (1974) first came to the conclusion that oceanic front can be generated by tidal mixing in the Irish Sea, more and more papers have turned out on the tidal mixing-caused front (Fearnhead, 1975; Simpson et al., 1977; Garrett et al., 1978; Griffiths et al., 1981 and Zhao, 1985). Hu (1986a and 1986b) proposed a mechanism for front-caused upwelling on the basis of measurements of temperature in the Yellow Sea. As he described, the Yellow Sea cold water mass in summer in the S Yellow Sea, the bottom of which is of concave shape, is located in the central part of the sea and formed with seasonally heating from the surface. In addition, the tidal current in the Yellow Sea is much stronger in the shallower part than in the deeper part and then in the shallower portion the strong tidal current causes water well mixed vertically down to the bottom, so that for the surface layer colder water occurs somewhere in between coastal water and offshore water, both of which are warmer. As long as the cold zone (front) is formed due to tidal mixing, two thermally-driven currents in opposite directions must be induced, which causes divergence in the upper layer and hence upwelling. The whole process can be summarily described as follows. Tidal mixing (kinetic energy converted into potential energy) → cold front in surface layer thermally-driven currents (potential energy converted into kinetic energy) divergence and upwelling. The present paper is aimed at mathematical simulation of the process with a two dimensional numerical model.

7.2 Mathematical model

1. Cartesian coordinates are used with x-axis positive from the centre of the basin to the coast and z axis positive upward from the sea surface. As an idealized model, the topography of the sea floor is simplified as the following function:

V. I. Ilyichev and V. V. Anikiev (eds.), Oceanic and Anthropogenic Controls of Life in the Pacific Ocean, 109–115.

$$H(x) = h\left(1-\frac{x}{x_0}\right),$$

where h is the largest water depth at the centre, which is assigned as 80 meters; xo is half width of the basin, which is assigned as 100 kilometres.

2. The turbulent coefficients of heat diffusion and the kinematic viscosity are of much importance to any models, which, however, are usually hard to be determined. Previous studies (Ding, 1985; Zhao, 1985) showed that the nearer to the coast the water is, the stronger is the tidal mixing. And obviously then, the stronger is the tidal mixing, the larger the coefficients are. On the other hand, vertical distribution of these coefficients is most likely an exponential function, and much smaller in the middle layer water than in the upper layer and bottom layer waters.

In the present paper the functions of the coefficients with respect to z are assumed as follows, which are basically agreed with what have been obtained by Guan (1963) and Yuan (1979).

Vertical heat diffusion coefficient Kz=1O(exp(8z/h)+exp(-4(h+z/h)) (cm^2 sec)
Horizontal heat diffusion coefficient Kx=30000 Kz (cm^2/sec)
Vertical kinematic viscosity Az=2Kz (cm^2/sec)
Horizontal kinematic viscosity Ax=50000Az (cm^2/sec)

3. Heating (cooling) rate during warming (cooling) seasons is assumed as follows:

$$Q = Q_0 \sin\left(\frac{2\pi t}{3600\times 24\times 360}\right) \ (cal/cm^2.sec)$$

where Q_0 is assigned as 0.002 cal/cm^2 sec.

4. Equations
Through comparison of order of magnitude of various terms in N-S equations, the governing equation system, including equation of continuity, equation of motion, and T-equation is obtained as follows, where the velocity in y direction, which is perpendicular to X-Z plane, is in geostrophic balance, and the equation of motion in z direction is of hydrostatic approximation.

$$\frac{\partial U}{\partial x} + \frac{\partial W}{\partial z} = 0$$

$$fv = \frac{1}{\rho_0} \cdot \frac{\partial P}{\partial x}$$

$$fu = \frac{\partial}{\partial z}(Az\frac{\partial v}{\partial z}) + \frac{\partial}{\partial x}(Ax\frac{\partial v}{\partial x})$$

$$\frac{\partial P}{\partial z} = \rho_0 g \ (1-\alpha T)$$

$$\frac{\partial T}{\partial t} + U\frac{\partial T}{\partial x} + W\frac{\partial T}{\partial z} = \frac{\partial}{\partial z}(Kz\frac{\partial T}{\partial z}) + \frac{\partial}{\partial x}(Kx\frac{\partial T}{\partial x})$$

where U, V and W are the velocities in x, y and z directions, respectively. T-water temperature, T-T_0+ Δ T. where T_0-an arbitrary constant and Δ T temperature deviation from T_0.
α = 0.00018/°C, f = 0.0000727/sec, g = 980 cm/sec^2, ρ_0 = 1.02 gram/cm^3 average density/

5. Boundary conditions:
With the assumption of the basin symmetry about z-axis (the centre), sufficient to examine what would happen in only half of the basin - x-positive area. And then we have

x=0 (at centre of the basin):

$$\frac{\partial U}{\partial x} = \frac{\partial V}{\partial x} = \frac{\partial W}{\partial x} = 0, \quad \frac{\partial T}{\partial x} = 0$$

z=O (on the sea surface, where there is no wind and only surface heating is taken into account):

$$\frac{\partial U}{\partial z} = \frac{\partial V}{\partial z} = \frac{\partial W}{\partial z} = 0, \quad \rho_0 = C_p Kz \frac{\partial T}{\partial z} = Q$$

where C_p is referred as 0.934 cal/gram °C

z = H(x) (on the bottom, where there is no heat exchange and completely viscous condition is adopted):

$$U = V = W = 0, \quad \frac{\partial T}{\partial x} \cdot \frac{\partial H}{\partial x} + \frac{\partial T}{\partial z} \cdot \frac{\partial H}{\partial z} = 0$$

The initial condition is given as follows:

$t = 0; \qquad U = V = W = 0, \quad T = T_0$

7.3 Numerical results and conclusions

Figure 7.1: Temperature Section, July, 1983 (From Hu, 1986b).

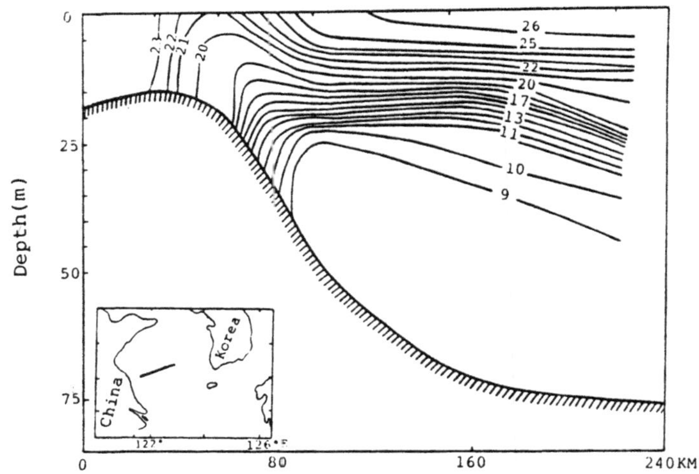

An alternative implicit finite difference scheme is used. The calculated numerical results are depicted in Fig. 3 - 5, and summarized as follows:

1. On Fig. 3 the calculated sectional distribution of ΔT, water temperature deviation, looks well consistent with the distribution of measured temperature (Fig. 1). Near the coast isotherms are vertically oriented, which means that the water is vertically homogeneous there because of strong tidal mixing, and temperature gets lower as it goes from the coast to the deeper sea because of water column length being horizontally different. Near the central part, however, due to weak tidal mixing,the water is stratified, the upper homogeneous layer is shallower and hence temperature is higher than in the surroundings. Between the central part and the coastal part a cold front is found, in which thermally-driven upwelling is expected to take place.

Figure 7.2: Temperature Distribution at 10 m level, July, 1983 (Heavy-arrow-deduced circulation) (From Hu, 1986b).

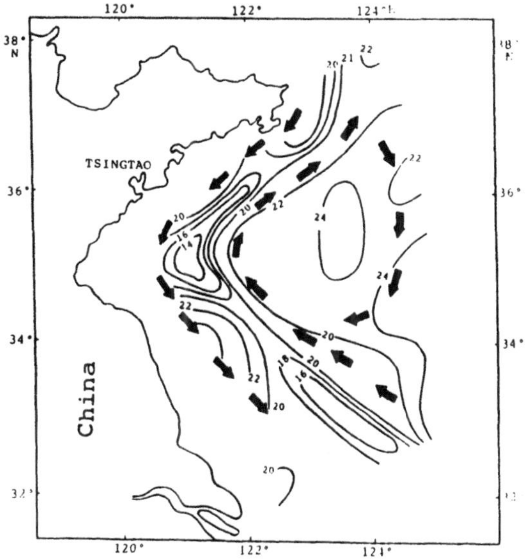

2. Calculated velocity in y direction is depicted at Fig. 4. There are two zero-velocity-lines, one of which is around 20 km from the centre of the basin, another one is around 60 km from the centre. Compared with Fig. 3 the right zero-velocity-line in Fig. 4 is corresponding with the front centre in Fig. 3. Right of the zero-velocity-line, flow is inward with maximum velocity of 13 cm/s, while to the left from it, the current is flowing outward with less speed and maximum velocity of 11 cm/s. Left from the left zero-velocity-line, a very weak inward flow seems to exist. In summary, if we consider the basin as a whole, a cyclonic gyre near the coast and a anticyclonic gyre in the inner part of the basin should exist, which causes upwelling in between and cold front is maintained. Around 20 km from the centre of the basin a downwelling is found, which corresponds to zero-velocity-line and near the centre of the basin an upwelling again ambiguously appears. The whole picture of the circulation is quite consistent with Hu's conceptual model (Fig. 2) with an exception of the cyclonic circulation near the centre.

Figure 7.3: Calculated Sectional Distribution of Temperature Deviation ΔT.

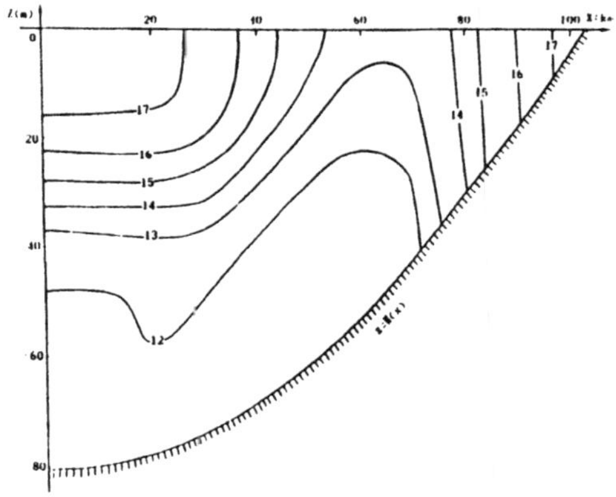

Figure 7.4: Calculated sectional distribution of velocity in Y direction (cm/s).

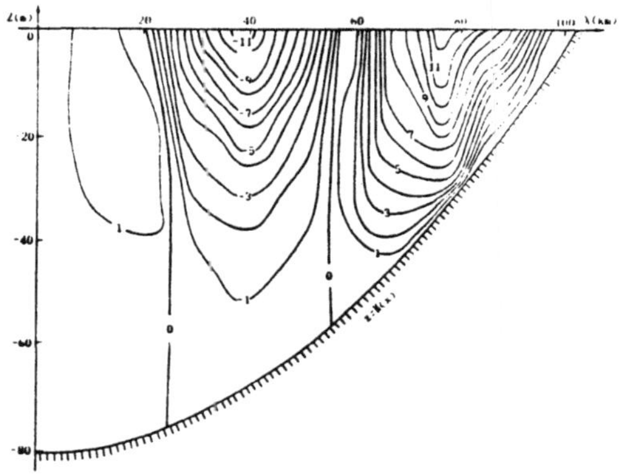

Figure 7.5: Calculated vertical circulation.

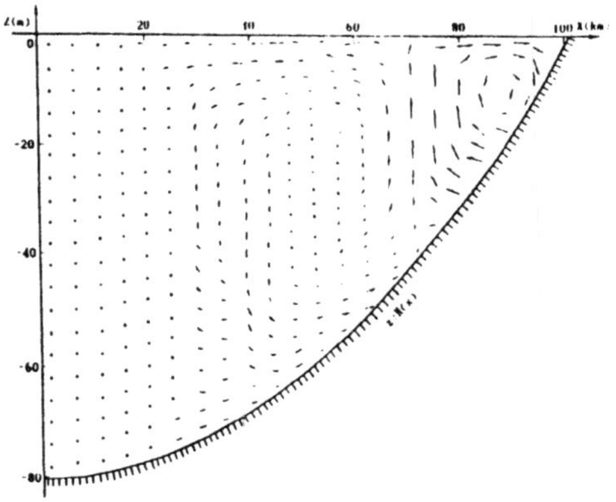

3. Fig. 5 is the vertical circulation in the X-Z plane. It can be easily seen that an upwelling exists in the area between 60-80 km from the centre and a downwelling is found between 20-10 km, which are roughly consistent with the zero-velocity-lines in Fig. 4 and the cold front in Fig. 3. The upwelling near the centre inferred from Fig. 4 is ambiguous.

All the results, which quite agree with the conceptual model given by Hu (1986a,1986b), show the presented idealized model basically works for simulating of the formation of cold front and its derived circulation on the shelf. Of course, the present model should be refined with to describe in details the features and processes relared to front-caused circulation on the shelf.

References

1. Ding, W.L.: The characteristics of the tides and tidal currents in the Bohai and the Yellow Seas. Studia Marina Sinica 25, 27-40 (1985)
2. Fearnhead, P.G. : On the formation of fronts by tidal mixing around the British Islands. Deep-Sea Res. 22(5). 311-322 (1975)
3. Garrett, C.J.R., Keeley, J.R., Greenbery, D.A. : Tidal mixing versus thermal stratification in Bay of Fundy and Gulf on Maine. Atmosphere-Ocean 16(4). 403-423 (1978)
4. Griffiths. D.K., Pingree. R.D., Sinclair, M. : Summer tidal fronts in the near-arctic regions of Foxe Basin and Hudson Bay. Deep Sea Res. 28(8). 865-873 (1981)
5. Guan B.X. : A preliminary study of the temperature variations and the characteristics of the circulation of the cold water mass of the Yellow Sea. Oceanogia et Limnologia Sinica 5(4), 255-284 (1963)
6. Hu, D.X. : Front-caused upwelling in the Yellow Sea. In: Proceedings of SCOR Symposium on Interfaces in the Ocean. CSIRO Marine Lab., Hobart, November, 24-26, (1986a).
7. Hu, D.X. : Some striking features in shelf circulation in the Yellow Sea and East China Sea. In: Tseng, C.K., Tomczak, M. (ed.) , Oceanography in the Bohai Sea. Huanghai Sea and East China Sea. (Manuscript under review) (1986b).
8. Simpson, J.H., Hunter. J.R. : Fronts in the Irish Sea. Nature 250, 104- 406. (1974)
9. Simpson, J.H., Hughes. D.G., Morris, N.C.G. : The relation of seasonal stratification to tidal mixing on the continental shelf. In Angel. M. (ed.) A Voyage of Discovery, Suppl. to Deep Sea Res ., 327-310 . (1977) .
10. Yuan, Y.L. : A preliminary study on the circulation related to the cold watermass of the Yellow Sea. 1. the thermal structure and the characteristics of the circulation in the central part of the Yellow Sea. Oceanologia et Limnologia Sinica 10(3) 187-199 (1979)
11. Zhao. B.R. : The fronts of the Huanghai Sea Cold Water Mass induced by tidal mixing.
Oceanologia et Limnologia Sinica 16(6), 452-460 (1985)
12. Zhao. B.R.: A preliminary study of continental shelf fronts in the western part of southern Huanghai Sea and circulation structure in the front region of the Huanghai Cold Water Mass (HCWM).Oceanolgia et Limnologia Sinica 18(3). 218-226 (1987)

Tangdong Qu and Dunxin Hu
Institute of Oceanology
Academia Sinica
Qingdao
China

8 HYDROLOGICAL FRONTS OF THE NORTHEAST PACIFIC OCEAN IN AUTUMN SEASON

G.I. Yurasov

Abstract

On the basis of STD measurements performed in SeptemberOctober 1986 at five meridional sections from 180-140°W to 50-35°N, the analysis of the water structure and isolation, classification and determination of front positions was carried out. The main types of water structure in the regions are subarctic and subtropical ones separated by the subtropic front which belongs to the area near 40°N. Zoning of the water area according to 6 types of water structures was made and the main and 5 intermediate thermohaline fronts were isolated as zones of elevated horizontal gradients.

8.1 Introduction

The area of the Pacific Ocean between 30-50°N east of 180°, except for a region adjoining the N America, still remains the least investigated. By 1961 the total number of observations within 1O-degree squares had been 30 to 60, mostly in summer season (Leontieva, 1968). In the subsequent period (Favorite et al., 1976), up to 1970 the data was not replenished much, especially for autumn season. In the recent years this situation did not change either, although a series of surveys to small water areas was carried out.

In relation to this, analysis of the quasi-synchronous hydrological survey data of the immense water region (30 to 50°N, 180-140°W), conducted aboard the R/V Akademik Lavrentiev from September 16 to October 16, 1986, may apparently be of scientific interest in the study of this dynamic and contrasting region of the Pacific. Measurements made with MARK-III Sounder in the 0-1200 m layer on transects along 180°,170°,160°,150°,and 140° W, made enabled to examine the global features of structure, identification, and position of hydrological fronts and as well as to approach the rearrangement of the related water.

8.2 The recent views on the region examined

Zonal distribution of hydrological characteristics in high ocean areas is highly affected by considerable meridional variability of solar radiation fluxes, arriving through the surface, as well as by evaporation and precipitation. In their turn large-scale features of wind stress, oceanic currents and bottom topography (Levine, White, 1983) make these meridional gradients concentrate in the corresponding narrow zones, forming thereby oceanic fronts. Reaching the depths of several hundreds of metres, the latter divide large stretches of the ocean which differ in slight variations of temperature-salinity characteristics, inherent in certain structure types having uniform dependence by vertical (Dobrovolsky, 1961; 1958).

At present, the structure and position of frontal zones have received a detailed study based on vast material of long-term measurements for the NW Pacific only (Bulgakov, 1972). As to the NE region, inadequate data of measurement of thermohaline water characteristics have not enabled one, even until recently, to make similar generalisations, despite the attempts to analyze the material available

V. I. Ilyichev and V. V. Anikiev (eds.), Oceanic and Anthropogenic Controls of Life in the Pacific Ocean, 117–128.

(Favorite et al.,1976). It should be noticed that both works mentioned above were focussed on the analysis of water mass distribution features for summer and winter seasons only, whereas the fronts were represented as lines fringing an immense area of interaction of waters of subarctic and subtropical structures from N and S.

In addition to monographic works (Leontieva, 1968; Favorite et al., 1976) water structure and dynamics in the N Pacific were the subject of numerous publications (La Fond, La Fond ,1977; Levine, White, 1983; Lynn, 1986; Niiler, Reynolds, 1984; Saur, 1980; Roden, 1971, 1974, 1977, 1980a, 1980b) discussing different aspects of these phenomena. Proceedings from these publications which are based on the data of separate expeditions in restricted areas in different seasons of the year, the issues of structure, identification, classification and position of fronts, as well as of circulation and water structure, have been partly considered.
According to those studies, one of the major hydrological fronts in this area is Subarctic Front (SAF), setting a S limit (40-43°N) to subarctic waters of low temperature and salinity in surface layers and with a halocline at 150 depth. At the surface the SAF coincides with isohaline 33.88‰. Another major front is subtropical (STF), separating the water of subtropical structure from S and located at 30.5°N. It can be identified in winter at the surface by isohaline 34. 8‰ and isotherm 18°C (Roden, 1977).

More detailed surveys of the last years enabled identification (Roden, 1930a) of the subtroprical front between meridians 152-158°W at 31. 5°N and the second front at 34°N which is referred to by Roden as "Front 34.5°N and can be identified at the surface in winter by isohalines 34.4-34. 5‰ and isotherm 16°C. Analysis of the mass XBT measurements in the area 30° to 45°N, 160°E to 160°W (Levine, White, 1983) has shown that in this area the thermal fronts form as a result of interaction of the Kuroshio extension with bottom topography which forms the front branch at first over the Shatsky Rise and then another branch in the area of Emperor Mountains. Position of those fronts between the meridians 140-135°W south of 40°N was investigated with the help of a series of successive surveys in June 1972-1975 (Roden, 1977). In this study the author proposes a new name "the north subtropical front" instead of front 34°N" (Roden, 1980) and surmises that it exists on the water surface of the entire N Pacific.

Thus, analysis of even incomplete list of works above enables us to draw a conclusion that the recent concepts on the frontogenesis and structure of waters in the central and eastern areas of the N Pacific have not yet been formed and any attempt to develop them will be useful.

In such light the analysis of hydrological features of the NE Pacific, tryed in the present paper, should be regrded.

8.3 Hydrological fronts according to the data of expeditionary research in September-October, 1988

Hydrological fronts are characterized by drastic changes in thermohalinic characteristics, when crossed, and hence they can be isolated either by absolute values or by standard deviation in horizontal gradients (Levine, White, 1983). The first criterium is applied gfor analyzing the data of a particulate survey, while the second one presupposes the use of data file of measurements within a given water area over a specified and sufficiently large time period.

We will consider a front to be a zone of increased horizontal gradients of temperature and salinity with respect to their mean values (over the area measured) which can be taken as a climatic standard. Calculation of climatic gradient over the water area covered by experimental operations have shown that it fluctuated in the 0-500 m layer and decreased with depth within the range of (1.26-0.446).10^{-2}°C/mile and (2.60-0.12).10^{-3} ‰/mile for temperature and salinity, respectively. Calculations for each pair of neighbouring stations on transects enabled us to reveal areas where zonal gradients exceeded significantly the climatic ones. Each transect showed five such sites which were subsequently identified as major and intermediate fronts. At the fronts, the values of temperature and salinity gradients in the layer considered (see Table 8.1) are (8.0-0.7).10^{-2} ‰ and (14.2-0.2)‰ per mile respectively, decreasing both with depth (similar to the climatic front) and along the fronts eastwards. At 150-140°W frontal temperature contracts are at their minima to indicate that Californian Current breaks up the zonation in the distribution of hydrological parameters specific to a region at large.

The largest horizontal gradients are associated with the front of 40°N. They exceed the climatic gradients almost by an order of magnitude. At the rest fronts the gradients are less though exceeding the climatical ones by two to four times.

Since at present the classification of frontal zones is rather conventional even for the major fronts (e.g. the NW Pacific Front which divides the waters of subarctic and subtropical structures is referred to as "polar front" (Suda, 1936), "subarctic front" (Dobrovolsky, 1958), "subarctic convergence" (Sverdrup et al., 1960), "Kuroshio front" (Levine, White, 1983), etc.) we shall advance classification of our own which takes into account both the qualitative rearrangement of temperature and salinity fields and geographical characters.

On this basis, six hydrological fronts have been isolated in the region studied (see Figure 8.1b):

- Northern Subarctic Front, NSAF
- Intermediate Subarctic Front, ISAF
- Quasi-Subarctic Front, QSAF
- Subarctic Front, SAF
- Northern Subtropical Front, NSTF
- Intermediate Subtropical Front, ISTF

Table 8.1: Maximum values of horizontal gradients of temperature and salinity at hydrological fronts according to measurements on transects.

Front name	Temperature gradients/x10^{-2} °C					Salinity gradient/x10^{-3} $^{0}/_{00}$/mile					Climatic values	
	180°	170°W	160°W	150°W	140°W	180°	170°W	160°W	150°W	140°W	Grad.T	Grad.S
NSAF	5.5	-	-	-	-	9.1	-	-	-	-		
ISAF	4.1	4.0	2.6	3.3	3.7	9.4	6.4	4.5	6.5	5.4		
SSAF	6.0	3.6	2.8	3.5	3.4	14.2	8.0	8.2	7.9	5.4		
SAF	7.9	4.5	6.2	3.9	3.9	13.2	12.0	14.1	9.6	9.4	0.5-1.3	0.1-2.6
NSTF	7.3	5.2	4.5	2.4	1.6	7.9	7.6	3.1	10.4	8.2		
ISTF	-	1.8	3.4	2.9	4.3	-	3.0	14.3	5.5	12.9		

Figure 8.1: Structural zones (a) and hydrological fronts (b) in the area of study. 1- boundaries of different types of water structure; areas of propagation of 2 - cold intermidate layer, 3 - subsurface salinity maximum, 4 - subsurface salinity minimum.

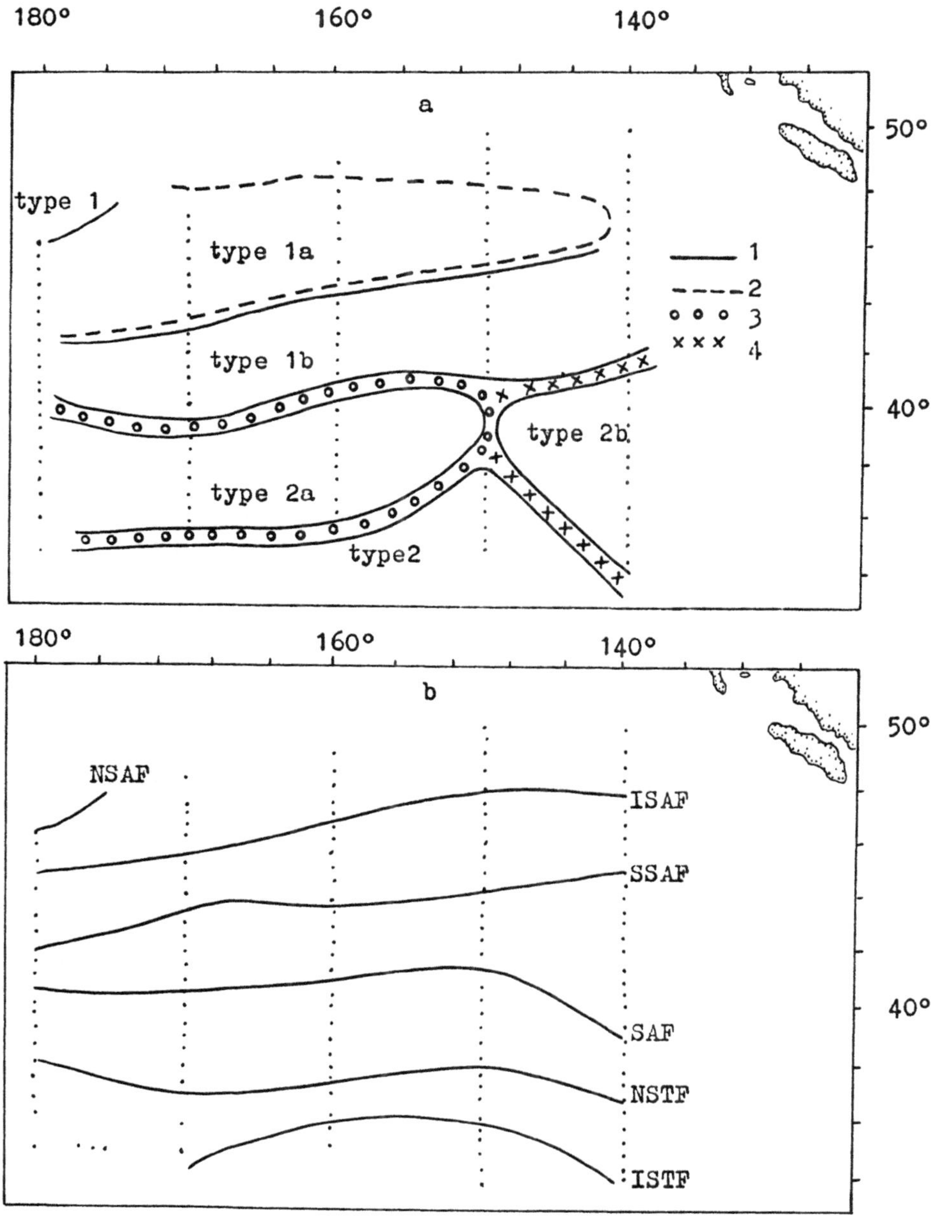

Isolation of so many fronts within a restricted water area of the high ocean may confuse as even for the most dynamically active NW Pacific (up to 160°E) the detailed analysis of the data of both long-term measurements and individual expeditions (Bulgakov,1972) enabled to isolate two fronts only. However, the recent research (Levine, White, 1983) showed that east of 160°E, under the control of bottom topography in the area of the Shatsky Rise and next in the vicinity of the Emperor Mts the Southern Subarctic Front (see Bulgakov, 1972, for nomenclature) branches to form four fronts, identified within lat. 30°-40°N, the positions of which fit well with the frontal divides (SAF,NSTF,ISTF) of the S part of the area discussed in the work. The similar effect of bottom topography on the 'Northern Subarctic Front" in the area of the Emperor Mountains seems also to contribute to its branching and formation of zones of additional fronts N of 40°N (QSAF, ISAF, NSAF).

To illustrate the rearrangement of the vertical water structure related to the availability of fronts, in Figure 8.2 we give a T,S-field transect along 170°W, built from the data of our measurements. As follows from the Figure, the fronts are well recognizable in the 0-500 m layer. In this case to the north from the QSAF one can reliably distinguish all intermediate cold layer at the depth of 100 to 150 m against the monotone increase in salinity with depth which is specific to the subarctic type of water structure. At the same time, to the S from the SAF where the vertical distribution of temperature is monotonous intermediate salinity extremes are found; a subsurface maximum at 50 m and the intermediate minimum as a distinctive feature of the subtropical type of water structure, at 400-550 m, between QSAF and SAF, vertical temperature and salinity distribution does not show features of the type of water structure mentioned above which is typical for the mixing zone.

Thus, the main front, isolated in the area of the expedition research, is subarctic. It is characterized by the biggest horizontal gradients and at the same time is closely connected with the structural transformations in the water column, dividing waters of subarctic and subtropical structures. The position of the SAF (see Figure 8.1) coincides completely with the boundary contouring the subarctic subtypes of water structure from the S (Figure 8.3, types 1, 1a, 1b) and subtropical ones from the N (Figure 8.3, types 2, 2a, 3b). A similar close relation to water structure is observed for intermediate fronts occupying the northernmost (NSAF) and southernmost (ISTF) positions. The first isolates waters of subarctic structure of the W variety (Type 1) and the second of the subtropical structure (Type 2). Other fronts pass within the ranges of propagation of different subtypes of the main structures and are not closely connected with their boundaries. Division into the major and intermediate fronts agrees well with the recent concept for this problem (Fyodorov, 1983).

Figure 8.2: Vertical transect through the field of temperature (a) and salinity (b) along 170°W.

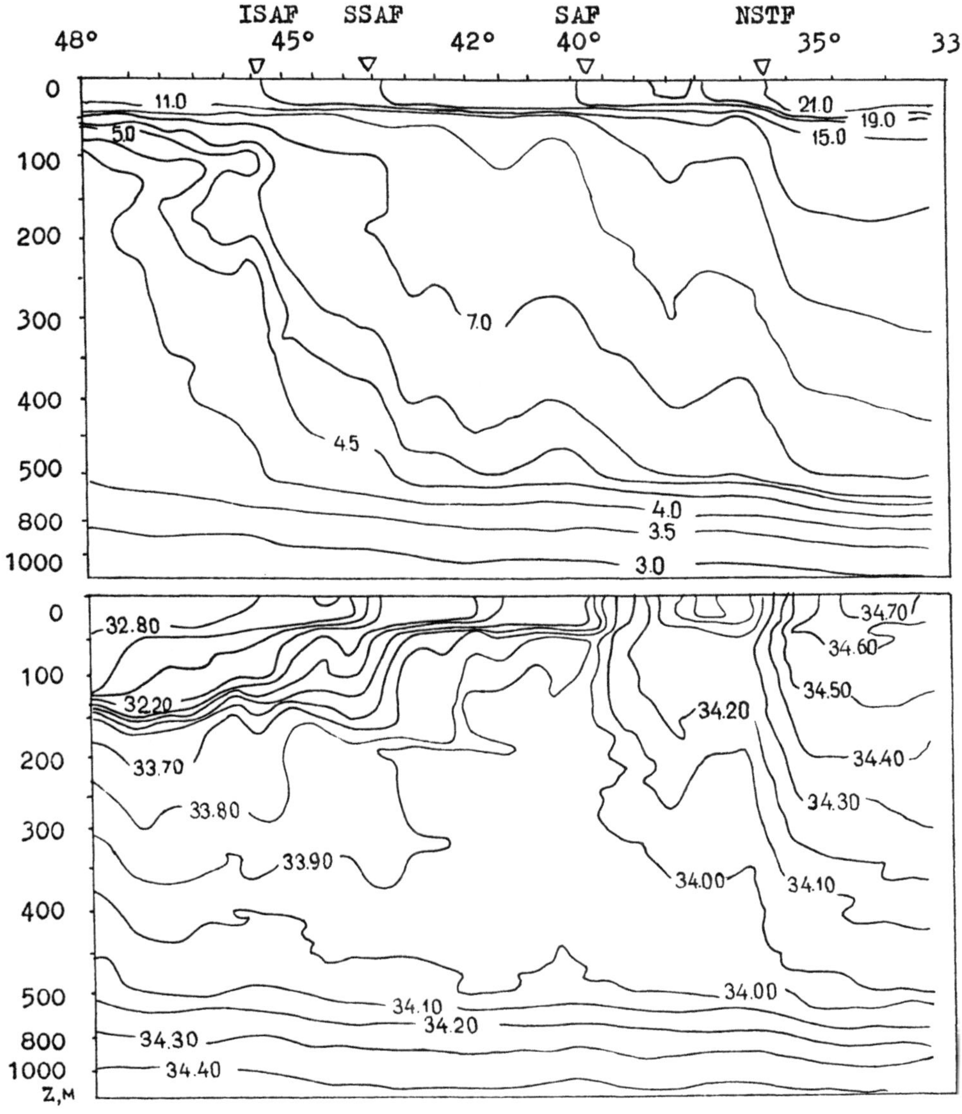

8.4 Thermohalinic water structure

T,S curve method is used widely to discriminate between the different types of water structures. With a joint examination of vertical distribution of temperature T(Z) and salinity S(Z) this method enables to determine the boundaries of propagation and characteristic indices of separate water mass cores, constituting the given structure. T(Z) and S(Z) curves having similar characters for each of the stations of a fixed transect, were united in separate groups. After that we collected distribution types for all transects and identified the field of their propagation. Then the boundaries between the fields were compared with the position of hydrological fronts.

According to publications (Leontieva, 1968) the entire area discussed is characterized as a wide zone of mixing of subarctic and subtropical waters which interact to form unstable varieties of these structures which are difficult to regionalize. Even in the NW Pacific Ocean, though using large amount of measurement data, we failed to isolate water structure in the mixing zone situated between the N and S subarctic fronts (Bulgakov, 1972). Nevertheless, as shown by the analysis of the expedition measurements, the chaotic transition from the subarctic zone to the subtropical one, specific of the zone of subarctic front of the NW Pacific, becomes ordered east of 180°. In the area discussed we succeeded in isolation of extensive water ranges with similar T,S-characteristics.

Thus, analysis of T(Z) within the region revealed the two main types of it:
- nonmonotonic one, characterized by the intermediate extreme values associated with the cold (CIL) or warm (WIL) intermediate layers;
- monotonous one, characterized by a constant (with variable gradient) decrease of temperature with increasing depth.

The areas of propagation of the T(Z) types observed are shown in Figure 8.1a, from which follows that the CIL is most pronounced in the NW portion of the region where the temperature in the core is 2.8°C and temperature difference between the CIL and the deeper situated WIL is 1.2°C. The area of location of the CIL in the W part of the region (at 180°) is confined to 41°N in the S, and is open from the N. Next, eastwards, it has the N boundary at 45°N. Even at 140°W, the CIL vanishes. Within the core area, temperature rises up to 7.1°C in NS and WE directions which is associated with water mixing at the depth of temperature extremes. The observed regularities and the configuration of the zone where the CIL manifests itself as a wedge with a summit at 140°W indicate that in the area discussed the cold intermediate layer (CIL) is caused by water advection from NW areas due to the N link of a large-scale anticyclonic circulation of the N Pacific Ocean, referred to as the Subarctic Current.

The S boundary of the CIL propagation for which the core temperature minor to 7°C was chosen as a criterium (Roden, 1974), is limited by the latitude 43-45°N. Southwards, up to 39°-41°N, temperature extremes vanish and isothermia is observed within the limit depth of their occurrence below the seasonal thermocline. More southwards, the vertical temperature distribution is patently monotonous with values increasing, throughout the layer down to 500 m. In the entire area of investigations, a thoroughly mixed quasihomogeneous layer occurs from the surface down to 30 m.

Analysis of the vertical salinity distribution shows that its main two types are:
- monotonous with salinity increasing from the surface to the bottom, having a seasonal jump and a halocline at the depth of the CIL occurrence;
- nonmonotonic, characterized by the intermediate layer of salinity decrease.

Within the first type mentioned, a subtype with quasi-isohalinic layer may be isolated at the depth of 200 to 500 m which is associated with the region situated N of 43°-45°N.

For the second S(Z) type three subtypes can be isolated: the first (between 180-150°W and 35-40°N) having a subsurface maximum, the second (east of 150°W between 38-35°N) with a subsurface minimum and intermediate maximum; the third (S of 35°N) having the maximum salinity at the surface. The regions where all the above-mentioned types S(Z) are located, are shown in Figure 8.1a. The quasi-isohalinic surface layer up to 30 m thick is also widespread over the entire area studied.

The unification of vertical distribution types of temperature and salinity in a joint analysis using T,S-curve method within the area considered against the background of two main types of water structure, i.e. the subarctic (type 1) and subtropical type 2) permitted isolation of four modifications: the subarctic E (type 1a), quasi-subarctic (type 1b), quasi-subtropical (type 2a) and E subtropical (type 2b). T,S-curves of the structural types isolated from the averaged values of temperature and salinity are shown in Figure 8.3. As seen from Figure, subarctic subtypes of water structure are characterized by a monotonous halinic and nonmonotonic thermal distribution of characteristics, whereas the subtypes of subtropical structure, by monotonous thermal and nonmonotonous halinic distribution. In the mixing zone (type 1b) the distribution of both T(Z) and S(Z) is nonmonotonous. On all curves of the vertical distribution one can discern a quasihomogeneous layer and seasonal thermo- and haloclines. Below, characteristics are changing slightly except for waters of subarctic structure which have, in the vicinity of the CIL, a nearly constant salinity and intermediate halocline at depths 140 to 180 m.

The boundaries between the water structure types isolated are shown in Figure 8.1a. The main dividing here between the subarctic and subtropical waters is associated with the latitude 40°N. To the N of it structural subtypes were isolated on the basis of thermal features related mainly to the value of temperature gradient between the CIL and WIL and hence the boundaries are rather conventional. To the S of 40°N the boundaries of structural subtypes of subtropical waters are delineated from the extreme values of S S(Z). It should be noted that most effective isolation of hydrological fronts is possible from halinic features.

Let us consider the features of water masses forming each of the structures isolated. In this case we will apply the following rule of the T,S curve (Stokman, 1944), "the number of water masses, determined from the shape of T,S-curve, is equal to the number of points of the given curve extremum "plus two", the words "plus two" implying two end points of a T,S-curve in which the tangent to the T,S-curve changes the direction of its inclination with respect to a right line, connecting its ends". Based on this approach, in Table 8.2 we present characteristics of water masses for each of the isolated water structures within the 0-1200 m layer in which measurements are made (water mass core indices are given from the average T,S-curves, shown in

Figure 8.3). It should be noted that though conventional, the classification of water structures and water masses mirrors all the principal features of the spatial structure of thermohalinic fields of this oceanic area.

Figure 8.3: T,S-curves for different types of water structure: 1- west, subarctic; 1a - east subarctic; 1b - quasi subarctic; 2 - subtropical; 2a - quasi subtropical, west; 2b - subtropical.

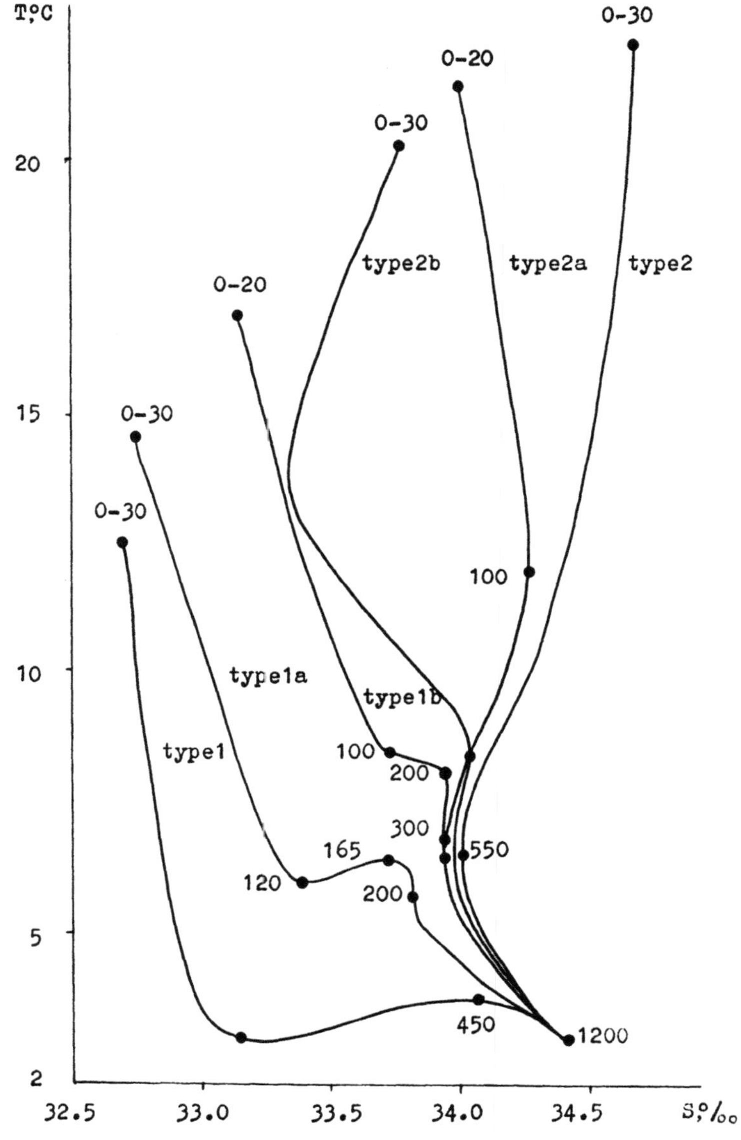

Table 8.2: Classification and generalized characteristics of water masses

Structural watertype	Water mass denomination	Depth range m	Core incedices		
Subarctic,	Surface mass	0-40	0	13	32.68
west variety	Cold intermediate WM	40-200	120	2.8	33.06
(type 1)	Warm intermediate WM	200-600	450	4.0	34.10
	Deep-seated WM	600-...	à-	-	-
Subtropical,	Surface WM	0-40	0	14.6	32.74
east variety	Cold intermediate WM	40-140	120	5.6	33.31
(type 1a)	Warm intermediate WM	140-200	165	6.5	33.82
	Isohalinic WM	200-460	220	5.4	33.82
	Deep-seated WM	460-...	-	-	-
Quasi-	Surface WM	0-40	0	17.0	33.12
subarctic	Isothermic WM	40-140	100	8.3	33.60
(type 1b)	Intermediate, high salinity WM	240-500	200	8.3	34.00
	Intermediate, low salinity WM	240-500	300	6.5	33.90
	Deep-seated WM	500-...	-	-	-
Subtropical	Surface WM	0-100	0	22.1	34.70
(type 2)	Intermediate, low salinity WM	400-700	550	6.0	33.94
	Deep-seated WM	700-...	-	-	-
Quasi-	Surface WM	0-40	0	21.4	34.00
subtropical,	Subsurface WM of increased salinity	40-200	100	11.5	34.33
(type 2a),	Intermediate WM of lowered salinity	200-600	475	6.3	33.98
west variety	Deep-seated WM	600-...	-	-	-
Subtropical,	Surface WM	0-40	0	20.3	33.80
east variety	Subsurface WM of lowered salinity	40-170	65	13.4	33.22
(type 2b)	Intermediate WM of increased salinity	170-400	300	8.3	34.07
	Intermediate WM of lowered salinity	400-600	460	5.5	33.94
	Deep-seated WM	600-...	-	-	-

8.5 Discussion

A detailed classical analysis of the structure of waters and hydrological fronts enabled isolation of six types of water structures in the region discussed (three in the subarctic zone and three in the subtropical zone). The boundary between the subarctic and subtropical zones is closely associated with the subarctic front, meandering slightly between 38°30'N and 41°30'N and standing out against the general background by the highest values of horizontal temperature and salinity gradients.

Of certain interest is the problem of identification of fronts by the values of thermohalinic indices of the surface water layer. With the exception of the NSAF and SSTF, with positions identified only on one of the transects, other fronts, isolated in autumn, may be defined from the following values of temperature and salinity: ISAF - 14.2°C,32.7 ‰; SSAF - 16.4°C, 33.1 ‰; SAF - 19.1°C, 33.6‰; NSTF - 21.5°C, 34.2‰; ISTF - 22.8°C, 34.7‰.

Hydrological fronts, isolated in the autumn period, provide support for the event of branching of the main front of the NW Pacific Ocean under the effect of bottom topography, revealed for the spring period (Levine, White, 1983) and also indicate a quasi-stationary position of these branches.

G.I. Yurasov
Pacific Oceanological Institute
USSR Academy of Sciences
Far East Branch
Vladivostok
USSR

PART III
FORMATION AND DISTRIBUTION OF LIFE

9 A UNIQUE NITROGEN BALANCE AT THE ANAEROBIC FLOOR OF THE BRACKISH LAKE NAKANOUMI

Takasi Kimoto, Hiroshi Hashitani and Taitiro Fujinaga

Abstract

In continuous observations of the anaerobic, H_2S generating floor of the brackish lake Nakanoumi (35°3O'N by 133°10'E, Japan), we found that ammonia alone decreased remarkably without the decrease of total nitrogen and the increase of nitrite and nitrate in the first observation in October 1986. In the second observation in September 1987, particulate amino acids concentration was increased with the decrease of ammonia without the decrease of phosphate concentration. This newly observed phenomena can not be responsible to explain some conventional physical or biological reactions, possibly suggesting new chemical synthesis of particulate amino acids at the anaerobic sea bottom in the presence of hydrogen sulphide.

The marine environments of coastal regions are characterized by their physical, chemical and biological phenomena, therefore, the relative concentrations of nutrients are significantly influenced by location, depth, transportation of water mass, pH, oxidation-reduction potential, photosynthesis and respiration of the planktons, bacteria activity etc.
To survey such a natural mechanisms, it is important to continuously measure the chemical compounds both horizontally and vertically.
Recently we developed a new continuous monitoring system for measuring temperature, salinity, pH, dissolved oxygen (DO), turbidity (Tur), chemical oxygen demand (COD), phosphate (PO_4-P), total phosphorus (T-P), ammonia (NH_4-N), silicate (Si) and chlorophyll-a (Chl) continuously (hourly measurements possible) (Fujinaga and Kimono, 1986).
In this system, two 2kW capacity diesel electric generators, automatic sampling equipment, a wireless telemetry system and a data logger were installed in a transportable barge the size of which was 5.4 x 2.8 x 2.em as shown in Figure 9.1. A specially designed automatic washing system for the sampling and measuring equipment overcame the problems caused by the growth of algae on the equipment. The sampling intake was automatically controlled and was set to definite sampling depths for each one hour sampling period.

The details of the automated measuring apparatus are as follows. The automated measuring apparatus for inorganic phosphate is based on the formation of phosphomolybdate, reconcentration and separation of phosphomolybdate by column and measurement with a voltammetric detector. Total phosphorus is digested by an acidic persulphate solution to inorganic phosphate in an autoclave at 12O°C.
Nitrate is measured with a specially designed dual wavelength spectrophotometer. The absorbance difference between 223 nm and 232 nm has a good correlation with the concentrations of nitrate plus nitrite. Automated measuring apparatus for ammonia is based on preconcentration and separation of ammonia with a gas permeable micro-porous membrane, followed by measurement of the ammonia-enriched acid solution with a conductivity sensor.
Total nitrogen is digested to nitrate by a basic persulphate solution in an autoclave at 120°C. Chemical oxygen demand is measured by a permanganate oxidation method in an alkaline solution. Silicate is measured though the formation of

V. I. Ilyichev and V. V. Anikiev (eds.), Oceanic and Anthropogenic Controls of Life in the Pacific Ocean, 131–137.

silicomolybdate and subsequent measurement with a voltammetric detector. Chlorophyll-a concentrations are determined using a spectrofluorometric method. Temperature, pH, dissolved oxygen, salinity and turbidity are measured with conventional sensors.

Figure 9.1: Map of Lake Nakanoumi.

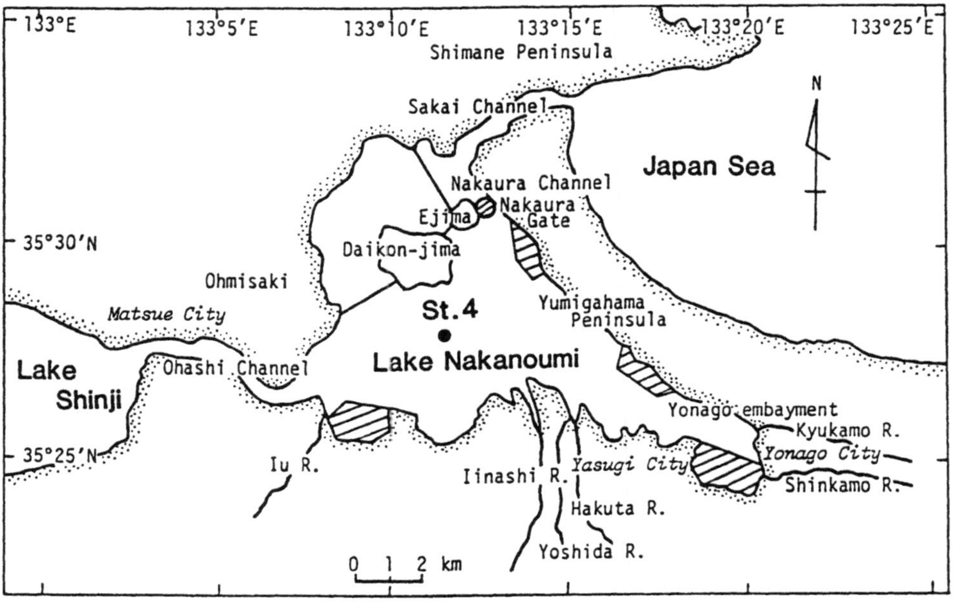

Lake Nakanoumi is connected with the Japan Sea only via the Sakai Channel, which it approximately 0.3 km wide and 7.5 km long as shown in Figure 9.2. This lake has a surface area of 97.5 km2 and total storage volume of 5.2 x 10^8 m3 at mean water-level of 0.2 m. The mean depth of the lake is 5.4 m and about 70% of this area is shallower than 7 m. Salinity decreases toward the interior of Lake Nakanoumi from 30-35 per-mil at Sakai Channel to 5-10 per-mil at Ohashi Channel (Ohtake et al., 1984).

The depth of the measuring site (St.4) was 6.5 m. A strong halocline was found at the depth of 4 - 4.5 m. Salinity below halocline was about 30 per-mil and that above halocline was about 30 per-mil. In summer dissolved oxygen had increased at the surface more than 100% due to the photosynthesis of phytoplanktons, while it had decreased at the bottom to nearly 0% through the biodegradation of organic matters. This anaerobic condition accelerates elution of phosphate, silicate and ammonia from the sediment. It also creates suitable conditions for hydrogen sulphide

Figure 9.2: Continuous water quality monitoring system.

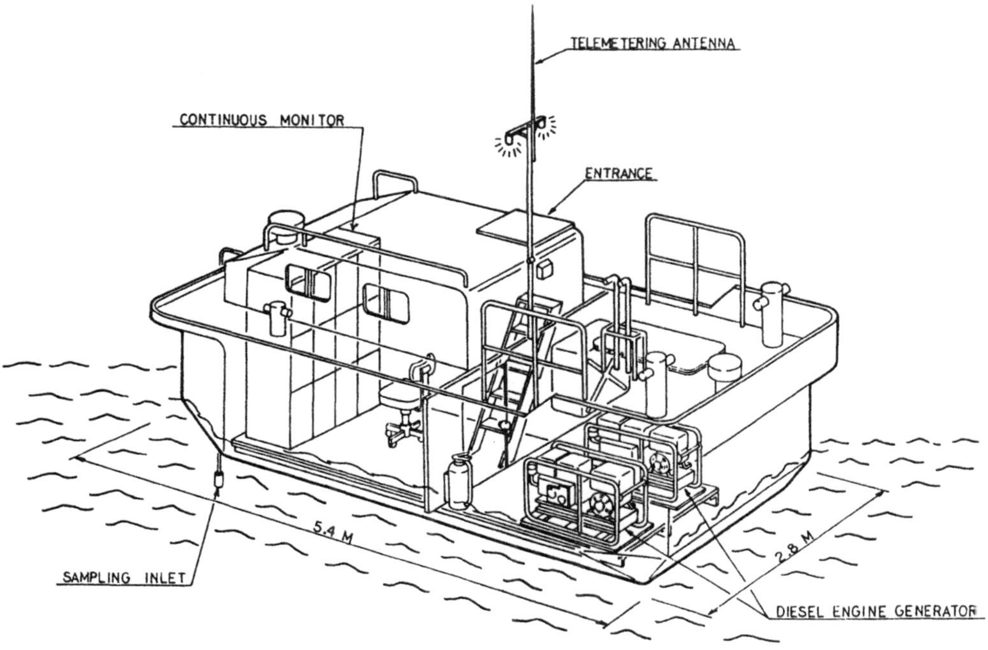

generation via bacteria sulphate reduction.
Sampling was conducted at two depths, 1 m below the surface and 1 m upwards from the bottom.
Simultaneous manual observations were performed at every one meter depth on such items as water temperature, salinity, dissolved oxygen, phosphate, nitrite, nitrate and total nitrogen. A weather station for measuring wind direction, wind speed, ambient temperature and atmospheric pressure was also installed.
Table 9.1 shows the correlation coefficients for data obtained at the surface and the bottom sampling points during Oct. 16th to 29th in 1986. The numbers of each data are 168. The concentration of phosphate, silicate and ammonia at the bottom layer were inversely correlated with the concentration of dissolved oxygen and these results can explain that anaerobic condition accelerates the elusion of those items from the sediment.

Phosphate and silicate have a high correlation at 0.90 while the correlation coefficients between the ammonia concentration and that of phosphate and silicate respectively were low at 0.78 (vs. PO_4-P) and 0.66 (vs. Si), this is because the ammonia at the bottom decreased more quickly than the phosphate and silicate concentrations.
Figure 9.3 shows a typical vertical distribution of nitrogen containing compounds normalized by the total phosphorus concentration because the variation caused by the diffusion should be cancelled.

Table 9.1: Correlation coefficients of continuous monitoring data obtained at the surface and the bottom sampling points.

Table I. Correlation coefficients of continuous monitoring data obtained the surface and the bottom sampling points.

	Upper Layer (1m depth from water surface)										
	W.Temp.	Salinity	pH	DO	PO_4-P	T-P	NH_4-N	Si	Chl.	COD	Tur.
W.Temp.		0.42	-0.08	-0.52	0.85	0.71	-0.01	0.51	0.17	-0.19	-0.31
Salinity	-0.02		0.01	-0.41	0.32	0.29	0.02	-0.06	0.36	-0.04	-0.28
pH	-0.53	0.04		0.53	-0.42	-0.33	0.09	-0.19	-0.19	0.13	-0.39
DO	-0.73	-0.34	0.64		-0.60	-0.52	0.22	-0.50	-0.15	0.46	0.14
PO_4-P	0.77	0.09	-0.72	-0.85		0.80	-0.05	0.64	0.22	-0.20	-0.13
T-P	0.75	-0.09	-0.73	-0.76	0.96		-0.04	0.51	0.31	-0.15	-0.09
NH_4-N	0.56	0.30	-0.46	-0.70	0.77	0.69		-0.17	0.07	0.25	0.08
Si	0.82	-0.08	-0.65	-0.82	0.90	0.91	0.65		-0.12	-0.38	-0.15
Chl.	-0.10	-0.74	0.00	0.36	-0.11	0.11	-0.36	0.04		0.29	0.20
COD	-0.20	-0.44	0.14	0.42	-0.25	-0.08	-0.34	-0.17	0.56		0.24
Tur.	0.31	-0.05	-0.61	-0.45	0.66	0.71	0.60	0.49	0.10	-0.05	
	Bottom Layer (1m upwards bottom)										

W.Temp.: Water Temperature, Chl.: Chlorophyll, Tur.: Turbidity

Figure 9.3: Vertical nitrogen balance.

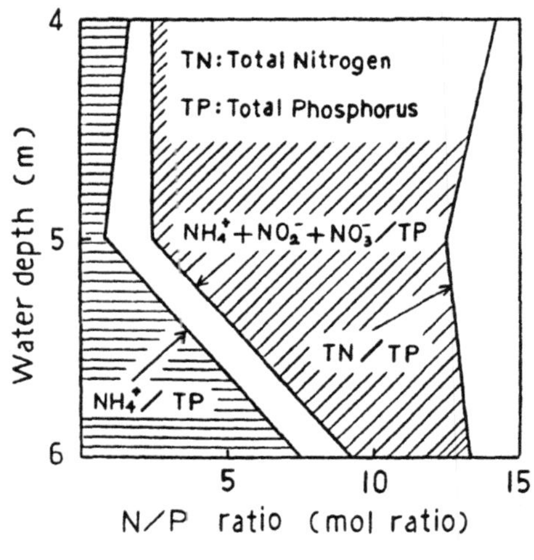

Concerning Figure 9.2, only ammonia decreased vertically between 6 m water depth (within 50 cm upwards from the bottom) and 5 m water depth.
It has been reported (Seiko et al., 1986) that the decrement of ammonia might be caused by nitrification and denitrification by bacteria. If these biological reactions are the major reason of the decrement, nitrite and nitrate concentration should be increased or the total nitrogen concentration should be decreased. Actually, however, these concentrations changed very little as can be seen in Figure 9.3.
In September 1987, we performed the continuous observation again at the same site in Lake Nakanoumi in order to find more detailed information about the characteristics near sediment boundary. In this observation, the amino acid concentrations were measured at 10 cm, 25 cm, 50 cm, 100 cm, 150 cm and 200 cm upwards from the bottom. Ammonia, phosphate and silicate concentrations in the interstitial water were also measured with the core samples collected there.
Dissolved oxygen in this observation decreased almost to 0% saturation from 10 cm to 50 cm upwards from the bottom and the strong smell of hydrogen sulphide was detected in this anaerobic zone.
Figure 9.4 shows the vertical distribution of phosphate, silicate, ammonia and total amino acid concentrations from the interstitial water over the range from 50 cm below the sediment surface to 50 cm above the sediment surface.

Figure 9.4: Concentrations near water-sediment boundary.

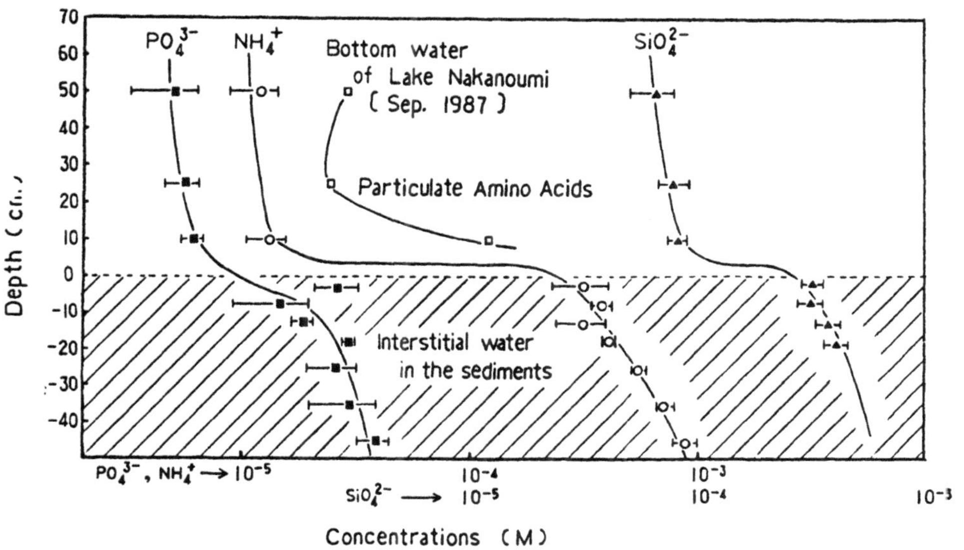

As shown in Figure 9.4, only the ammonia concentration decreased vertically across the water-sediment boundary in a manner similar to the previously observed result in Figure 9.3. The samples were prepared by three pretreatment procedure before amino acid analysis: a) filtered by 0.2 μm Nucleopore filter, b) filtered and hydrolyzed (1 ml sample with 1 ml distilled HCl was heated at 105°C for 24h after vacuum degassing), and c) directly hydrolyzed.

For amino acid analysis, the samples were treated with ortho-phthalaldehyde and N-acetyl-L-cysteine to produce fluorescent amino acid derivatives which were efficiently resolved and detected on a reversed-phase high performance liquid chromatography and the fluorometric detector system (Buck and Krummen, 1984).
Only method (c) was found to yield detectable amounts of amino acids. From this result we could conclude that particulate amino acids such as peptides existed at the bottom.
Figure 9.5 shows the distribution of various particulate amino acids concentrations at 10 cm and 25 cm upwards from the bottom.
The particulate amino acid concentrations plotted against the depth are shown in Figure 9.4. It is clearly seen that the particulate amino acids concentrations have increased with the corresponding decrease of ammonia concentration.

Figure 9.5: Amino acids composition.

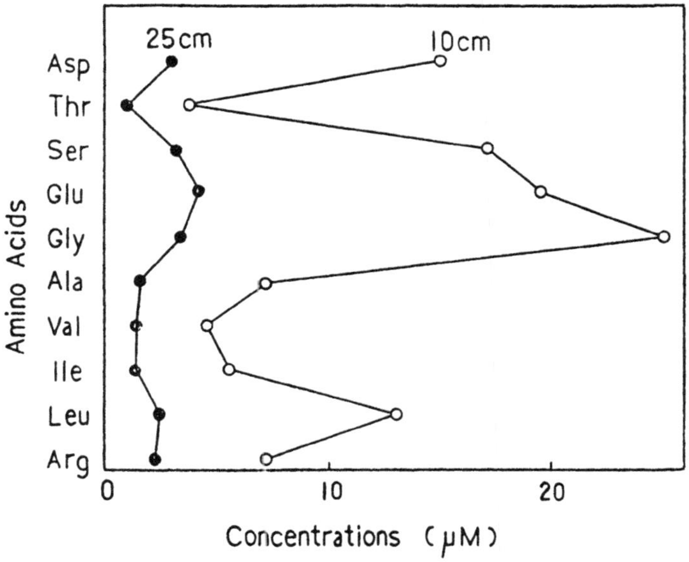

We conclude that a new non-biotic synthesis of particulate amino acid can take place at anaerobic sea bottom.
Although some anaerobic chemosynthetic bacteria (*Thiobacillus* etc.) could also be responsible for similar synthesis, the decrease in phosphate concentration may be too small to prove it, as can be seen in Figure 9.4.

We thank Drs. Yoshio Date, Kunio Rondo, Yasushi Seiko, Minoru Okumura, and Kaoru Fujinaga for their assistance.

References

Buck, R.H., Krummen, K. : Resolution of amino acid enantiomers by hiqh-performance liquid chromatography using automated pre-column derivatisation with a choral reagent.
J. Chromatography,31, 279-285 (1984).
Fujinaga, T., Kimono, T., : Development of continuous monitoring system for marine environment. Transactions of the Research Institute of Oceanochemistry, 4-20 (1986).
Ohtake, H., Katz, S., Date, Y. : The role of tides in removing phosphorus from Lake Nakanoumi, a coastal lagoon in Japan. Arch. Hydrobiol. 99, 443-462 (1984).
Seiko, Y., Rondo, K., Kimura, T., Haihitanl, H.: Nitrogen Metabolism in the brackish lake Nakakoumi. Abstract for International Symposium on New Sensors and Methods for Environmental Characterization (SMEC), S5-05 (1986).

Takasi Kimoto
Kimoto Electric Co., Ltd
3-1 Funahashi-cho
Tennoji-ku
Osaka 543
Japan.

Hiroshi Hashitani
Shimane University
Nishikawazu-cho
Matsue-shi
Shimane 690
Japan.

10 THERMODYNAMICALLY NONEQUILIBRIUM NATURE OF THE OCEAN/ATMOSPHERE INTERFACE AND THE ORIGIN OF LIFE

V.A. Tverdislov, G.G. Khundzhua, L.V. Yakovenko and V.V. Anikiev

10.1 Introduction

A hypothesis proposed few years ago by one of the authors of this paper (V.A.T.) it further corroborated. The hypothesis suggests that asymmetric (and non-equilibrium) distribution of cations between biological systems and surrounding milieu could have originated from the non-equilibrium distribution of ions between the bulk phase and thin cold layer on the sea surface. The process of a photocell formation might have included a stage when aerosol droplets usually present near the sea surface spontaneously evolve into sealed vehicles, the droplets having entrapped potassium-enriched sea water from the cold layer on the surface of the "primary broth".

Physico-chemical properties of all biological systems are strongly dependent on the particular behaviour of matter in the neighbourhood of their boundaries which are at the same time interfaces of different thermodynamic phases. Traditionally, such interfaces are considered as equilibrium ones. The interfaces of biological systems (firstly, cells) and their surroundings, however, play an active role in the processes of energy and substance exchange between the phases. Thermodynamically non-equilibrium nature of the interfaces may be the main reason for components redistribution and transport between and in the phases. We believe that this approach to the problem of the origin of ionic asymmetry of all living systems may be fruitful.

Most elaborated hypotheses of the life origin consider cell precursors (coacervates or clay particles) as ion-exchangers which selectively take certain mineral ions from the surrounding sea water /1,2/. The partition of ions thus obtained must he an equilibrium one. The existing hypotheses can not explain the evolution of the ion-homeostasis system of a photocell from a steady-state to the non-equilibrium one. We try in this article to develop further the hypothesis in accordance to which photocells were from the very beginning formed as non-equilibrium systems and the main role in the process of their origination was played by the non-equilibrium ions (and other substances) distribution between the thin cold layer on the sea surface and the bulk phase of sea water.

10.2 Thermodynamics of the ocean/atmosphere interfacial processes. The cold layer

Ocean and atmosphere comprise is an open thermodynamic system in the sense that it receives energy from the sun and loses it by different ways. Both high degree of transparency of atmosphere and insignificant albedo of the ocean result in the temperature of the ocean being on the average 0.1 - 1.4 K higher than that of the air. Such disbalance causes spontaneous processes of heat- and mass-exchange consisting of intensive IR-radiation from the sea surface, water evaporation and contact heat exchange through the (imaginary) boundaries of the system. All these processes take place mostly in a thin layer on the very surface of the ocean (the so called radiation layer). Generally this layer loses so much heat that the loss can not be counterbalanced by absorption of the sun radiation. Therefore, one may expect to find rather high gradients of temperature in the boundary layer between the

V. I. Ilyichev and V. V. Anikiev (eds.), Oceanic and Anthropogenic Controls of Life in the Pacific Ocean, 139–144.

ocean and atmosphere.
In fact, at the beginning of 1970s, when a highly responsive thermoprobe was used in studies of the sea surface, the temperature profiles of the boundary layer were obtained for the first time under the conditions of the open sea with the roughness up to 4. It was clearly shown that the temperature gradient is steepest just on that thin (200-500 u) boundary layer called "the cold layer" because its temperature is lower than that of the underlying water and the air above it. The gradient may be as high as 800-1000 K/m /3/ (Cf. Figure 10.1). The cold layer proved to be rather stable in storms and is usually restored in a few seconds after destruction of a wave.

Figure 10.1: Temperature difference between the bulk phase and the cold layer (T,K) in dependence on the distance (Z, mm) from the free surface.

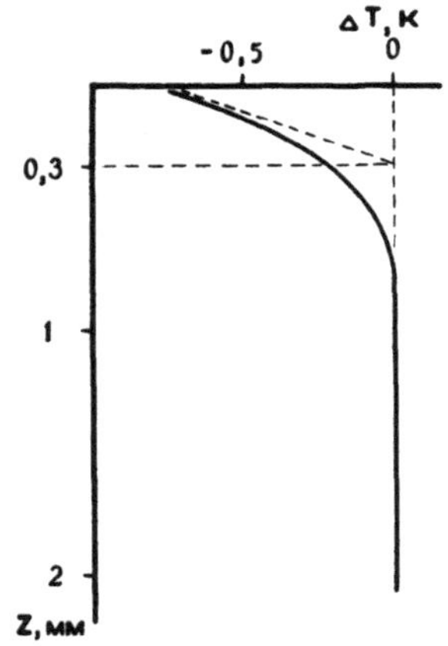

At present it is well known that sea water components are distributed nonuniformly between the cold layer and the bulk phase of water. Thus, in the cold layer the concentration ratios of K^+ and Cal + to Na^+ and Mg^{2+}, reap., are much higher than in the bulk phase /5/. Concentrations of some trace elements in the cold layer may be a few hundred times higher than their concentration in the bulk phase, and vice versa. The existing explanations of these effects are based on an assumption that water as a solvent should have some peculiar properties near the surface (to the depths not deeper than a few hundred A^0 /4/. Therefore, the vertical distribution of sea water components is to be considered as an equilibrium one. We suppose that this distribution of components is caused by non-equilibrium processes taking place on the ocean/atmosphere interface and iq, therefore, a non- equilibrium one.
This supposition was indirectly confirmed by data obtained in our laboratories. Samples of sea water were taken with capillary nets (micro nets) in different parts of the Caspian and the White Seas under well defined conditions. Concentrations of ions in the samples were assayed with the help of Atomic Absorption Spectrofometer. The concentration ratio of K^+ to Na^+ aws found to be higher (in some cases more than twice that in the bulk phase) when evaporation of water is more intensive.
By microelectrode measurements in laboratory conditions an electric potential difference was found out between the bulk phase and cold layer on the surface of

sea water in a bath. It is important that in all cases when heat and mass exchange between the solution and air was prevented by special arrangements of the experiments, and, therefore, the cold layer could not be formed, the potential difference was not detected.

Generally, an explanation of the observed effects can not be given not only in the framework of concepts of equilibrium adsorption, but also by of linear approximation commonly used in descriptions of such phenomena as thermodiffusion, thermoelectricity, etc. /6/. It seems plausible that non-linear thermodynamic processes should play an important role in bringing about the described effects in the extremely non-equilibrium system comprising the ocean and atmosphere.
Now we can assert that one of the effects of the luminous energy flux from the sun to the system under consideration consists of formation of a dissipative structure with vertically non-uniform distribution of temperature and concentrations.
The question arises whether the discrete structures so characteristic for all the living matter can originate from a 2-D distributed system. The answer lies in the process of aerosol formation in the layer of atmosphere just above the sea surface. It is known that at the destruction of a wave some quantity of air is trapped and dragged down under the sea surface where it forms numerous bubbles of various sizes.
The diameters of the bubbles was shown to be are averagely smaller than 0.4 mm /4/. When a bubble reaches the surface it collapses, producing a gush of water from the bubble's bottom and a great number of micro droplets from the bubbles edges (Figure 10.2).

Figure 10.2: Aerosol formation at an air bubble breaking up on the seasurface: (1) jet-stream droplets; (2) microdroplets.

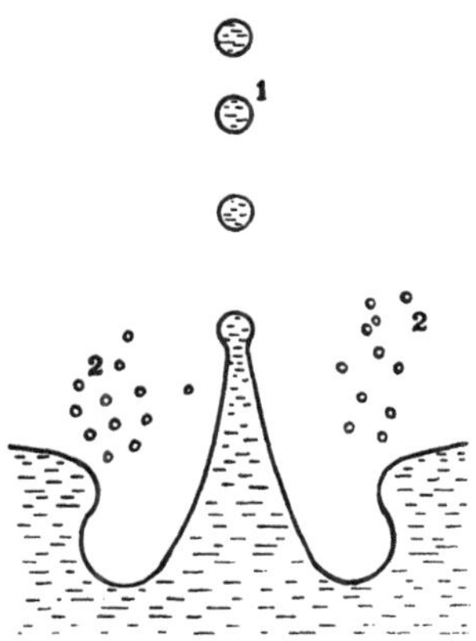

The gush breaks up into jet-stream droplets. The droplets of both kinds rise into the air up to a few cm. The jet-stream droplets are bigger and fall back to water quickly. The micro droplets originate from the cold layer and have therefore the same "inverted" chemical composition. In some cases the K^+ to Na^+ concentration ratio in micro droplets may exceed five times that in the bulk phase /4/ . If a monolayer of a surface-active agent on the sea surface exists, the micro droplets formed are also covered with a layer of the agent /7,8/.

10.3 Discrete prebiotic structures with inverted ionic composition

Initial stages of the prebiotic evolution could have taken place in the surface boundary layers of estuaries of the primeval ocean where abiogenically formed organic molecules had been concentrated /1/. One of the principal questions, which we are going to provide an answer to, is whether there were zones in the sea where discrete structures with non-equilibrium ionic asymmetry could have come into being, and what biological consequences could it have.
In the first part of this article we showed that there are systems in nature in which ions of biological interest are distributed nonuniformly, the distribution is non-equilibrium and similar to that in the case of living cells. The systems in question are the cold layer and aerosol formed of it, the aerosol consisting of discrete micro droplets. The problem is that how could these micro droplets evolve into photocells. One of the possible ways of such evolution could have been spontaneous formation of lipid vehicles entrapping sea water from the cold layer which, in turn, may be enriched with potassium ions and organic substances. Abiogenically formed lipids, as surface active agents, may have formed disperse monolayer on some parts of the ancient ocean surface /9,10,11/. Then micro droplets of the aerosol must have been covered with the same disperse lipid monolayer. As micro droplets are very small they fly in streams of air for a long time. During the flight of a micro droplet a some amount of water evaporates, the micro droplet shrinks and the disperse monolayer of lipids becomes condensed. When a micro droplet falls back to the sea surface it may adsorb another monolayer of lipids complementary to the first one and become, therefore, covered by a lipid bilayer (Figure 10.3).

Figure 10.3: Aerosol droplet transformation into a liposome.
(1) - the ocean/atmosphere interface with a disperse monolayer of amphiphylic lipid molecules; (2) - a micro droplet covered with a monolayer of lipid molecules; (3) - a liposome with a lipid bilayer membrane contains sea water from the cold layer which has 'inverted' chemical composition (high potassium -low sodium concentrations).

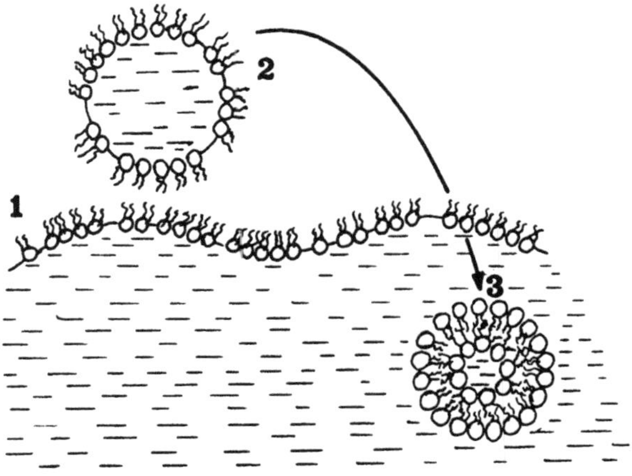

This new system which is usually called a liposome, is rather stable and can retain its content without significant changes for hours and even weeks depending on the circumstances /12/. In our opinion the lifetime of the liposome is sufficient for primary processes of metabolism, biosynthesis and transport of organic and inorganic substances could have started. The described system satisfies general requirements to the objects under biological evolution:

- discreteness of organic matter is the principal condition of the natural selection /13/.
- it seems to be vital that the photocell should be covered by a lipid membrane which is almost impermeable for hydrophillic biomacromolecules and water-soluble low molecular compounds because of its internal hydrophobic region.

Existence of non-equilibrium and asymmetric ion distribution between the photocell and the surroundings was of principal significance for the cell survival. The partition of ions is the main constituent of the cell's safeguarding system. Most usual components of sea water, inorganic ions, are used in this system because any leakage in the membrane or an unfavourable change of circumstances would result in a redistribution of ions and changes in the transmembrane electric potential difference, both of which may be used as distress signals.

The asymmetry of the ion distribution concerns mostly cations. This corresponds to the anionic nature of most biomacromolecules. Among widely spread cations potassium and sodium are less active chemically than calcium and magnesium and do not form water-insoluble compounds. Cations of these two classes are distributed between the cell and its surroundings very nonuniformly and the two pairs of them are used in closely related ion regulation systems in contemporary living cells.

Nonuniform and non-equilibrium distribution of cations between a cell and the medium could have been achieved by various ways. However, some physico-chemical factors made the range of possibilities rather narrow. First of all, the principle of electroneutrality must be kept inviolate in the cell's volume. Therefore, any current of cations out of the cell can occur only together with an equivalent current of cations into the cell or of anions out of the cell. This leads to the situation when some cations must be accumulated by the cell while cations of other sorts must be removed from the cell's interior.

The final separation of cations must be such that transmembrane concentration ratios are around 10. This figure is based on the two following facts: i) as a rule, enzymes' activities change appreciably with appr. 10-fold changes in ion concentrations; ii) lipid membranes as liquid-crystal-like structures are controlled my electric fields with the potential gradients about 10^5 V/cm. Divalent cations do not suit the purpose well because their concentrations are much lower than those of monovalent cations. Moreover, intracellular calcium and magnesium are mostly in wound states. The chemical composition of the ocean remained, apparently, constant during the whole period of biological evolution /2/. As sodium ion concentration in the sea water is much higher than that of other ions, sodium accumulation by a cell to yet 10-times higher concentration would cause a great osmotic disbalance. Therefore, in order to keep osmotic pressure in an appropriate range a cell can provide itself with the required ion concentration differences on its membrane only by accumulation of potassium ions in exchange for sodium ions. Similar partition of ions occur when a liposome containing sea water from the cold layer is formed as is described above. Thus, the chemical composition of the ocean and its surface cold layer

predetermined the cations distribution between cells and their surroundings peculiar to all living beings.
Functioning of metabolic and biosynthetic systems in a potassium enriched protocell demanded that the ionic composition of the cell interior should have been kept constant. This was achieved after special systems of active ion transport appeared on the later stages of biological evolution.

The authors are grateful to Prof. L.A.Blumenfeld, Prof. S.E.Shnol, and Drs. A.N.Zaikin, V.I.Lobyshev, I.L.Tverdislova, E.V.Karavayeva for fruitful discussions and help in experimental work.

References

1. Oparin A.I., Life: Its Nature, Origin and Development. Academic Press, New York, 1962.
2. Rutten M.G. The Origin of Life by Natural Causes, Elsevier Publishing Company, Amsterdam-London-New York, 1971.
3. Khundzhua G.G., Andreev E.G. Doklady AN USSR, No.4, p.208, 1973,(in Russian).
4. Home R.A., Marine Chemistry. The Structure of Water and Chemistry of the Hydrosphere. Wiley Press-Interscience Division of Wiley and Sons, New York-London-Sydney-Toronto, 1969.
5. Yeremeyeva L.V., Bezborodov A.A. Geochemistry of the ocean/ atmosphere interface. Reprint by Ukrainian Acad.Sci., Institute of Geological Sciences, Kiev, 1981. (in Russian).
6. Garrett W.D., Limnol. Oceanogr., v.1O, p.602, 1965.
7. Riley G.A., Limnol. Oceanogr., v.8, p.372, 1963.
8. Wangersky E.C., Am.Scientist, p.358, 1965.
9. Kenton D.H., Steinman G., Biochemical Predestination. Mc Grow Hill Book Company, New York-London, 1969.
10. Goldacre R.J., In: J.F.Danielli, K.G.A. Pankhurst and Ridford (Eds.) Surface Phenomena in Chemistry and Biology, Pergamon Press, New York, p.276, 1958.
11. Haggis G.H., In: G.H.Haggis (Ed.), Molecular Biology, Wiley, New York, N.Y., p.151, 1964.
12. Ivkova M.H., Pechatnikov V.A., Gracheva O.A., Pechatnikova E.V., Ivkov V.G., Gen.Physiol. and Biophys., v.6, p.45, 1987. (in Russian).
13. Shnol' S.E., The Physicochemical Factors of Biological Evolution, Soviet Theor.Rev., Supply. Ser.Bill. (Ed. V.P.Skulachev), v.I, Harvard Acad.Publ., New York, 1981.

V.A. Tverdislov
Lomonosov University
Moscow USSR

G.G. Khundzhua
Pacific Oceanological Institute
USSR Academy of Sciences, Far East Branch
Vladivostok USSR

11 MIGRATION MECHANISM OF THE AYU

Katsumi Tsukamoto and Kazuo Uchida

Abstract

Mechanism of fish migration was investigated based on (1) ecological, (2) endocrinological and (3) behaviourial aspects of the upstream migration of juvenile ayu *Plecoglossus altivelis*, amphidromous salmonoidei fish.
(1) Age and size increased with the time of migration,i.e., the timing of migration did not depend on a fixed age and size. Fish with earlier birth dates and larger or moderate growth rates migrated earlier at younger ages and smaller sizes. As the first step for migration, fish requires a fixed age and size peculiar to each timing of migration, which are determined by its own birth date and growth rate. Cyclic switching of size and migratory pattern in successive generations was also observed in landlocked ayu of Lake Biwa, i.e. since small lake residents apparently spawn earlier than large stream migrants, their progeny should be early born and migrate early to become large migrants in the inlet streams. Conversely, the progeny of large migrants would be born late, would likely remain in the lake as small residents, and would in turn produce early born, early migrating progeny.
(2) Thyroid activity and plasma thyroxine level were lower in migrants than in premigrants just before migration. Fish with the higher thyroxine level showed the stronger tendency to migrate upstream in an artificial river. Endocrinological conditions must be satisfied as the second step before initiation of migration. (3) Lastly, environmental factors trigger the migration. Increases in fish density, water temperature and light intensity, and starvation facilitated the migratory behaviours of ayu such as jumping activity and swimming upstream. These factors also changed spacing of ayu in an experimental tank. To interpret the triggering mechanism of upstream migration by environmental factors, a behaviourial model is presented by placing an endogenous drive for migration between stimuli (current or waterfall) and response (upstream migratory behaviour). Environmental factors elevate the level of drive through nervous or endocrine system, lowers the threshold of migratory behaviour, and thus, migration is initiated in response to a stimulus which has never before been effective.

11.1 Introduction

What drives fishes upstream? What is it that calls them far off the ocean? Fish migration has long been one of the most interesting and still mysterious subjects of biological research. Diadromous fishes must migrate between the sea and the freshwater encountering risks of the drastic changes in food condition, osmotic pressure and other environmental factors. They should pass the mouth of river and the coastal areas at least twice in their life and these areas are apt to be polluted. Study on fish migration would make some contribution not only to biology itself but also to the problems of human impact on the environment and natural resources. The objective of this study is to know the mechanism of fish migration by examining the priming conditions for migration and the triggering mechanism of migratory behaviour.

V. I. Ilyichev and V. V. Anikiev (eds.), Oceanic and Anthropogenic Controls of Life in the Pacific Ocean, 145–172.

11.2 Experimental design

In this study, we used the amphidromous ayu *Plecoglossus altivelis* Temminck et Schlegel, which is categorized in diadromy but migrate between the fresh water and the sea without any relation to spawning. Most diadromous fishes such as anadromous salmons or catadromous eels migrate in relation to spawning, which must be a high light of their life cycle. Their migration is accompanied by various physiological and ecological changes such as maturation, social behaviour, feeding habit, etc. These apparent changes related to spawning may mask the nature of the endogenous drive for "movement." Therefore, amphidromous species such as the ayu seems to be good material for understanding the essence of fish migration.

The ayu is a salmonoidei fish with a life span of only one year. This is an important species for game fishing as well as fish culture. Ayu spawn adhesive eggs in gravel along the lower reaches of rivers in autumn. After about two weeks, newly hatched larvae, 6 mm long, drift downstream to the sea and spend the winter months there (Figure 11.1).

Figure 11.1: Migration patterns of ayu

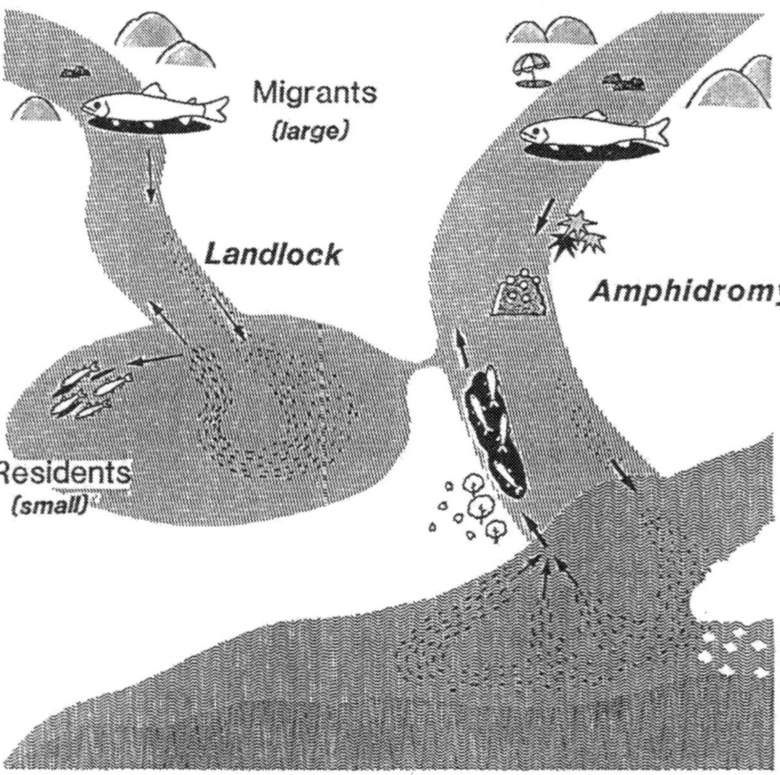

The juveniles (about 60 mm long) start an upstream migration in spring and grow to a length of 150-300 mm by the end of summer, feeding on algae adhering to rocks. In autumn, mature fish spawn and then die. This typical amphidromous pattern is preserved in landlocked populations, that are known from Lake Biwa and several other lakes in Japan (Figure 11.1).

In the landlocked population of Lake Biwa, however, there are differences among fish groups with respect to body size, migration pattern and reproductive characteristics (Azuma 1970, 1973a, 1973b, 1973c, see Figure 11.1). Such differences in migratory pattern also can be seen in Arctic charr *Salvelinus alpinus*, (Nordeng 1961), Dolly Varden *Salvelinus malma* (Maekawa 1984), brown trout *Salmo trutta* (Jonsson 1985), sockeye salmon *Oncorhynchus nerka* (Ricker 1938), and masu salmon Oncorhynchus masou (Kubo 1980). Differentiation among the ayu in Lake Biwa varies from that in other species since the habitat is restricted to the lake and its inlet streams. If we distinguish Lake Biwa ayu by time of their upstream migration, they may be roughly divided into three groups: (1) spring migrants, which go upstream in spring (March-May) and grow large (about 150 mm) in the inlet streams of the lake; (2) summer migrants, which start the migration in summer (June-August); and (3) residents, which remain in the lake throughout their lives until their terminal spawning migration to lower stream reaches in late summer or early autumn. Residents mature at small sizes (about 80 mm, called dwarfism by Azuma 1973c) and spawn in August-September earlier than large spring migrants which spawn in October-November. Both migrants and residents spawn in the same inlet streams but in different reaches. Examination of the differences between migrants and residents in the landlocked population may help to clarify the migration mechanisms of amphidromous fish. Furthermore, the Lake Biwa fish do not have the problem of changing osmoregulation faced by their marine counterparts, which may simplify interpretation of migration patterns. For commencement of migration, fish must pass through the two processes: (1) satisfying the priming conditions for migration which need comparatively long term; and (2) elevating the endogenous drive for movement which occurs in a short time. These processes are examined in the study based on the three aspects: (1) ecological, (2) endocrinological, and (3) behavioral aspects. Finally, the mechanism of fish migration is discussed by synthesizing these results.

11.3 Ecological aspects

Many biological developments are size dependent. Size is, however, a function of age, and we also should examine age as a factor controlling the motivation for migration. In this chapter, we examine the age of migratory fish as well as the size in order to determine the fundamental necessary conditions for migration.

Methods

Juvenile ayu which ascended from Lake Biwa to inlet streams, River Ado, River Chinai and River Uso, were trapped with the commercial fish weir about 200-500 m upstream from the river mouth (Figure 11.2). From April to July in 1982-1984, about 30 fish were collected monthly and preserved in 70% ethanol for age determination and examination of body size and meristic characters. Residents which remained in the lake till summer were examined and compared with summer migrants caught at the same time in an inlet stream, River Uso. Amphidromous stocks were also examined in three rivers; River Shinano, River Tenryu and River Yabe (Figure 11.2). Juvenile ayu which migrated upstream from the sea were collected about once every 2 weeks in spring 1982, and 1984-1986.

Based on daily growth increments on sagitta otoliths (Figure 11.3), the ages of ayu could be easily and accurately determined within a 2% error even for fish 300 days old (Details in Tsukamoto and Kajihara 1987). After standard length was measured, sagitta otoliths were extracted and were mounted in euparal on a glass microscope slide. All rings outside a 15 μm radius were counted through a light microscope with two sets of arrow image projectors (Nikon Teaching Head) or with a camera lucida at 600 X magnification. Sagittae have radii of about 15 μm at hatching with 2-7 embryonic increments (Tsukamoto and Kajihara 1987). The mean of the counts by two readers was accepted as the age when the counts agreed within 5%. Birth dates and catch dates were expressed as Julian dates, which were used in all statistical analyses.

It could be assumed that the growth of ayu was linear during the larval and juvenile stages (Fukuhara and Fushimi 1986). Thus we simply calculated the growth rate of each fish from the estimated age and the difference between body length at sampling and mean body length at hatching.

We examined four factors, i.e. age, size, birth date and growth rate, and determined the role of these factors in the upstream migrations of the ayu. We also applied these results to the landlocked fish of behaviorally and morphologically differentiated groups in order to clarify the mechanisms by which migratory patterns vary among landlocked ayu.

Age and Size of migration

Both age and size of juvenile ayu caught by the weirs in the inlet streams were linearly correlated with the time of upstream migration ($P<0.001$; Figure 11.4). Successive ages at migration time were different one from another ($P<0.005$) except between two March samples and between May and June ($P>0.1$). Size showed a similar tendency except for the March 1983 datum (Figure 11.4). The minimum age and size at migration were 193 days and 69.6 mm. Fixed age and size peculiar to the time of migration should be required for the initiation of upstream migration.

Birth date and growth rate

Clear correspondence between mean birth date and the time of upstream migration was observed (Figure 11.5). The earlier a fish was born in autumn, the earlier it started upstream in spring next year ($P<0.001$). Growth rate decreased with the time of migration ($P<0.01$). Examination of the relationship between birth date and growth rate of migrants in the inlet streams and residents in the lake showed that the migrants were early born with high or intermediate growth rates (about >0.25 mm/day), whereas the residents were late born fish or early born with extremely poor growth (<0.25 mm/day). It may be concluded that birth date and growth rate determine the timing of migration, or the migratory patterns.

Figure 11.2: Study area. Asterisks show the sampling locations of migrants of Lake Biwa aye (top) and amphidromous ayu (bottom). Solid circles indicate lake residents, and an asterisk in solid circle is premigrants.

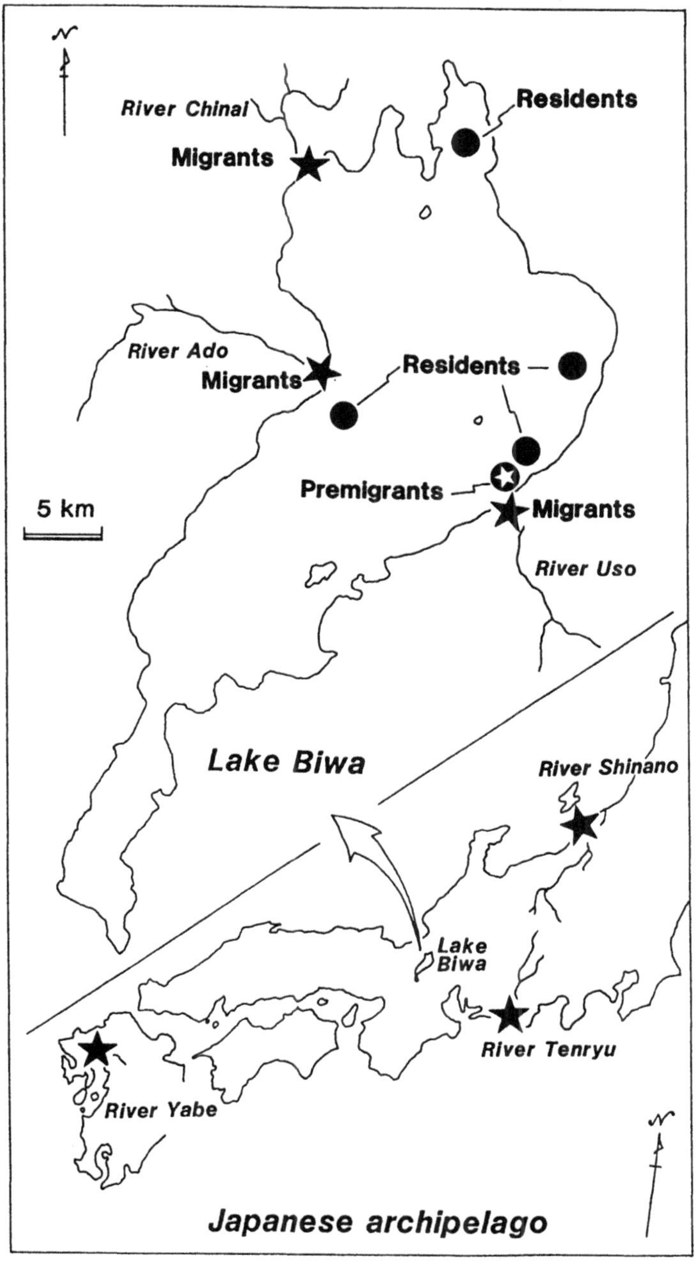

Figure 11.3: Sagitta otolith of ayu. A: a field collected sample of 18 mm in SL and 42 days old (estimated). B: reared fish of 35.5 mm and 90 days old.

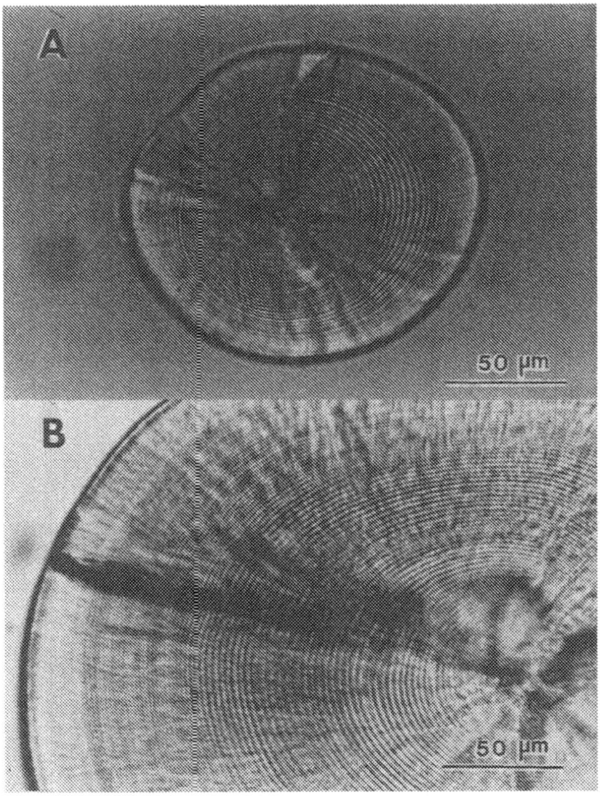

Figure 11.4: Changes in age (top) and body size (bottom) with the time of upstream migration in the landlocked population caught by the weir. Dot in parenthesis indicates the earliest fish going upstream in March of a different year (1983).

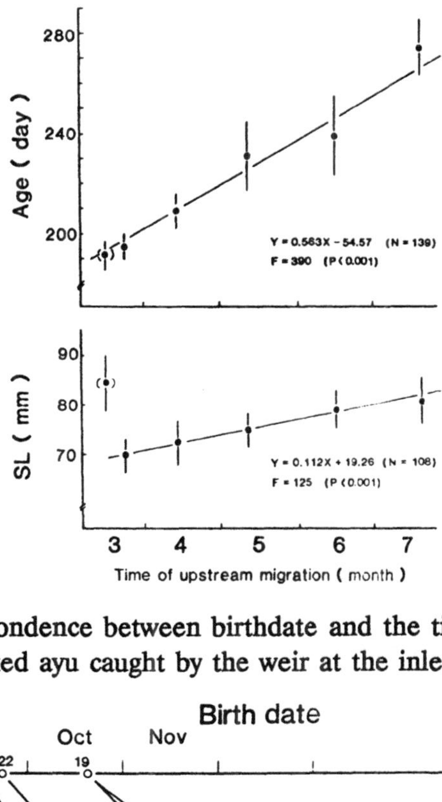

Figure 11.5: Correspondence between birthdate and the time of migration in the landlocked ayu caught by the weir at the inlet stream of Lake Biwa.

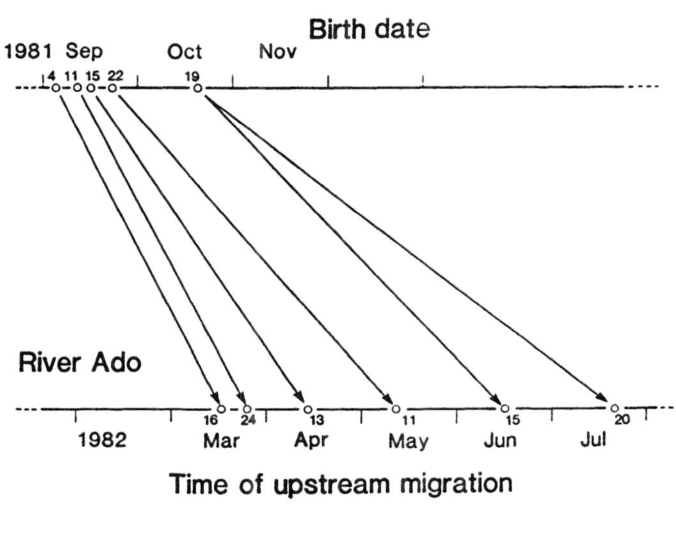

Amphidromous population

In the amphidromous marine populations of ayu associated with the Rivers Shinano, Tenryu and Yabe, linear correlations similar to the landlocked fish were obtained between the timing of migration and age ($P < 0.001$), growth rate ($P < 0.05$) and birth date ($P < 0.005$), although in neither river did size increase with the migration time ($P > 0.05$), which might have been due to the short sampling period of about one month. The minimum ages for migration were about 150-180 days and the minimum sizes were about 55-70 mm, both of which were smaller than those of the landlocked fish.

Principles in migration

During a part of 6-months marine life in winter, larval and juvenile ayu distribute in the surf zone as a nursery habitat (Tsukamoto 1988). They use this habitat as transient. Analysis of sequential samples at fixed surf zones during winter months shows similar results to those obtained in the upstream migrants: both age and size of ayu larvae at a surf zone increase with catch date ($P < 0.001$); growth rate decreases with catch date, and as a whole, it also decreases with birthdate, i.e. the earlier born, the better growth: the relationship between birth date and catch date is significant ($P < 0.001$), showing that the earlier a fish hatches, the earlier it recruits to surf zone. These results all agree with the the landlocked ayu larvae sampled at the larval habitat in the lake. Thus, we obtain some commonalities between larval recruitment and upstream migration of juvenile fish in both landlocked and amphidromous marine ayu.

By synthesizing all the above results on the relationships between the timing of migration (recruitment) and age, size, growth rate and birth date, we obtain the principle of migration of the ayu (Tsukamoto et al. 1987, Tsukamoto 1988): the earlier born fish with the faster growth rate migrate earlier at the younger age and the smaller size, whereas the later born, or the fish with the lesser growth rate migrate later at the older age and the larger size. Thus, the age and size peculiar to the migration time seem to be necessary conditions for upstream migration by ayu. Birth date and growth rate determine the age and size at migration. This process could be a fundamental first step in the mechanism of migration.

Migrants or residents?

The process of differentiation in migratory pattern among landlocked ayu might be summarized as follows. Most of the early hatched fish recruit early into the larval habitat in the lake and go upstream early as spring migrants (Figure 11.5). Late born fish and early born fish with poor growth recruit late to the larval habitat, remain in the lake until the end of spring and afterwards, separate into summer migrants and residents according to their endocrinological conditions (see below).

The spawning season of small residents is earlier (one month or more) than that of large migrants. Thus, the progeny of small residents must be born early and go upstream early to be large migrants. That is, small residents would produce large migrants in the next generation (Figure 11.6). On the other hand, large migrants mature late and their progeny, with late birth dates, have the potential to remain in the lake and become small residents.

Figure 11.6: Cyclic switching of the size and migratory pattern in successive generations of the lanklock ayu in Lake Biwa.

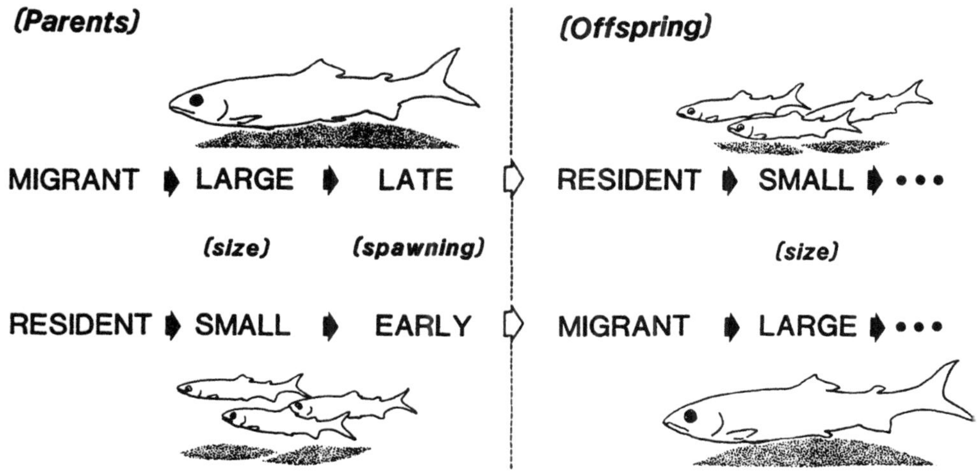

Thus, there would be cyclic switching in size and migratory pattern in successive generations (Tsukamoto et al. 1987). Bilton (1971) presented a similar hypothesis on alternation of age and size at return in successive generations of sockeye salmon. Azuma (1970, 1973a, 1973b, 1973c) classified the landlocked ayu in Lake Biwa into four groups (group A-D) based on the differences in morphological and ecological features. He suggested that the Lake Biwa fish were undergoing intraspecific differentiation, although its degree was not advanced. However, we could not find significant difference among residents, spring and summer migrants, for three meristic characters, the numbers of vertebrae and dorsal and anal fin pterygiophores ($P > 0.05$), although the fish showed significant differences in body forms which agreed with Azuma's (1973a) distinction between standing water type (residents in this paper) and running water type (migrants). It is known that body form changes easily with environmental conditions (Kafuku 1966, GFES 1978). Maruyama and Ishida (1978) showed that the rate of body hight increase of the ayu diminishes among fish inhabiting shallow water. Moreover, enzyme variation was not significant between residents and spring migrants (Nishida and Takahashi 1978; SFES 1982). Variation in body form might merely result from differences in environmental conditions, and not from genetic isolation. Although segregation of spawning seasons exists and spawning areas of migrants and lake residents are located in different reaches of a stream, these factors do not maintain spatial isolation of larvae in the lake. A mark-recapture study, in which an isozyme was used as a genetic marker, showed that the released fish spread widely through the lake and seemed to mix completely with the wild larvae (SFES 1983).

We consider that the differences in migratory patterns of the loandlocked ayu, migrants or residents, stemmed mainly from the difference in environmental temperature during larval stages. Eggs and larvae can experience quite different temperatures at the beginning and the end of a spawning season; temperatures can decline about 13 °C over this interval both at the stream spawning ground and at the lake surface (Tsukamoto et al. 1987). Temperature affects not only somatic development but also food abundance, and thus might also change the ontogeny of behavioral characteristics such as feeding, schooling, and vertical migration. We find that the growth rate of ayu, which integrates all these factors, affects the time of the recruitment. In addition, it is apparent that the chance match or mismatch of ayu larvae to the patchiness of food also influences the growth rate regardless of birth date.

Precisely speaking, however, there is an intermediate type of summer migrant between small residents and large migrants of the ayu. Early born fish with poor growth rates remain in the lake. Growth, and thus migratory pattern, might be determined by chance availability of food. Therefore, the genes of some small residents and large migrants would be well mixed through successive generations, contrary to the isolation between even-year and odd-year spawning populations of pink salmon Oncorhynchus gorbuscha (Ishida 1966). In conclusion, it is not genetic isolation but the differences in birth date, temperature-dependent growth rate, and endocrinological condition (see below) that cause the size and migratory pattern differences in landlocked ayu in Lake Biwa.

11.4 Endocrinological aspect

In the previous chapter, we found that 'birth date' and 'growth rate' determine the timing of migration; and that fixed 'ages' and 'sizes' peculiar to the time of migration are needed for the migratory ayu. Then, is the migration immediately initiated if only these factors are satisfied?

Four factors, age, size, birth date and growth rate, were compared among behaviorally different fish groups of Lake Biwa ayu collected at the same season, June 1984: summer stream migrants, lake residents, and 'premigrants' just before upstream migration which were caught by set net near the coast of lake. Although these three groups had roughly the same conditions of age, size, birthdate and growth rate, some went upstream and the others remained in the lake. Some of the fish remaining in the lake had greater size and growth rate than summer migrants, but did not themselves migrate. This result suggests that the age, size, birth date and growth rate are not sufficient conditions for the initiation of migration, although they are apparently necessary ones. Then, what causes such behavioral differences among migrants, premigrants and residents?

It is apparent that endocrine glands are involved in fish migration (Hoar 1953, Fontaine 1954, Baggerman 1960, Woodhead 1975, Leatherland (1982). Thyroid gland is the most famous among them for several aspects of migration including salinity preference (Folmar and Dickhoff 1980), oxygen consumption, temperature tolerance, parr-smolt transformation (Folmar and Dickhoff 1980), sexual development, locomotor activity (Hoar 1953), and metabolism of protein, lipid, carbohydrate and vitamin. We focus on the thyroid and thyroid hormone in this chapter and discuss their role in the upstream migration of ayu.

Methods

Thyroid gland activity was compared among three fish groups of Lake Biwa ayu:

summer stream migrants, lake residents, and premigrants. The height of epithelial cells of a thyroid follicle was measured histologically (Details described in Tsukamoto et al. 1987).

In order to determine the relationship between blood thyroxine (T_4 concentration and the upstream migratory behaviour in the ayu, five groups of juveniles (landlocked fish, amphidromous fish and reared fish from three different hatcheries; 500 fish each, 60-70 mm long) were released in an artificial river (length 250 m, width 1 m, depth 0.3-0.7 m, velocity 0.3-0.8 m/s), and plasma T_4 levels, which were determined by radioimmunoassay (RIA) procedures (Dickhoff et al. 1978, Aida et al. 1984), were compared between fish going upstream and downstream in each group (Tsukamoto et al. 1988).

Thyroid gland

The mean height of follicular epithelial cells (HFEC) in the thyroid gland of summer migrants was significantly lower (8.4 μm; SD, 3.3) than it was in the premigrants (11.4 μm; SD 2.1) and the residents (11.6 um; SD 1.9) that remained in the lake ($P<0.05$). The large variance of summer migrants stemmed from the frequency distribution of HFEC measurements, which had a mode less than 7 um and a tail that placed 20% of the means above 11 μm. Fish in the latter 20% had follicles with very scanty colloid and high epithelium, whereas colloid volumes of the modal follicles did not differ greatly from those of other groups. These results showed that fish going upstream had
follicles that were hypofunctioning, or were hyperactive with little hormone reserve, in contrast to the groups remaining in the lake, which had both high thyroid activity and much reserve. Honma and Tamura's histological observation (1963) showed that the thyroid presented hyperfunction just before the upstream migration in the juenile of landlocked ayu. Similar result was obtained in amphidromous ayu by Iwai (1962). Woodhead (1975) demonstrated that thyroid activity of the cod and the spurdog was not primarily associated with reproduction, but was closely related to their oceanodromous migration.

Thyroxine and migration

Plasma T_4 concentration of fish swimming upstream in the artificial river was 1.3-4.0 times higher in each group (mean of 5 fish groups, 11.2 ng/ml) than that of each counterpart going downstream (4.4 ng/ml; Figure 11.7). A great variation of plasma T_4 concentration was also found among the initial samples of different fish groups (Figure 11.7). Juvenile ayu of different groups show different patterns of dispersion when stocked in natural rivers (Tsukamoto et al. 1979, Masuda et al. 1981). Close correspondence was observed between plasma T_4 level and the tendency to swim upstream in rivers: i.e. the landlocked ayu of the Lake Biwa, which had the highest T_4 concentration among five groups, showed a strong tendency to go upstream in the mark-recapture experiment in a natural river, whereas hatchery-reared fish A (HR-A), with the lowest T_4 level, had tendency to distribute downstream and settled there. It can be roughly concluded that fish with the higher T_4 level have the stronger tendency to migrate upstream.

Figure 11.7: Plasma thyroxine concentrations (mean±SE, N at the base of the bar) of juvenile ayu of five groups: lanklocked fish of Lake Biwa (LAKE), amphydromous river fish (SEA) and reared fish from three different hatcheries (HR-A, HR-B, HR-C). Thyroxine levels of fish going upward and downward after released in the artificial river together with the initial value of each fish group are shown.

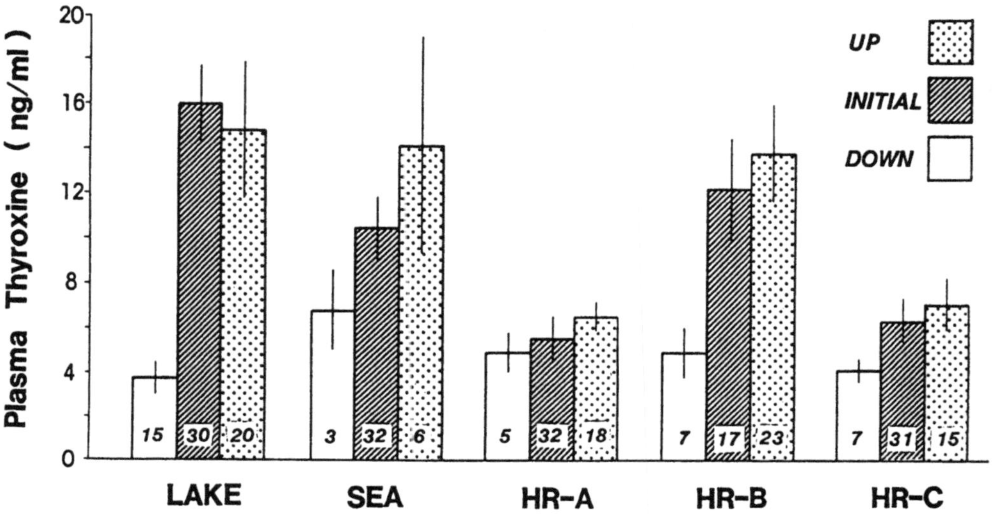

These results demonstrated undoubted evidence or the positive correlation between plasma T_4 and the upstream migratory behaviour itself.

Recent RIA study (Tsukamoto et al. unpublished) revealed that plasma T_4 level of the landlocked ayu were different among behaviorally different groups: premigrants just before upstream migration showed a high value (10.1 ng/ml), lake residents, a low value (3.3 ng/ml), and stream migrants, an intermediate one (4.8 ng/ml). T_4 surge seemed to occur in fish just before migration, and excreted T_4 had already been metabolized in migrants. Sower and Schreck (1982), Ueda et al. (1984), Yamauchi and Takahashi (1987) obtained similar results in coho salmon, chum salmon and Japanese smelt, respectively, suggesting that thyroxine is involved in determining the readiness for the movement of fish from the sea to fresh water.

Comparisons between the histological results and plasma T_4 levels show that premigrants and residents are in quite different states although they have similar heights of follicular epitherial cells. Residents might have reached a preparatory stage for migration with high activity of the thyroid but without the T_4 surge. Premigrants seem to be in a state just before or during migration with the T_4 surge in the blood, whereas migrants show a state of exhaustion both histologically and in actual plasma T_4 level. Thus, the endocrinological condition must be satisfied as the second step for migration.

11.5 Behavioral aspect

What was different among lake residents, premigrants and stream rate but the difference in activity of thyroid gland and its hormone level in blood? Then, does the aye start the upstream migration immediately after these endocrinological conditions are satisfied together with specific age and size? It is well known that rainfall, sudden change in temperature and other special environmental changes have the potential to cause the migration of various animals. It might be also environmental cues that act as a trigger of the upstream migration of aye which finishes preparing all priming conditions for migration and crowd to a mouth of river.

In this chapter, we examine the influences of environmental falters on the upstream migration of aye through laboratory techniques by measuring the behavioral indices which represent the tendency to migrate upstream: (1) jumping behaviour, (2) swimming upstream, and (3) spacing. Behavioral model involving the 'drive' (motivation) for migration is also presented in order to clarify the triggering mechanism of the upstream migration of aye.

Methods

Five groups of juvenile aye (landlocked fish, amphidromous fish and reared fish from three different hatcheries, 60-70 mm long) were examined in experiments of the above three behavioral indices (Tsukamoto et al. in press). Behavioral characteristics were also compared for three groups of Lake Biwa aye: summer stream migrants, lake residents, and premigrants (Uchida 1986).

Figure 11.8: An apparatus to measure the jumping activity of ayu.

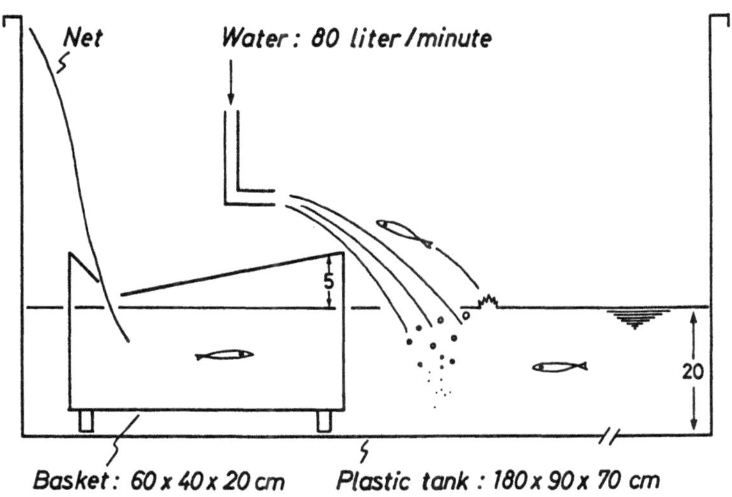

'Jumping behaviour' was defined as jumping response toward a waterfall which was given as a stimulus (Figure 11.8). Strength of tendency to jump was estimated for a fish group as % fish which jumped during a fixed period. The strength of tendency of an individual fish was also measured in ladder-type experimental tank (Uchida 1986) and was expressed as the number of jumping that the fish performed for a period.

'Swimming upstream' is a behaviour to go upward for a fixed distance in an artificial stream (20-50 cm/sec) which is induced by a water inlet of experimental fish container (a stimulus). Strength of tendency to go upstream was estimated in the same manner as in jumping behaviour.

'Spacing' is a schooling behaviour of 100 juvenile aye which are contained in an observation tank (diameter 80-120 cm, water depth 30 cm; Figure 11.9). Each fish group presents a peculiar type of spacing: aggregation, pod or school, loose or tight, and slow or fast. Although some fish group disperse all over the tank, they are considered to gather into a loose school or aggregation when contained in a larger tank with sufficient space. Therefore, under a fixed environmental and physiological conditions, a fish group has its own 'Optimum Distance to the Nearest Neighbour (ODNN)", and forms a school or aggregation so that fish may keep their own ODNN in a sufficiently large space. The value of ODNN can be given as the mean of the distances between nearest neighbours in a school or an aggregation.

Behavioral indices

Close relationship was observed between the results of jumping test in laboratory and the migratory behaviour observed in the mark recapture study in a river, i.e. a fish group with the higher jumping rate (e.g. landlocked stock) showed the stronger tendency to migrate upstream when released in a natural river (Tsukamoto et al. in press). Jump-fish of each group, which jumped in the jumping test before the release, actively swam upstream in a river, whereas non-jump-fish, each counterpart of a group, showed a less tendency to migrate upstream (HFES 1985). Fish stock with the stronger tendency to swim upstream in an artificial stream presented the more active upstream migration in a released river (Tsukamoto et al. in press). Correlation was also obtained between the results of spacing and mark-recapture study in a field: fish with the larger ODNN, which widely dispersed in a tank or formed a loose school (see Figure 11.9 A), showed the stronger tendency to swim upstream in a river, whereas fish with the small ODNN which formed a pod or a tight aggregation (see Figure 11.9 B) dispersed downstream or remained at the point of release.

Behaviorally different three groups of the Lake Biwa aye also showed different responses in each test of jumping behaviour and spacing (Uchida 1986). Stream migrants presented the largest ODNN and the highest jumping rate among three groups, and premigrants, lake residents in order. Thus, it is apparent that laboratory tests of jumping, spacing and swimming upstream are available as behavioral indices to represent the tendency to migrate upstream in nature.

Figure 11.9: Spacing of juvenile ayu. A: landlocked ayu showing a wide dispersion (less intensive aggregation with large ODNN). B: reared fish presenting a tight pod (intensive aggregation with small ODNN). Aggregation sometimes changes into circular schooling with a peculiar diameter.

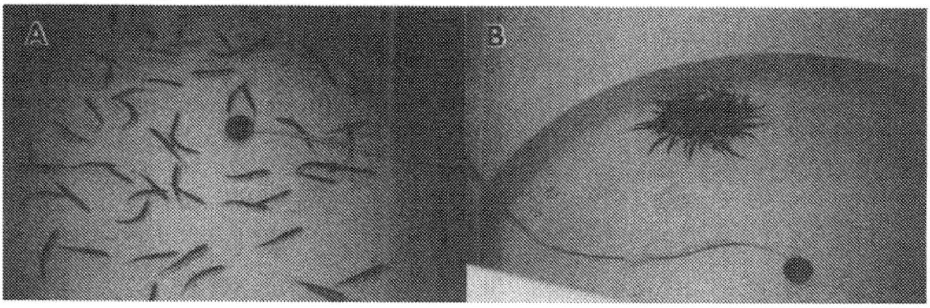

Environmental factors

Several environmental factors influence the behavioral indices, and thus the migratory behaviour of the aye. Increase in fish density facilitated both jumping activity and swimming upstream (Tsukamoto et al. 1985; Figure 11.10). Rise of water temperature accelerated jumping behaviour of the aye and contrarily, a fall suppressed it (Figure 11.11). Fish acclimated at the higher temperature presented the more active jumping behaviour and the larger ODNN in spacing (Uchida 1986), i.e. temperature level as well as its change had effect on behaviour. Sudden changes of light intensity, both rise and fall, facilitated jumping behaviour. Water depth also influenced the behaviour of aye: shallow water stimulated both jumping behaviour (Uchida 1986) and swimming upstream (Ujiie et al. 1983).

Internal circadian rhythm seems to control jumping activity of the aye (Uchida 1986). Jumping was observed only during the hours that corresponded to daytime, when fish were kept for more than a week under a constant temperature and light condition (complete light or dark condition).

Feeding condition also affected both jumping and spacing (Uchida 1986). Fish with deprived food jumped more actively than those at full stomach. Hungry fish dispersed widely in a tank, presenting the large ODNN (Figure 11.9 A), but they showed aggregation with the small ODNN (Figure 11.9 B) immediately after they were fed. This phenomenon was a reversible change easily caused by feeding condition.

Figure 11.10: Density effect of juvenile ayu on jumping activity and swimming upstream.

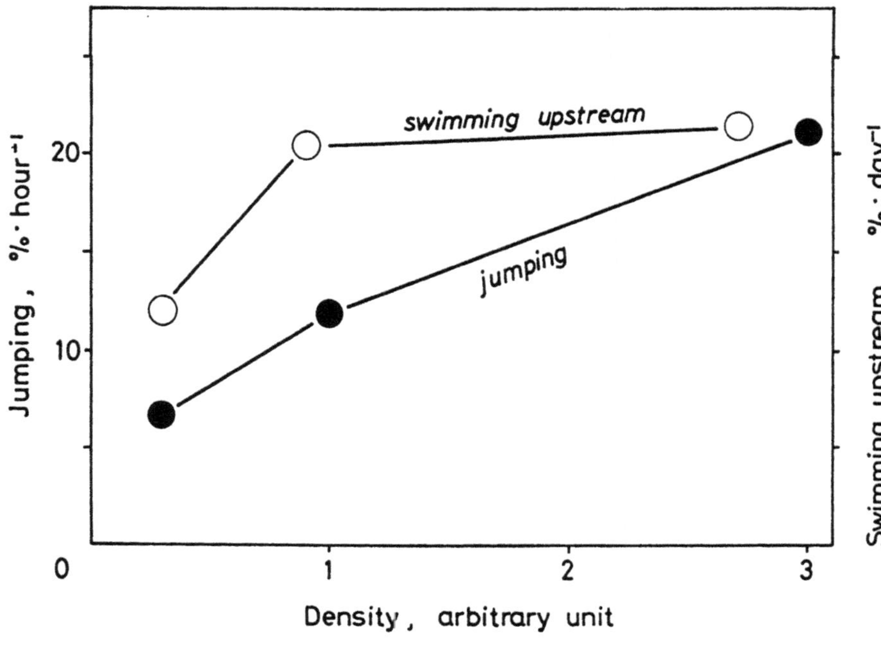

Figure 11.11: Effect of water temperature changes on jumping activity of the juvenile ayu. Two experimental groups (A and B) were simultaneously exposed to the symmetrical changes of water temperature.

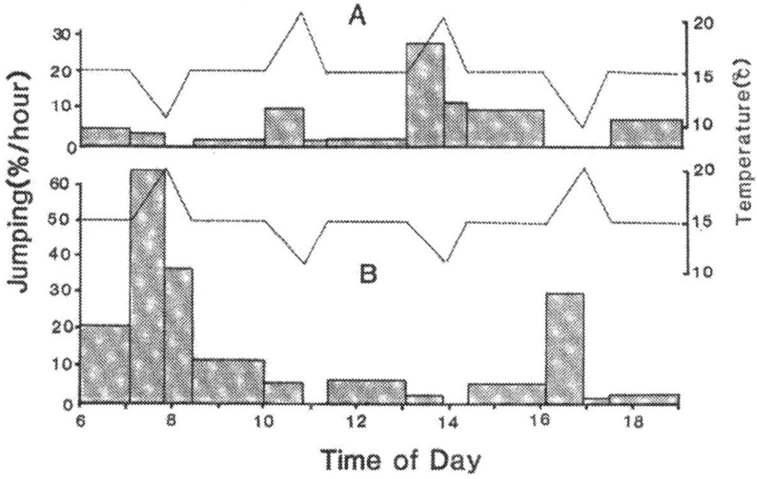

These laboratory data coincide well with the previous field observations. Juvenile aye crowd into a river mouth one after another in early spring. But low temperature of riverine water suppresses their upstream migration, and fish density increases rapidly at the mouth of river. Thus, a rise of temperature triggers their upstream migration. The number of migratory fish increases with a rise in temperature of riverine water (Cotta 1952, Kusuda 1963; Tanizaki 1966, Fushiki et al. 1978). Sudden change in light intensity also facilitates the upstream migration of aye at fish way of a dam (Koyama 1980). Aye show upstream migration only during daytime (Fujioka et al. 1980), presenting a clear diurnal rhythm in a natural environment.

In addition, age and size also affect the jumping activity of the aye (Uchida 1986). The larger fish of same age group, or the older fish of same size group presents the more active jumping behaviour, An individual fish that shows a strong jumping activity in the first test does jump actively in the second or third test, and vice versa for a less active jumper. Each fish has each individual character in jumping behaviour.

Stimulus and drive

Stimuli for jumping and swimming upstream are waterfall and current, respectively. Sound and current as well as waterfall can induce jumping (Uchida 1986). Aye responded most to a 200 Ho sound and the jumping rate increased with sound pressure (Konagaya 1980). Although a rise in temperature or an increase in fish density can also induce jumping without a stimulus of waterfall in a tank, such jumping occurs randomly and has no orientation. This is considered to be a vacuum activity, which is released without any kinds of stimuli by increasing drive level and thus by decreasing behavioral threshold to the minimum level. Upstream migration is not a random movement but an orientated travelling, i.e. from the sea to a river, or from lower reaches to upper ones. What induces only random movements is not a stimulus for upstream migration. Therefore, a rise in temperature or increase in fish density can not be regarded as a stimulus but a promotor for jumping, or upstream migration of the aye.

Even if a stimulus (waterfall or current) is a fixed one, environmental factors modify the responses of behaviour (jumping or swimming upstream), and the response is also different according to fish stocks. Consequently, we must put an intermediate function between a stimulus and a response d a response (Figure 11.12). We call this function as 'drive', which is a cause to drive and urge an animal to do something. Drive is a wide concept which includes all physiological and psychological mechanisms connecting between stimulus and response. Motivation for the upstream migration of the ayu is also included by this drive.

Interrelationship among behaviours

Interrelationship among the above indices can be summarised as follows: fish with large ODNN show active jumping and strong tendency to swim upstream, whereas fish with small ODNN have less tendency to jump and swim upstream. Activity rhythms coincided well between jumping behaviour and swimming upstream when these two indices were measured at the same time in an experimental tank (Uchida 1986). An individual fish that showed the stronger jumping activity in ladder type experimental tank demonstrated the greater tendency to swim upstream in a similar tank. Fish showed a jump toward the current direction (stimulus) when they were prevented by an obstacle of wire net from swimming upstream in an experimental stream. These results suggest a close interrelationship between the indices, and therefore, it is quite possible that jumping and swimming upstream are released through a common behavioral mechanism.

Repulsive force

Different fish stocks had different values of ODNN in spacing, which also corresponded with the tendency to swim upstream in a river. Since fish form a school or aggregation so that they may keep their own ODNN value constant, 'repulsive force' occurs between individuals when they come close to enter the range of their ODNN, and then, they go away from each other (Parr 1927, cited in Radakov 1972). Contrarily, when they are far apart beyond the ODNN, 'attractive force' acts between neighbours, and they approach each other to the distance of ODNN.

Figure 11.12: Behavioural model for the triggering mechanism of the upstream migration of ayu. An endogenous drive of migratory behaviour is located as an intermediate function between stimulus and response, and the drive of ayu is assumed as repulsive force which acts between individuals when they come close to enter the range of their own Optimum Distance to the Nearest Neighbour (ODNN).

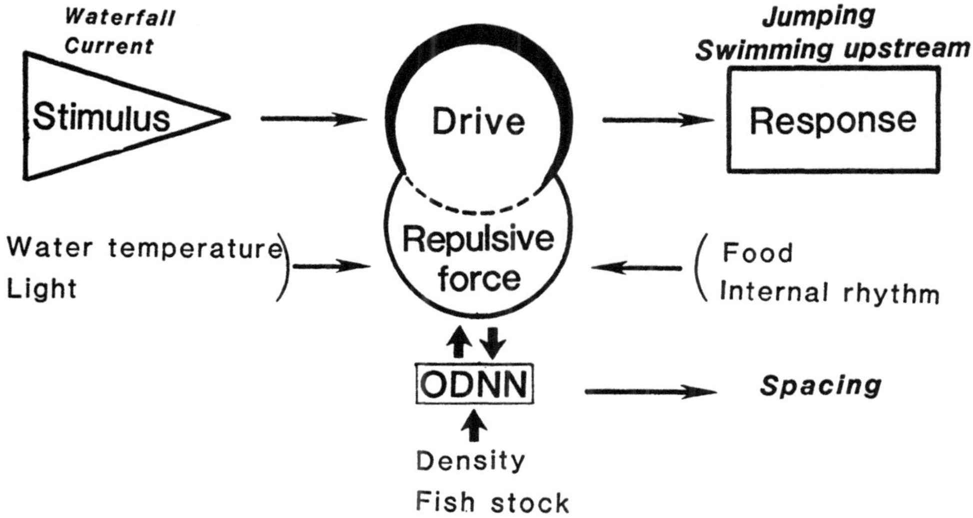

Behaviourial model

We assume that a drive of the migratory behaviours of ayu is a repulsive force itself that acts between neighbours in a tight school, and obtain a behaviourial model for the triggering mechanism of upstream migration of ayu (Figure 11.12). Various phenomena above described as well as the interrelationships among indices can be explained by the model. Differences in jumping activity and tendency to migrate upstream among fish stocks have different values of ODNN, repulsive forces occurring between individuals differ among fish stocks when they are contained in a limited space or released at a point of river; thus, different repulsive forces, that is, different levels of drive cause the different responses even if a stimulus is constant. When fish are compressed into a narrow space and density rises, the distance between neighbours diminishes than ODNN and a resulting repulsive force increases with a rise in fish density. Accordingly, a level of drives elevates, or a threshold to stimulus falls, and thus, a rise in fish density causes an increase in jumping activity and swimming upstream even if to a fixed level of waterfall or current. Furthermore, environmental factors such as temperature, light, and water depth as well as feeding condition and internal rhythm have a direct effect on a level of repulsive force. A rise in water temperature, for example, increases repulsive force and rises a level of drift, and thus, jumping is facilitated. Repulsive force increased ho temperature rise expands the ODNN and accordingly, a change occurs in spacing: school loosens, or fish aggregation disperses.

11.6 Migration mechanism

Migration model of ayu

Migration model is presented by synthesizing all ecological, endocrinological and behavioral results obtained in the upstream migration of the aye (Tsukamoto 1988, Figure 11.13). Drive and priming conditions which are arranged in the left side of the figure are preparatory conditions required for the initiation of migration, i.e. adaptive functions and abilities for a new habitat such as smoltification, salinity preference, temperature tolerance and excitation of sensory receptors. Factors affecting these conditions are divided into three categories in the right side (1st-3rd steps in Figure 11.13). First, migratory fish must be provided with a fixed age and size peculiar to a time of migration, which are determined by birth date and growth rate. Through these processes, priming conditions such as 'form' and 'ability' are prepared as the first step. Next, contribution of endocrine system is needed for migration. Endocrinological factor affects osmoregulatory 'function' and salinity 'preference' and endows fish with priming conditions for migration as the second step. Lastly, environmental factor triggers migration as the third step. It acts central nervous system through receptors and directly elevates the level of drive. On the other hand, environmental factor such as rainfall, river discharge or a sudden rise in water temperature induces hormone surge via nervous system. Hormone surge also has the potential to elevate the level of drive. Elevated drive by nervous or endocrine system, or by both, lowers the threshold of migratory behaviour to an environmental stimulus, and thus, migration is initiated in response to a stimulus which has never before been effective (Tsukamoto 1988).
Environmental factor in the third step is, of course, involved in both the first and

second steps, although it is explained in the third step of this model. We conventionally categorized the factors affecting migratory conditions into three steps and put emphasis on their order, but their processes, in a strict sense, would sometimes advance simultaneously and it is difficult to divide clearly these steps in time.

Upstream migration of the aye can be divided into two phases, i.e. approaching phase to a river mouth and ascending phase in a river. It might be interpreted that the first and the second steps in the model correspond to the the former phase from the sea to a mouth of river, while the third one is involved in the latter phase including the switching point from the former to the latter phase.

Bag German (1960) presented a model on the triggering mechanism of migration with emphasis on endocrinological aspect. In her model, priming factors, e.g. photo period and water temperature, affect endocrine system and provide with necessary conditions for migration such as osmoregulatory ability, metabolism and perception of receptors. Releasing factor, e.g. temperature, river discharge or weather, finally triggers migration in the animal that complete the preparation of migratory conditions. However, the model lacks the behavioral interpretation on the triggering mechanism of releasing factor. Releasing factors such as water temperature and weather conditions are not stimuli of the migratory behaviour but the factors that elevate the level of drive, since they do not directly cause an orientated response by themselves in fish contained in still water.

Age and size

Various biological phenomena depend on age or size, such as development of various organs, metamorphosis, maturation (Aim 1959), and smoltification (Katz 1983). Various environmental conditions weaken the close relationships between age or size and biological developments by modifying the developmental time table. On the other hand, size is closely related to age as a function of time. Differentiation of the aye stomach depends on both age and size (Yamazaki 1986). Development of jumping behaviour by aye, which is closely related to migration behaviour, seems to be controlled by both age and size (Uchida 1986). Age and size cannot be separated in biological development. As in the case of maturation (Aim 1959), it appears that there must be minimum necessary conditions of both age and size for the upstream migration by aye. These conditions seem to be about 130 days and 50 mm for amphidromous marine populations and about 180 days and 60 mm for the landlocked population. Although the cause of the age and size differences between these two kinds of populations is still unclear, the landlocked fish spawn 1-2 months earlier than the amphidromous ones. Furthermore, environmental conditions such as water temperature, salinity and food abundance may affect the age and size at migration.

Figure 11.13: Migration model of ayu. Migratory drive and priming conditions in the left side of the figure are prerequisites for the initiation of migration. Factors which control these conditions are divided into three categories in the right side (1st-3rd steps).

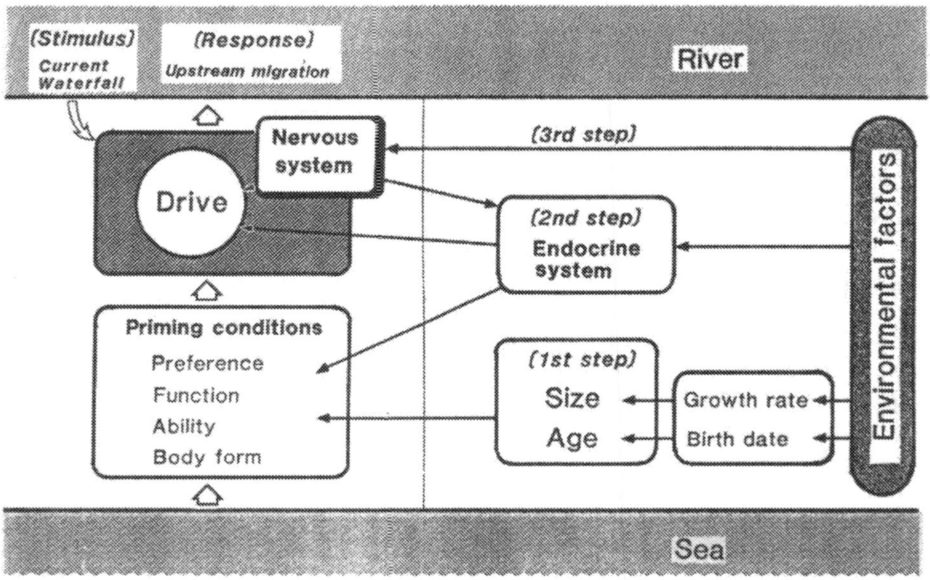

Some of the results obtained for the aye agree with those from studies of salmonid fishes. The age and size at return varies greatly in species such as chum salmon *Oncorhynchus zeta* (Seiko 1954; Sanyo 1966), sockeye salmon (Milton 1971), Arctic char (Nordeng 1961) and brown trout (Jonsson 1985). Among fish born in the same year, it is apparent that the age that salmon or trout return to spawn increases with the year they migrate from the nursery stream. The correspondence is 1 : 1 on the yearly scale with which these fish typically are aged. A closer comparison with the aye results cannot be made because of the differences in time scale. Sockeye salmon returning at age 5 were smaller and showed a faster growth rate than those returning at age 6 (Egorova, unpublished: cited in Hanamura 1966). This finding is consistent with our results; aye size increased with the migration time and fish with larger growth rate started their migrations earlier.

Hormonal control in fish migration seems to have two kinds of actions, first and slow processes. Clear coincidence of T_4 surge at new moon and the initiation of downstream migration of coho salmon *Oncorhynchus kisutch* (Grau et al. 1981) suggests an existence of fast process in the endocrine system. It is apparent that salinity preference, osmoregulatory function are obtained through a slow process according to somatic growth, various ontogenic changes and metamorphosis smoltification.

We found that thyroid activity and plasma T_4 concentration have close relationships with the migratory behaviour of the aye. Thyroid drugs have been shown to increase locomotor activity in salmons (Hoar et al. 1952, Hoar et a. 1955), goldfish (Hoar et a. 1955), trout (Wood head 1975), guppy (Sage 1968), and cod (Wood head 1970). Hoar et a. (1955) observed that T_4 treatment increased the general level of excitability of the goldfish and activated the jumping behaviour of yearling salmon. The schooling behaviour of chum and coho salmon fry was less intensive in the treated fish than in the control, while thiourea treatment increased the tendency to form schools (Hoar et a. 1952). These changes induced by T_4 are generally advantageous for the migratory fish and may explain the process of migration to some extent. However, these laboratory observations on the salmons such as locomotor activity, schooling and jumping behaviours have never been shown to bear any direct relationship with migratory behaviour in a field. Grau et a. (1981) confirmed that the timing of downstream migration of cono salmon. coincided with a T_4 surge at new moon. Yamauchi Et al. (1985) observed in an artificial river that the downstream migration of the juvenile mast salmon was caused by a rainfall with turbidity following the new moon. Next step would be an examination of relitionships among T_4 surge, drive and migratory behaviour itself. This is quite possible in birds, that prolactin plays an important role in premigratory fattening and at the same time increases, so called, 'night restlessness' itself, which seems to be an endogenous drive for bird migration (Meier et al. 1965, Meier 1972). In fish, however, it is still unclear whether or not thyroid hormone has a direct relationship with the drive for migration itself. We observed that the artificial elevation of plasma T_4 level by the intraperitoneal injection of T_4 or the immersion in T_4 solution had no effect on jumping behaviour and schooling (ODNN) in the juvenile aye (Otaka et al. unpublished). It is probable that plasma T_4 might have no direct relationship with both the migratory behaviour and its endogenous drive. We can not deny the possibility that plasma T_4 may play a subsidiary role which supports the function and ability required for the migration, e.g. the elevation of metabolism, or swimming ability. It also seems necessary to examine the relationships between migratory behaviour and the other hormones such as sex hormone, growth hormone, prolactin and cortisol.

Drive for migration

We assumed a repulsive force as an endogenous drive for upstream migration of aye. The level of drive is roughly estimated by the value of ODNN. However, we have little knowledge on the substance of drive or repulsive force, or on its neurophysiological and endocrinological background. Behavioral mechanism of downstream migration is also unknown in smelts of salmons or adult eels, where a stimulus for the behaviour has been equivocal. These are subjects remaining for the future.

It is easy to categorize fish migration into spawning migration, feeding migration or wintering migration according to their objects. It is also superficial interpretations that the juvenile chum salmon undertakes its migration for the N Pacific in order to avoid a rise in temperature of coastal water, or that adult salmons ascend a river since they cannot live in sea water any more. Various conditions such as osmoregulatory function, smoltification or premigratory fattening are the functions or the abilities that have been gradually obtained through long evolutional histories of fish migration. It is not a species which is completely provided with all these priming conditions that begins the first migration. Furthermore, there are many species whose migration route is restricted only to fresh water or to the sea, and they do not need to change the osmoregulatory mechanism dramatically. To be able to live in the sea/fresh water, to select the sea/fresh water, and to intimate different things. If examining only migratory functions or abilities, we can not get to the ultimate goal, why fishes do migrate. Migration merely implies moving from one habitat to another. The essential factor of fish migration must be an endogenous drive for movement, like "Zugunruhe" or "night restlessness" in birds, which drive an animal and urge it to perform a certain action. Behavioral and psychological approaches as well as neurophysiology should be involved in the studies on fish migration.

References

Aida, K., T. Kato, and M. Awaji. 1984. Effects of castration on the smoltification of precocious male mast salmon *Oncorhynchus mason*. Bullet.in of the Japanese Soc.of Scientific Fisheries 50:565-57.

Alm, G. 1959. Connection between maturity size and age in fishes. Institute of Freshwater Research, Drottningholm, Sweden, Report 40:5-115

Azuma, M. 1970. Studies on the variability of the land-locked ayu-fish, *Plecoglossus altivelis T. and S.*, in Lake Biwa. I. On the mode of distribution and some body form variation at early phases of development. Japanese J. of Ecology 20:63-76.

Azuma, M. 1973a. Studies on the variability of the land-locked ayu-fish, *Plecoglossus altivelis T. and S.*, in Lake Biwa. II. On the segregation of population and the variations in each population. Japanese J.of Ecology 23:126-139.

Azuma, M. 1973b. Studies on the variability of the land-locked ayu-fish, *Plecoglossus altivelis T. and S.*, in Lake
Biwa. III. On the differences in the process of maturation, spawning habits, and some morphological features of each population. Japanese J. of Ecology 23:147-159.

Azuma, M. 1973c. Studies on the variability of the land-locked ayu-fish, *Plecoglossus*

altivelis T. and S., in Lake Biwa. IV. Considerations on the grouping and features of variability. Japanese J.of Ecology 23: 147-159.
Baggerman, B. 1960. Factors in the diadromous migrations of fish. Pages 33-60 in I. C. Jones, editor. Hormones in fish: Proc.of Symposia of the Zool.Soc. of London, No.1. The Zool.Soc.of London, London, U.K.
Milton, H. T. 1971. A hypothesis of alternation of age of return in successive generations of Skeena River sockeye salmon (*Oncorhynchus nerka*). J.of the Fisheries Res.Board of Canada 28:513-516.
Dickhoff, L. C. Folmar, and A. Gorbman. 1978. Changes in plasma thyroxine during smoltification of smoltification of coho salmon, *Oncorhynchus kisutch*. General and Comparative Endocrinology 36:229-232.
Fontaine, M. 1954. Du determinism physiologique des migrations. Biological Review 29 : 390-418 (in French).
Folmar, L. C., and W. W. Dickhoff. 1980. The parr-smolt transformation (smoltification) and sea water adaptation in salmonids. A review of selected literature. Aqua culture 21:1-37.
Fujioka, Y., S. Fushiki, K. NASA, Y. Oh no, S. Tazawa, and S. Satoi. 1980. Effect I. The fish ways of anadromous fishes in the ayu-I. Scientific Report of Shiga Prefectural Fisheries Experimental Station 33:16-22.
Fukuhara, O., and T. Fushimi. 1986. Development and early life history of the aye reared in the laboratory. Bull.of the Japanese Soc.of Scientific Fisheries 52(1):75-80.
Fushiki, S., S. Tazawa, and H. Matoba. 1978. On the migrating upstream of ko-ayu (*Plecoglossus altivelis T. & S.*) at the artificial spawning channel in spring and summer season. Scientific Report of Shiga Prefectural Fisheries Experimental Station 30:15-19.
GOES (Gift Prefectural Fisheries Experimental Station). 1978. The effect of stocking of reared juveniles of the aye. Freshwater Fisheries Laboratory and Research Division of Japan Fisheries Agency, Final report of 1972-1976, Tokyo, Japan.
Grau, E. G., W. W. Dickhoff, R. S. Nishioka, H. A. Bern, and L. C. Folmar. 1981. lunar phasing of the thyroxine surge preparatory to seaward migration of salmonid fish. Science (Washington, D. C.) 211:607-609.
Hanamura, N. 1966. Salmon of the north Pacific Ocean, part III. A review of the life history of north Pacific salmon, 1 Sockeye salmon in the Far East. International North Pacific Fisheries commission Bulletin 18:1-27.
HUES (Hiroshima Prefectural Fisheries Experimental Station, Freshwater Fish Branch). 1985. The effective utilization of the river as a fishing ground. Report of 1984 Research by grant of Japan Fisheries Agency, Shyobara, Hiroshima, Japan:1-24.
Hoar, W. S. 1953. Control and timing of fish migration. Biological Reviews of the Cambridge Philosophical Society 28:437-452.
Hoar, W. S., M. H. A. Keenleyside, and R. G. Good. 1955. The effects of thyroxine and gonad Al steroids on the activity of salmon and goldfish. Canadian J.of Zoology. 33:428-439.
Hoar, W. S., D. MacKinnon, and A. Redlich. 1952. Effects of some hormones on the behaviour of salmon fry. Canadian J.of Zoology 30:273-286.
Honda, Y., and E. Tamura. 1963. Studies on the endocrine glands of the salmon id fish, the aye, *Plecoglossus altivelis* Temminck and Schlegel. V. Seasonal changes in the endocrine of the land-locked form, the koayu. Zoological, New York Zoological Society 48:25-31.
Cotta, H. 1952. On the change of the number of the anadromous "Aye' (*Plecoglossus*

altivelis T. & S.). J. of Ichthyology 2(3):113-116.
Ishida, T. 1966. Salmon of the North Pacific Ocean, part III. A review of the life history of North Pacific salmon, 2. Pink salmon in the Far East. International North Pacific Fisheries Commission Bulletin 18:29-39.
Iwai, T. 1962. Studies on the *Plecoglossus altivelis* problems: embryology and histophysiology of digestive and osmoregulatory organs. Bulletin of the Misaki Marine Biological Institute Kyoto University 2:1-101.
Jonsson, B. 1985. Life history patterns of freshwater resident and sea-run migrant brown trout in Norway. Tran.of the American Fisheries Society 114:182-194.
Hafuku, T. 1966. Morphological differences between domesticated common carp and wild ones. Bulletin of Freshwater Fisheries Research Laboratory (Tokyo) 16:71-82.
Katz, T. 1983. Studies on maturation and the smoltification in pond cultured mast salmon, Oncorhynchus mason. Pages 1-12 in Progress reports on mast salmon production 3, Japan Fisheries Agency, Marine Ranching Program, Tokyo, Japan.
Konagaya, Y. 1980. Jumping response of aye to sound. Bull.of the Japanese Soc.of Scientific Fisheries 46(1):31-34.
Koyama, N. 1978. Ecology of aye. Pages 1-176: Cyukoshinsyo. Chyuokouronsya, Tokyo, Japan.
Kubo, T. 1980. Studies on the life history of the "mast" salmon (*Oncorhynchus mason*) in Hokkaido. Scientific Report of the Hokkaido Salmon Hatchery 34:1-95.
Kusuda, R. 1963. An ecological study of the anadromous "aye", *Plecoglossus altivelis T. and S.*-II. Seasonal variations in the composition of the anadromous aye schools. Okumo, Kyoto. Bull.of the Japanese Soc.of Scientific Fisheries 29:822-827.
Leatherland, J. F. 1982. Environmental physiology of the teleostean thyroid gland: a review. Environmental Biology of Fish 7(1):83-110.
Maekawa, K. 1984. Life history patterns of the miyabe charr in Shikaribetsu Lake, Japan. Pages 233-250 in L. Johnson and B. Burns, editors. Biology of the Arctic char: Proc.of the International Symposium on Arctic char. University of Manitoba Press, Manitoba, Canada.
Maruyama, T, and R. Ishida. 1978. Effect of water depth of the pond on the growth and body form of the aye. Bull.of Freshwater Fisheries Research Laboratory (Tokyo) 28:211-219.
Masuda, N., M. Endo, K. Tsukamoto, T. Takeda, and R. Ishida. 1981. Influences of hatchery and lake origins of released aye juveniles *Plecoglossus altivelis* on their dispersion in the Tsubusa river. Bull.of the Japanese Soc.of Scientific Fisheries 47(8):1093.
Meyer, A. H. 1972. Temporal synergism of prolactin and adrenal steroids. General and Comparative Endocrinology, Supplement 3:499-508.
Meyer, A. H., D. S. Farber, and J. R. King. 1965. A possible endocrine basis for migratory behaviour in the white-crowned sparrow, *Zonotrichia leucophrys*. Animal Behaviour 13:453-465.
Nishida, M. and Y. Takahashi. 1978. Enzyme variation in populations of aye, *Plecoglossus altivelis*. Bull.of the Japanese Soc.of Scientific Fisheries 44:1059-1064.
Nordeng, H. 1961. On the biology of char in Salangen, North Norway. I. Age and spawning frequency determined from scales and otoliths. Nytt Magasin for Zoology (Oslo) 10:67-123.
Parr, A. E. 1927. A contribution to the theoretical analysis of the schooling behaviour of fishes. Oct. Pap. gingham Oceanogr., Colln.1.
Radakov, D. V. 1972. Schooling in the ecology of fish. Pages 1819: John Wiley and

Sons, New York, USA.
Rocker, W. E. 1938. "Residual" and kokanee salmon in Cult us Lake. Journal of the Fisheries Research Board of Canada 4:192-218.
Sage, M. 1968. Respiratory and behavioral responses of *Poecilia* to treatment with thyroxine and thyourea. General and Comparative Endocrinology 10:304-309.
Sanyo, S. 1966. Salmon of the North Pacific Ocean, part III. A review of the life history of North Pacific salmon, 3. Chum salmon in the Far East. International North Pacific Fisheries
Commission Bulletin 18:41-57.
Seiko, R. S. 1954. The stocks of West Kamchatka salmon and their commercial utilization. Izvestia TINRO 41:3-109.
(in Russian).
SUES (Shiga Prefectural Fisheries Experimental Station) 1982. Study on the propagation of the land-locked aye. SUES, Report of 1981 research by grant of Japan Fisheries Agency, Hikone,
Japan.
SUES (Shiga Prefectural Fisheries Experimental Station) 1983. Study on the propagation of the land-locked aye. SUES, Report of 1982 research by grant of Japanese Fisheries Agency, Hikone, Japan.
Sower, S. A., and C. B. Schreck. 1982. Steroid and thyroid hormones during sexual maturation of Cohn salmon (*Oncorhynchus kisutch*) in sea water or fresh water. General and Comparative Endocrinology 47:42-53.
Tanizaki, R. 1966. Catch of the aye and water temperature during the season of upstream migration. Physiology and Ecology 14:33-38.
Tsukamoto, K. 1988. Migratory mechanisms and behavioral characteristics in ayu. Pages 100-133 in T. Ueno and M. Okiyama, editors. Ichthyology Currents: Asakura Shot en. Tokyo, Japan.
Tsukamoto, K., A.Ida, and T.Take. 1988. Plasma thyroxine concentration and upstream migratory behaviour of the aye in an artificial river. Bulletin of the Japanese Society of Scientific Fisheries 53:1687-1693.
Tsukamoto, K., R. Ishida, K. NASA, and T. Kajihara. 1987. Switching of size and migratory pattern in successive generations of landlocked aye. Pages 492-506 in M. J. Dad swell, R. J. Klauda, C. M. Moffitt, R. L. Saunders, R. A.
Rulifson, and J. E. Cooper, editors. Common strategies of anadromous and catadromous fishes: American Fisheries Society Symposium 1. American Fisheries Society, Bethesda, Maryland, USA.
Tsukamoto K, and T. Kajihara. 1987. Age determination of aye with otolith. Bulletin of the Japanese Society of Scientific Fisheries 53(11):1985-1997.
Tsukamoto, K., N. Masuda, Y. Morn and K. Kajihara. 1979. Dispersion of aye juveniles in the Tsubusa river-III Comparison between hatchery-reared aye and wild aye. Bull. of the Japanese Soc.of Scientific Fisheries 45(11):1365-1370.
Tsukamoto, K., T. Take, M. En do, and S. Masuda. in press. Evaluation of some behavioral parameters as predictive indices of stocking effectiveness of aye. Bull.of the Japanese Soc. of Scientific Fisheries.
Tsuhamoto, K., K. Uchida, Y. Murakami, M. Endo, and T. Kajihara. 1985. Density effect on jumping behaviour and swimming upstream in the aye juveniles. Bulletin of the Japanese Society of Scientific Fisheries 51(2):323.
Uchida, K. 1986. Behavioral study on characteristics of the upstream migration of the aye, *Plecoglossus altivelis*. Pages 1-121, doctoral dissertation. University of Tokyo,

Tokyo, Japan.

Ueda, H., O. Hiroi, A. Hera, K. Yamauchi, and Y. Nagahama. 1984. Changes in serum concentrations of steroid hormones, thyroxine, and vitellogenin during spawning migration of the chum salmon, *Oncorhynchus zeta*. General and Comparative Endocrinology 53:203-211.

Ujiie, S., S. Fushiki, and Y. oh no. 1983. Effectiveness of the fish ways of anadromous fishes in the ayu-III. Scientific Report of Shiga Prefectural Fisheries Experimental Station 36:21-35.

Woodhead, A. D. 1975. Endocrine physiology of fish migration. Oceanography and Marine Biology Annual Review 13:287-382.

Woodhead, P. M. J. 1970. An effect of thyroxine upon the swimming of cod. J.of Fisheries Research Board of Canada 27(12):2337-2338.

Yamauchi, K., M. Ban, N. Kasahara, T. Izumi, H. Kojima and T. Harako. 1985. Physiological and behavioral changes occurring during smoltification in the mast salmon, *Oncorhynchus mason*. Aqua culture 45:227-235.

Yamauchi, K., and H. Takahashi. 1987. Migratory behaviour and hormone. Pages 156-171 in M. Morisawa, K. A Ida, and T. Hirano, editors. Biology of Migratory Fishes. Japan Scientific Societies Press, Tokyo, Japan.

Yamazaki, Y. 1986. Feeding habit of the amphidromous aye. Master's thesis, University of Tokyo, Tokyo, Japan.

Katsumi Tsukamoto and Kazuo Uchida
Ocean Research Institute
University of Tokyo
Nakano
Tokyo
164 Japan

12 SPATIAL DISTRIBUTION OF MESOPLANKTON IN KUROSHIO FRONTAL ZONE DURING "MEGAPOLYGON" EXPERIMENT

T. A. Zadonskaya

Investigation of space-time plankton distribution in the seas and oceans is one of the main tasks in solving the problem of ocean bioresources utilization. Numerous observations and fishery show that frontal and shelf zones are the most productive. However, answering the questions concerning place, time and a number of objects as well as attempting to make long-term predictions one should carry out complex and detailed investigation of the frontal zones and oceanic circulations related to them.

The basic questions which we aimed to answer conducting this research were as follows: 1) In what way the large-scale distributions of plankton are related to the Kuroshio frontal zone? 2) In what way mesoscale plankton distributions are related to mesoscale eddies? 3) In what way mesoscale plankton distributions might be related to vertical structures of hydrological properties?

Our research has been carried out as part of the "Megapolygon" program in the area with high dynamic activity of water masses, the frontal Kuroshio zone, the Kurils current in summer (June - August) 1987 on board the research vessel "Academician A.Nesmeyanov" (Figure 12.1). The area under observation was 270 x 333 miles, the coordinates being 38°N - 42°30'N and 150°-157°E. In June the stations were spaced 45' apart latitudinally and longitudinally, in all 49 stations on 8 meridional sections. In August the repeated survey has been performed in a similar way with the longitudinal spacing being shortened to 30' (23 miles). Mesoplankton samples have been collected following standard method by Judy net with a 0.1 m^2 inlet having filtering cone of nylon sieve No.38. The layers 0-50, 50-100, 100-200 and 200-300 m were sampled. Biomass and species composition were determined in net zooplanktonic samples. Temperature and salinity profiles have been determined by "Istok-4" probe.

To obtain picture of general spatial structure in the area under study it is reasonable to use integral biomass data in 0 - 300 m layer. Such distribution is shown in Figure 12.2. A broken line indicates the main fronts position which are subarctic (SAM), the polar (Pi) and the Kuroshio extension front (KEF). The eddies of different sign are presented in the figure, the cold cyclonic and the warm anticyclonic eddies.

While comparing mesoplankton spatial structure with the hydrological one we can observe that the highly productive Zone I corresponds to the SAF and PF fronts confluence, adjacent to it on the cold water side. It is meridionally extended and is considerably expanded in the N part of the polygon , which is under the cold N front influence.

Zone 2 in the NE is also adjacent to the subarctic front, and maximum zooplankton biomass (BM) also occurs on the cold water side.

S area of maximum values (Zone 3) adjoins the Kuroshio extension front. The central maximum corresponds to the inner part of the anticyclonic eddy in the centre of the inter frontal zone. Zone 5 of minimum values adjoins the N front on the warm water side, while minimum in Zone 6 corresponds to cyclonic circulation.

Thus, the comparison of our results with simultaneous hydrological data enables us to make a conclusion that zones of extreme mesoplankton concentrations are due to the position of the main large-scale fronts, mesoscale eddies and general water circulation in the area under research.

V. I. Ilyichev and V. V. Anikiev (eds.), Oceanic and Anthropogenic Controls of Life in the Pacific Ocean, 173–181.

Figure 12.1: A map of the area under study.

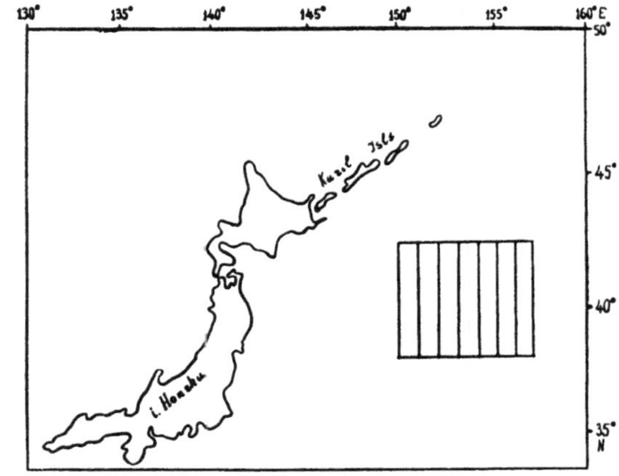

Figure 12.2: Spatial distribution of BM net zooplankton (in mg/m^3) in the integrated 0-300 m layer. June, 1987.

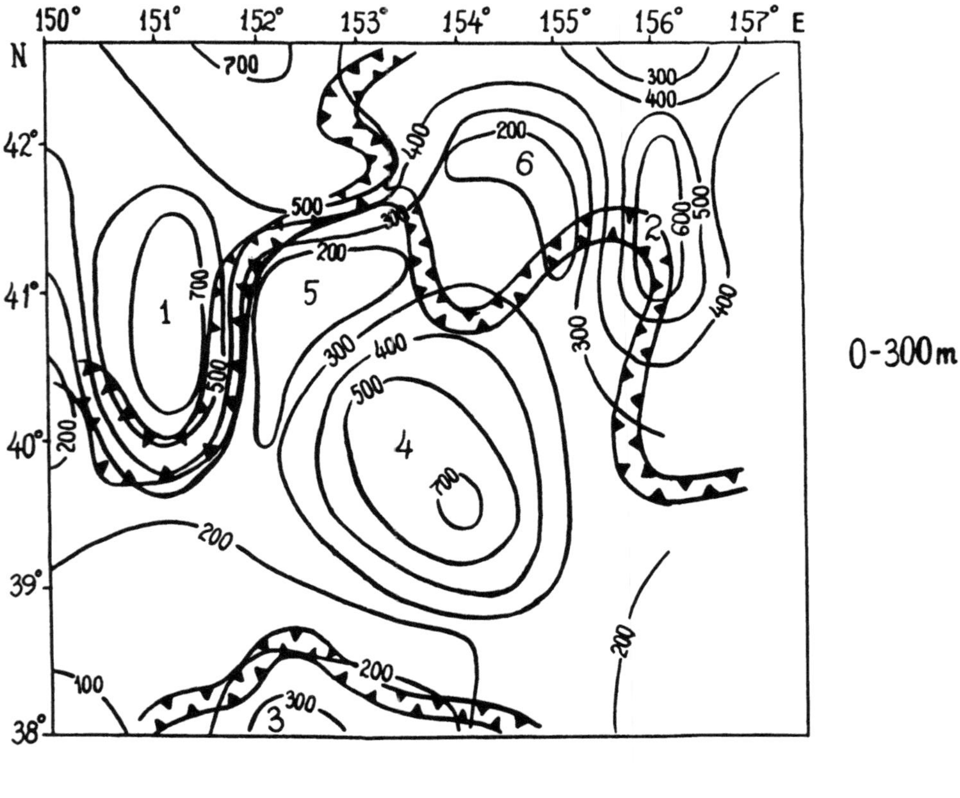

Hydrological data shows that NW situated SAM and PI are generally directed NE reaching 39°45'N in the S part near 151°N. The area rich with mesoplankton in this part of the polygon is situated to the NW from maximum hydrological characteristics. Sharp increase of mesoplankton biomass in the upper 0 - 50 m layer (Figure 12.3) corresponds to the surface isotherm of 14 -15° and the S front boundary. With the sharper temperature gradients the frontal zone is subject to further plankton increase, and 800-900 mg/m^3 biomass corresponds to 12° isotherm and NW boundary of the front. Sharp gradient zone of sexton biomass (400 - 900 mg/m^3) is 60 - 70 miles wide (150°30' and 153°E) and expands up to 200 Mile (42°N) coinciding in space with the main cold front.

Concentration changes on the frontal zone boundary occur simultaneously with changes of the zooplankton species composition. Temperate-cold waters in the NW with 8-14° surface temperature in June were characterized by a homogeneous plankton complex typical of the subarctic waters. lost frequent mass species occurred were as follows: *Calanus cristatus, C. plumhrus, Eucalanus bungii, Metridia pacifica, Pseudocalanus elongatus, Oithona similis,* as well as *Tomopteris spptentrionatus, Aglanta hemistoma, A.digitali, Euphasia pacifica, Sagitta sp., Hyperia galba, Conchoecia serruttata, C. borealis. Calanus plumhrus* is found to be a mass species in 50 - 100 m layer. Minimum zooplankton concentrations in 0 - 50 m layer of subarctic waters were 600 - 700 mg/m^3, while maximum values were 1500-1900 mg/m^3.

At the frontal zone boundary an overall plankton appearance changes simultaneously with the concentration. Side by side with typical bore Al species we have marked those of the mixed water zone like *Paracalanus parvis, Calanus tenuicornis, Mecynocera Clausi* as well as other representatives of warm water plankton *Calanis pacificus, Aetideus pacificus. Calanus pacificus* is found as a monoculture in 0 - 50 m layer (39°3O'N). Thus, sharp change of plankton between the zones with *Calanus plumchrus* and with *Calanus pacificus* which is accompanied by its biomass drop enables us to consider the temperate-cold and warm Kuroshio waters junction to be a boundary between different zoogeographical areas. In June 1987, based on our data, this boundary was found in 14-18°C water temperature (30°30' - 40°15'N, 150°-152°E). N boundary position of the subarctic front resulted in changing the position of various zoogeographical areas' boundary.

High sexton concentration which characterizes the adjoining NE areas of the frontal zone are due to abundant phytoplankton blooming, the latter factor lacking in the NE area. The blooming zone covers the entire area to the NE from the front and extends to 41°N in the S. Mass development of phytoplankton in this area was also observed in August, but unlike in June it was much weaker. In June mesoplankton in the NE area was characterized by boreal species typical of plankton complex of temperate-cold waters, while in August mixing of boreal and subtropical plankton complexes occured as well as penetration of warm water subtropical species *Candacia bradyi, Pleuromamma aiphias* up to 41°N and *C.pacificus* up to 42°30'N in 100 m layers. It is necessary to emphasize that the hydrological data in August do not show SAF in this area based on surface temperature and salinity distribution. Proceeding from the species composition of zooplankton in August the frontal zone in the NE was found to the N from 42°30'N.

Figure 12.3: The same in 0-50 and 50-100 m layers.

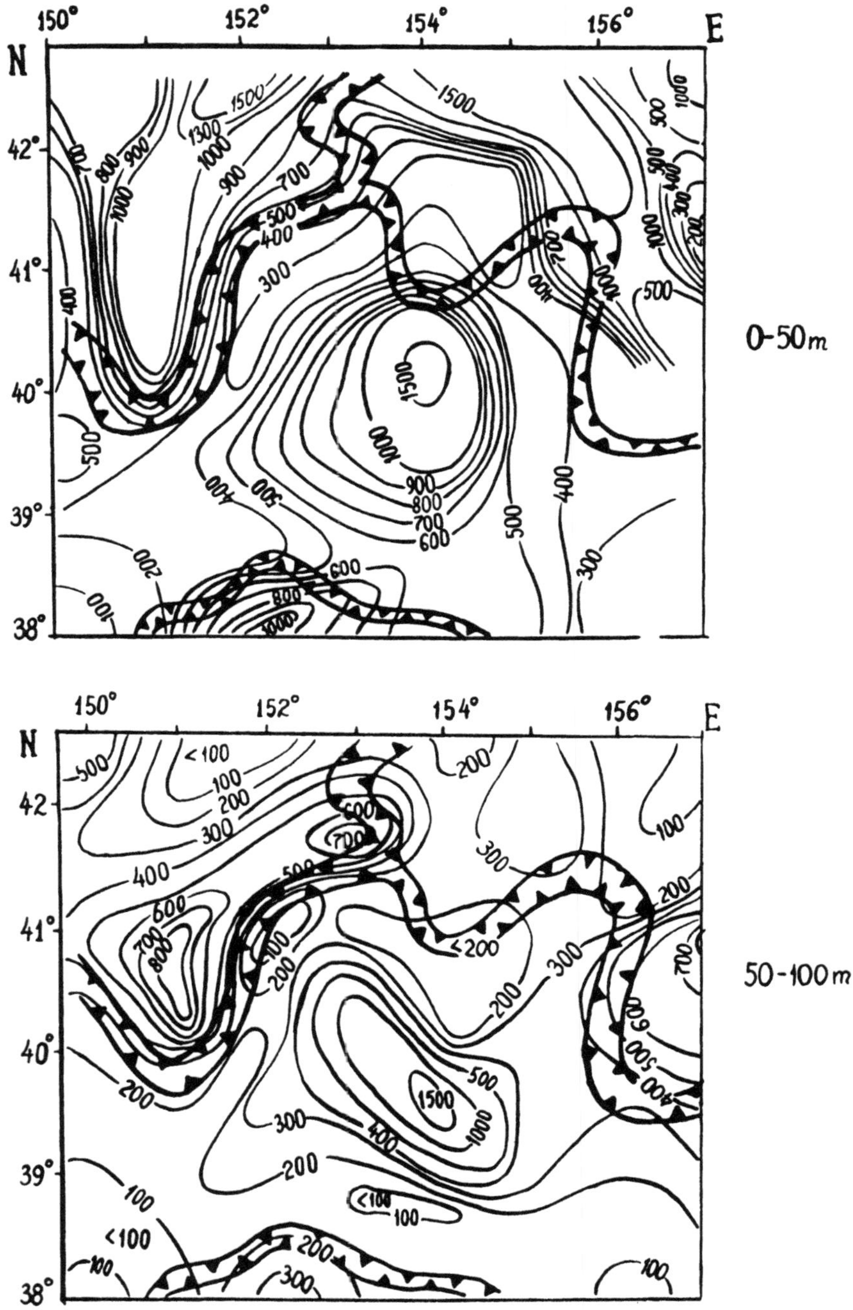

Figure 12.4: The same in 100-200 and 200-300 m layers.

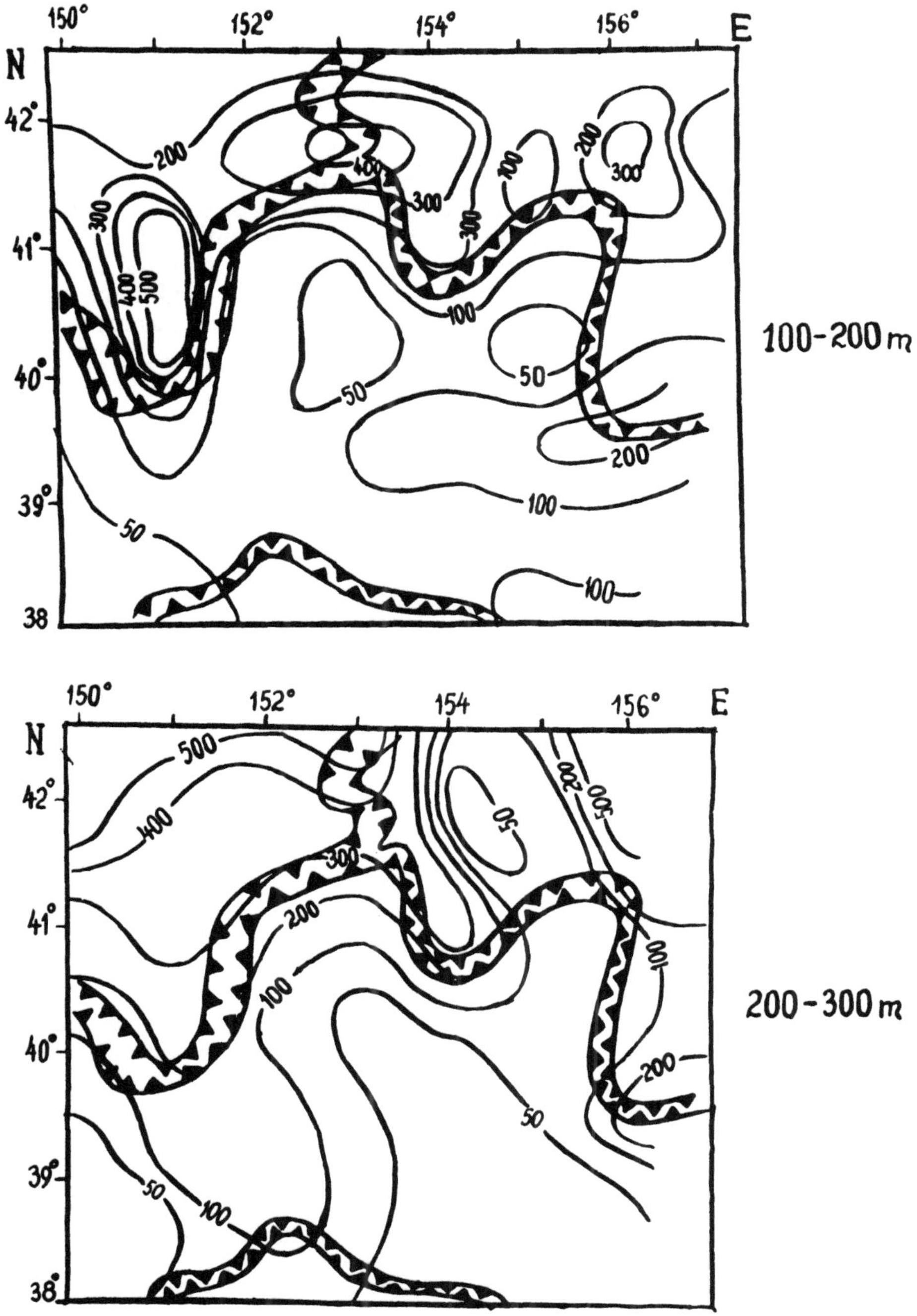

High plankton concentrations in the surface layer in the S frontal zone which represents the Kuroshio extension front (KEF) are due to abundant development of warm water tropical plankton complexes as well as mass phytoplankton development in a 0 - 50 m layer.
Plankton biomass changes versus depth, nevertheless, general distribution typical of the O - 5O m layer in plankton distribution remains. This can be illustrated by Figure 12.3 and 12.4.
Thus sharp concentration increase and species composition change in mesoplankton was observed in zones of largest horizontal temperature gradients - subarctic and polar fronts in the N as well as in the area of Kuroshio extension in the S. However vertical distribution shows differences in mesoplankton distribution of subarctic and subtropical waters. This is shown in the Table 12.1. In N near-frontal and frontal areas large BM mesoplankton values (up to 500 mg/m^3) maintained down to 300 m which we could observe while in S area of KEF high productivity was revealed only down to 100 m.

Table 12.1: Dependence of mesoplankton concentration (in mg/m^3) on depth for waters of various dynamic structure.

Layer (m)	SAF +PF	PF	SA waters	SAF	KEF	AC_1 Central	AC_2 Part,	AC_3 June	E-M Aug.	C
0-50	400-900	700-1500	900-1900	400-900	600-1000	1500	400-700	500	2000	200-300
50-100	200-700	300-700	500-800	200-500	200-300	1000	300	100-200	200-500	200-300
100-200	200-400	200-400	200-500	100-300	50-100	50	50-100	50	100-150	200-300
200-300	200-500	400-500	200-500	100-300	50-100	50	50-100	50	100-150	200-300

Mesoplankton distribution differs considerably in mesoscale eddies of opposite sign largely depending on characteristics of waters involved. On the one hand, decisive here might be the plankton transport processes as a passive admixture, and on the other - the lifetime of the eddy. The maximum BM mesoplankton corresponds to the anticyclonic eddy with surface layer converging (area 4, Figure 12.2), while the minimum relative to the adjoining high-productive subarctic waters corresponds to the cyclonic eddy (area 6, Figure 12.2) with surface layer diverging.
Vertical mesoplankton distributions in anticyclonic and cyclonic eddies are also different. In central parts of warm anticyclonic eddies the upper 100 m layer shows 2 - 3 times higher plankton concentration relative to the surrounding waters. In the 100 - 300 m layer the amount of mesoplankton sharply decreased to match the level of transformed subtropical water biomass. Mesoscale cyclonic eddy with waters of subarctic origin reveals uniform distribution both horizontally and vertically in the

studied 0 - 300 m layer, while the anticyclonic eddies showed BM drop 5 - 6 times as compared to the similar depths in cyclonic eddies.

In the anticyclonic eddy (area 4, Figure 12.2), subject to annual structural transformations, high BY values were due to phytoplankton blooming. In young anticyclonic formations whose lifetime seemed not to exceed several months the mesoplankton was represented by mass *Copepoda* aggregations, *Calanus Pacificus* in June, and *Salpa Doliolum sp., Talis sp.*, etc., in August (E-M, Figure 12.5).

Figure 12.5: The same in the integrated 0-300 m layer. August, 1987.

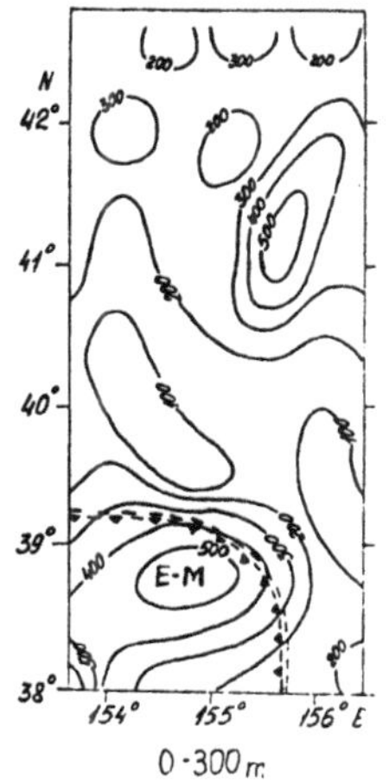

And in conclusion a few words on correlation of mesoscale plankton distributions with vertical structure of hydrological characteristics.

Figure 12.6 shows typical shapes of temperature vertical structure in frontal zones of different origin and eddies of opposite sign as well as mesoplankton biomass in these structural formations in the layers 0 - 50, 50 - 100, 100 - 200, and 200 300 m. As the Figure 12.6 illustrates, the patterns of temperature vertical profiles differ both in the areas of large scales (about 100 m) and small scales (that reveals in the presence of fine structure). The most productive layer in the area under study was the upper one, O - 50 m. The growing depths result in changes of both temperature profile pattern and mesoplankton biomass. In the zones of subarctic and polar fronts as well as cyclonic eddies with fine structure high mesoplankton concentrations (200 - 500 mg/m^3) remain down to 300 m.

Figure 12.6: Patterns of vertical temperature structure and BM mesoplankton (in mg/^{3}) in 0-50, 50-100, 100-200, and 200-300 layers.

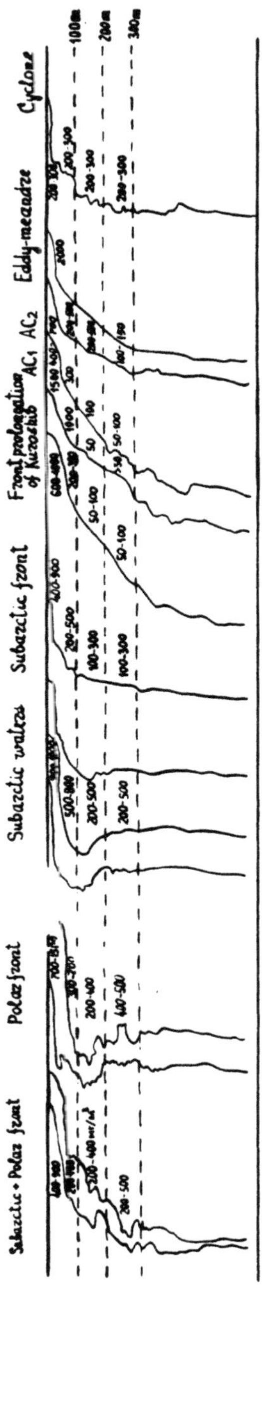

With fine structure temperature changes lacking in the zones of subarctic front in the NE as well as in the Kuroshio extension front and anticyclonic eddies, mesoplankton biomass sharply decreases versus depth and on the average is 50 - 100 mg/m^3. Thus high values of BM mesoplankton remain in the case of fine structure present and are lacking in the areas where temperature profiles are monotonous.

T. A. Zadonskaya
Pacific Oceanological Institute
USSR Academy of Sciences
Far East Branch
Vladivostok
USSR

13 SPAWNING HABITAT, CONTINENTAL SHELF AREA AND HERRING PRODUCTION IN THE NORTH PACIFIC OCEAN

D.E. Hay

Abstract

Abundance of Pacific herring (*Clupea harengus pallasi*) populations may be limited by (1) the available area of summer feeding habitat, mainly on coastal shelf waters with depths of 200 m or less, and (2) the area of suitable spawning habitat. Shelf areas may serve as summer feeding areas that account for most of the production of herring biomass. In offshore waters, adult herring rarely are found in waters exceeding 200 m depth and herring schools often are seen close (<5 m) to the bottom. Throughout the N Pacific, the largest herring populations occur in association with large continental shelf areas. Spawning and nursery habitat consists of shallow, near-shore areas protected from the open sea. The relative proportion of shelf and near-shore habitat varies among areas and populations. The Sea of Okhotsk has a broad shelf area, but maximal herring production often may not be realized because of limited available near-shore habitat suitable for spawning and rearing of the larval and juvenile stages. In contrast, in British Columbia there is a relative abundance of near-shore spawn and nursery habitat and a relative scarcity of suitable shelf areas.

13.1 Introduction

Pacific herring (*Clupea harengus pallasi*) occur throughout the N perimeter of the Pacific Ocean. The most S populations occur off the coast of California in the E Pacific and in the Sea of Japan and the Yellow Sea in the W Pacific. At higher latitudes, herring populations are found along both the N American and Asian shores N to the Arctic Ocean. The N and S limits to their distribution appear to be determined by temperature and salinity (Alderdice and Velsen, 1972) and herring are found only in relatively cold water masses. The developmental stages most restricted by temperature appears to be those of eggs and larvae. Based on the laboratory studies of Alder dice and Nelsen (1971) and Duenas (1981) optimal incubation temperature are about 8 degrees C. This temperature is consistent with the observed temperatures made during natural herring spawning although California herring spawn in slightly warmer temperatures and herring in the extreme north spawn in slightly cooler temperatures (Hay, 1985). In laboratory conditions juvenile and adult herring can survive for extended periods at warmer temperatures, up to 20 degrees C, although prolonged exposure to warmer water or sudden increases in temperature can lead to outbreak of diseases (Hay et al., 1988).

Pacific herring do not penetrate fresh water to any significant extent so low salinity is an obvious restriction to some habitats. In their extreme N ranges, both in N American and Soviet Arctic waters, spawning herring appear to be confined to partially estuarine environments found in the vicinity of large rivers. Within the latitudinal extremes, other factors appear to limit distribution and abundance. The area of the continental shelf to a depth of 200 m, and the topography of the coastal zone are factors that appear to be related to herring distribution and relative abundance. There are parallel clinal changes that occur both in the E and W Pacific such as a tendency toward smaller adult size, earlier spawning and shorter live spans

V. I. Ilyichev and V. V. Anikiev (eds.), Oceanic and Anthropogenic Controls of Life in the Pacific Ocean, 183–191.

in the S ranges (Tiller and Schmidtke, 1956; Hay, 1985).

It is well known that the biomass of single Pacific herring populations fluctuates over time (Hourston, 1980; Morita, 1985) within some limits posed by the environment. The purpose of this paper is to point out that such a limit, or the maximum observed herring biomass, is related to the area (up to 200 m) of the adjacent continental shelf. Herring spend the summer months feeding in continental shelf locations so the area of the shelf may provide a rough indication of the available summer feeding habitat. Therefore, in some locations, or in some years in all locations, the potential or maximal biomass of a herring population may be determined by the available summer feeding habitat. It is also clear, however, that in some locations, the survival of early life history stages may be impaired by inadequate near-shore spawning and juvenile rearing habitat.
Juvenile herring tend to be confined to near shore waters, at least during their first summer (Hourston, 1958).

13.2 Associations with the continental shelf

Although herring normally are regarded as pelagic, migratory species, their distribution is closely associated to coastal zones (for spawning) and continental shelves (for summer feeding). To my knowledge, there is no documented evidence of herring occurring in pelagic habitats in the open sea, far from shore (i.e. 1000 km). Also, herring sometimes are described as migratory (Taylor, 1964) but the extent of movement appears to be modest relative to some other species. Unlike the Pacific salmon species (*Oncorhynchus*) that regularly undergo widespread migrations in the open Pacific, herring tend to remain within the region of the continental shelf. In British Columbia, extensive fishing and hydroacoustic surveys of the coast have been conducted for more than 20 years (Kisser et al., 1987; McCarter et al. 1987, and references therein). In general the results of these surveys show that herring rarely are found at depths bigger than 200 m. Usually they are confined to locations with a depth of about 100 m or less. They frequently show distinct diurnal vertical migrations, forming widespread 'skimmers' near the surface or in shallow depths. During daylight hours they sometimes are some close to the bottom that sounders cannot distinguish them from the sediments. Therefore, some association with the bottom appears to be an ecological requirement of Pacific herring and the 200 m edge of the continental shelf appears to be the limit of their distribution in most offshore locations.

The maximum estimated biomass of Pacific herring populations and the corresponding area of the continental shelf are correlated (Table 13.1). The correlation is statistically significant ($r = 0.84$, $p < 0.05$). The areas of the continental shelves shown in Table 13.1 are only approximations and do not consider regional differences in the frequency distribution of depths or the mean depth of the shelf. This could be a deficiency in this analysis because some specific depth ranges of the continental shelf may be particularly important for herring. For instance, in Hecate Strait, off the coast of British Columbia, herring are most often found between the 50 and 100 m depth ranges and they seem to favour areas with slopes rather than flat bottom sites. Also, the estimated biomass of herring populations in Table 13.1 also are approximations. Total biomass estimates sometimes vary substantially. For instance, Laevastu et al. (1982) list an equilibrium biomass of nearly 2 million tones of herring in the E Bering Sea. The figures used in Table 13.1, taken from different sources are lower. The simplest biological explanation for the relationship between the potential size of herring population and shelf area is that the food availability is a function of shelf area. In recent surveys in the N British Columbia, we found that average densities of herring aggregations are lower in the summer, when they are feeding, than in the winter months when feeding activity is reduced (McCarter et al., 1988).

13.3 Coastal topography and spawning areas

Throughout their range in the N Pacific, herring spawning distributions are limited to locations that provide sheltered water for spawning and incubation of eggs and rearing of larval and juvenile stages. There are no documented cases of spawning sites in open, exposed water similar to those used by many Atlantic herring populations. In British Columbia, no major spawning occurs on shores facing an open sea. In California, spawning is confined to the few coastal indentations, the largest of which is San Francisco Bay. Similarly, on the outer coasts of Oregon and Washington States, spawning is confined to the few small coastal indentations, such as Tillamook Bay, Yaquina Bay, Umpqua Bay and Coos Bay (Butler 1980). Except for San Francisco, the spawning biomass of these small stocks is not significant relative to those In other parts of the north Pacific.

Small local herring populations are associated with many of the indentations of the British Columbia coast and the waters inside Vancouver Island and the Queen Charlotte Islands. The topographical complexity of the coast appears to have fostered the formation of a rich stock structure, with perhaps 40 or more separate spawning stocks (Hay 1986). In general, the size of the spawning stock is roughly related to the area of the spawning site (Hay and Kronlund 1987). The topographical complexly of the coastline and the diversity of herring populations in SE Alaska is roughly similar to that of British Columbia (Rounsefell, 1930; Carlson, 1980).
The herring populations in S Central Alaska, all occur in coastal indentations such as Yakutat Inlet, Prince William Sound, and inside waters around Kodiak Island (Rounsefell, 1930; Reid, 1971). There are no large populations that spawn in the Aleutian Island chain but there are several large herring population in the E Bering Sea. In the E Bering Sea, spawning is manly in sheltered areas of Bristol Bay and Norton Sound (Barton and Wespestad, 1980).

Table 13.1: Comparison of continental shelf area and maximum estimates of spawning stok biomass from diffrent herring populations in the N Pacific.

	location	Biomass (tonnes x 1000)	Shelf area (km^2)	Density (tonnes/km^2) or (g/m^2)
1.	Sea of Okhotsk	800 (a)	620,000 (b)	1.29
2.	Eastern Bering Sea			
	(a) 1960	1166 (c)	687,000 (d)	1.69
	(b) 1967-87	~150 (e)	687,000	0.21
3.	Sea of Japan (Hokkaido stock)	1000 (f)	250,000	4.00
4.	North and central California	50 (g)	25,000 (h)	2.00
5.	Outer Coasts of Washingtion and Oregon	1 (i)	25,000	0.04
6.	British Columbia Puget Sound S.E. Alaska	400 (j)	71,000 (k)	5.63
7.	Washington, Oregon, British Columbia, S.E. Alaska	401 (1)	96,000 (l)	4.17

(a) Based on an estimate by Benko et al. (1987)
(b) Chikuni (1985)
(c) Estimate based on analysis in Fried and Wespestad (1985)
(d) Gulland (1970)
(e) Approximation based on Fried and Wespestad (1985) and Lebida and Sandstone (1988)
(f) Minimum estimate based on catches which once approximated 1 million metric tonnes (Morita 1985)
(g) Based on Spratt (1981)
(h) Based on a mean continental shelf width of 25 km (Gulland, 1970) by a latitudinal range of 100 km
(i) Rough estimate of the biomass of several small stocks
(j) Estimate of 300,000 t for British Columbia, based on Hourston (1980) and a maximum of 100,000 t for Puget Sound (Washington) and SE Alaska that had a maximal catch of 71,000 t in 1929 (Reid 1971)
(k) Gulland (1970)
(l) Estimates summed from above.

13.4 Spawning habitat, biomass and egg mortality

Estimates of the maximum potential spawning area and maximum biomass have been estimated for three different regions of the N Pacific (Table 13.2). Compared to other locations, the British Columbia and SE Alaska coast have abundant spawning habitat relative to the biomass of the spawning stock. The estimate of the average density of spawning herring per square meter of available spawning habitat is provided only as a comparison, but it has little biological meaning because in British Columbia, herring rarely utilize more than one tenth of the available habitat in any single year. In contrast, it seems that when populations are in periods of abundance in California and the Sea of Okhotsk, much of the available spawning habitat is used.

Table 13.2: Comparison of potential spawning areas and spawning stock biomass among different areas.

		Potential Spawning area (km^2)	Maximum Biomass (tonnes)	kg/m^2
1.	Sea of Okhothsk	34.5 (a)	800,000 (a)	23.2
2.	British Columbia	93.0 (b)	300,000 (c)	3.2
3.	California	3.0 (d)	50,000 (d)	16.7

(a) Estimates from Benko et al. (1987)
(b) Estimate based on a total range of 3100 km of coast (Hay, 1985) by a mean spawning width of 30 m (Haegele and Schweigert, 1985; Hay and Kronlund (1987)
(c) Estimate of biomass for British Columbia is an approximation based on the probable biomass during the the reduction fishery in the 1960's when the total catch exceeded 220,000 tones (Hourston, 1980)
(d) Estimates from Spratt (1981)

The reported estimates of egg density in the Sea of Okhotsk are bigger, at 6 to 10 million eggs per square m (Bunko et al., 1987), than densities in British Columbia which rarely exceed 3 million per square m (Hay and Kronlund, 1987) and usually are less than 0.5 million per square m (for example, see Haegele and Schweigert, 1985, 1987; Schweigert and Haegele, 1988). Further, egg mortality estimates differ markedly between areas with Bunko et al. (1987) reporting that 90 % or more of dense spawns are lost but Haegele et al. (1981) stating that mortality in British Columbia egg depositions rarely exceeds 30 %.

13.5 Abundance of herring relative to other species

The mean biomass of the spawning population of herring in San Francisco Bay, California, (the largest stock in California) is about 24,000 tones between 1974 - 1980 (Pratt, 1981) and currently is estimated at about 50,000 tones. In contrast, the estimated size of the spawning population of northern anchovy (*Engraulis mordax*) is about 770,000 tones and that of Pacific mackerel was estimated at more than 450,000 tones (Cal. Dept. Fish. Game, 1987). As described above, there are no large herring populations and no substantial herring fishery off the coast of Oregon or the outer coast of Washington State but there is a large biomass of northern anchovy, up to 1,000,000 tones, that is geographically distinct from the California anchovy stocks (Richardson, 1981). In British Columbia the maximum total biomass of herring stocks is estimated at about 300,000 tones (Table 13.1). There are no large concentrations of anchovies but there are large populations of hake off the S British Columbia, with an estimated average abundance of 364,000 tones and a maximum of 605,000 tones (Ware and McFarland, 1987). Probably, herring are the single most abundant species off of the N British Columbia coast although the biomass of sand lance (*Amodytes hexapterus*) could be substantial, but the total quantity is unknown. In the E Bering Sea the biomass of herring has changed over time from an estimate of over 1 million tones to a current average of about 150,000 tones (Table 13.2). It is clear, however, that the Bering sea supports enormous stocks of other species such as the yellow fin sole (*Limanda aspera*), pollack (*Theracra chalcogramma*) and cottids that are considerably more abundant than herring (Gulland, 1970; Laevastu et al., 1982).

13.6 Management implications

The most important observations are (1) that herring spawning, and larval and juvenile nursery sites are confined to locations with sheltered water and (2) that the maximal biomass of herring stocks in the N Pacific is related to the area of the adjacent continental shelf. If these observations are valid then there are important implications for management and potential enhancement. If the biomass of a stock is limited by the available shelf area, then enhancement measures that attempt to promote better survival at early life history stages, are impractical. On the other hand, if production is limited during early life history stages and particularly during the egg stage, then the probability of success for enhancement efforts should be bigger. As an approximate guide, the ratios of shelf area to estimated spawning areas may be useful. If the ratio of spawning area to shelf area is high, then it is unlikely that efforts to enhance survival of early life history stages will be worthwhile. On the other hand, if the ratio indicate a relative abundance of shelf area, then enhancement efforts may have a better chance for success. In this regard, it appears that the use of artificial spawning reefs in the Sea of Okhotsk (Bunko et al., 1987) have the potential to increase production because the ratio of shelf area to spawning area is high in the Sea of Okhotsk. It is unlikely, however, that the same approach would succeed in British Columbia or SE Alaska.

References

Alder dice, D. F. and F. P. J. Nelsen. 1971. Some effects of salinity and temperature on early development of Pacific herring (*Clupea pallasi*). J. Fish. Reg. Board Can. 28: 1545-1562.
Bunko, Yu. K., Yu. N. Bogatkin and R.K. Farkhutdinov. 1987. Biological basis of the use of artificial spawning grounds for the reproduction of Okhotsk herring. Translation by Plenum Publishing Co. from Biologiya Morya, No. 1, p. 5661, Jan.-Feb. 1987.
Barton, L.H. and V.G. Wespestad. 1980. Distribution, biology and stock assessment of western Alaska's herring stocks. p. 2753. In B.R. Melteff and Y.G. Wespestad [en.] pro. Alaska Herring Symposium, Alaska Sea Grant Report 80-4.
Butler, J. 1980. Fish Commission of Oregon. p. 30. In D. Day [en.] Proc.of the Pacific Coast Herring Workshop. Wash. Dept. Fish. 32 p.
California Department of Fish and Game. 1987. Review of some California Fisheries for 1986. CalCOFI rep. 28: 11-20.
Carlson, H. R. 1980. Seasonal distribution and environment of Pacific herring near Luke Bay, Lynn Canal, Southeastern Alaska. Trays. Am. Fish. Sock 109: 71-78.
Chlkuni, S. 1985. The fish resources of the northwest Pacific. FRO Fish. Tech. Pap., 266: 190 p.
Duenas, C.E. 1981. Influence of incubation salinity and
tolerance of Pacific herring (*Clupea pallasi Valenciennes*) larvae. M.SC. Thesis. Univ. of British Columbia, Vancouver, Canada.
Fried, S.M. and V.G. Wespestad. 1985. Productivity of Pacific herring (*Clupea harengus pallasi*) in the eastern Bering Sea under various patterns of exploitation. Can. J. Fish. Aquat. Sci. 42 (Suppl. 1): 181-191.
Gull and, J.A. 1970. The fish resources of the oceans. Tech. Pap., 97: 425 p.
Hoist, V., J.F. Schweigert, and D. Foulnier. 1987. Stock assessments for British Columbia herring in 1986 and forecasts of the potential catch in 1987. Can. MS Rep. Fish. Aquat. Sci. 1929: 63 p.
Haegele, C.W. R.D. Humphrey and A.S Hours ton. 1981. Distribution of eggs and vegetation type in Pacific herring (*Clupea harengus Dallas I*) spawning in southern British Columbia. Can. J. Fish. Aquat. Sci. 38: 381-386.
Haegele, C.W. and J. F. Schweigert. 1985. Distribution and characteristics of herring spawning grounds and description of spawning behaviour. Can. J. Fish. Aquat. Sci. 42(Supply. 1): 39-55.
Haegele, C. W. and J.F. Schweigert. 1987. Herring stock estimates from diving surveys of spawn for the lower east coast of the queen Charlotte Islands in 1987. Can. MS Rep. Fish. Aquat. Sci. 1956: 41 p.
Hay, D. E. 1985. Reproductive biology of Pacific herring (*Clupea harengus pallasi*). Can. J. Fish. Aquat. Sci.42
(Suppl. 1): 111-126.
Hay, D. E. 1986. A stock hypothesis based on spawn and winter distribution. In C.W. Haegele [en.] Proceedings of the Fifth Pacific coast herring workshop, October 29-30, 1985. Can. MS Rep. Fish. Aquat. Sci. 1871: 145-148.
Hay, D. E., and A. R. Kronlund. 1987. Factors affecting the distribution, abundance, and measurement of Pacific herring (*Clupea harengus pallasi*) spawn. Can. J. Fish. and Aquat. Sci. 44: 1181-1194.

Hay, D. E., J. R. Brett, E. Bilinski, D. T. Smith, E. M. Donaldson, G.A. Hunter, and A. V. Solmie. 1988. Experimental impoundments of pre-spawning Pacific herring (*Clupea harengus Pallasi*): effects of feeding and density on maturation, growth and proximate analysis. Can. J. Fish.Aquat. Sci. 45: 399-406.
Hourston, A.S. 1980. The decline and recovery of Canada's Pacific herring stocks. rasp. P.-v. run. Cons. it. Explore. Mel 177: 143-153.
Hourston, A.S. 1958. Population studies of juvenile herring in Barley Sound, British Columbia. J. Fish. Reg. Bud. Can. 15: 90 9-9 6 0 .
Kieser, R., D. E. Hay, C. W. Haegele, D. C. Miller and T.J. Mulligan. 1987. Hydroacoustic herring survey results and trawl catches from Hecate Strait, October 15 to November 4, 1984. G.B. REED cruise GBR84C and M.V. CANADIAN # 1 cruise CAN84-1. Can. MS Rep. Fish. Aquat. Sci. 1917: 107 p.
Laevastu, T., F. Favorite, and H.A. Larking. 1982. Resource assessment and evaluation of the dynamics of fisheries resources in the northeastern Pacific with numerical ecosystem models, p. 70-81. In M. C. Merger [en.] Multi species approaches to fisheries management advice. Can. Specs purl. Fish. Aquat. Sci. 59.
Lebida, R.C. and G.J. Sandstone. 1988. Age, sex and size composition of Pacific herring from coastal Bering Sea spawning sites in Alaska, 1987. State o£ Alaska Technical Fishery Report 88-06. 75 p.
McCarter, P. B., R. Kisser, D. E. Hay, and D. C. Miller. 1987. Hydroacoustic herring survey results and trawl catches from Hecate Strait, November 26 to December 12, 1985. G.B. REED cruise GBR85E and M.V. SUNNFJORD cruise SUN85-1. Can. MS Rep. Fish. Aquat. Sci. 1940: 131 p.
McCarter, P. B., R. Kisser, D. E. Hay. 1988. Hydroacoustic herring survey results and trawl catches from Hecate Strait, July 14-27, 1987. W.E. RICKER cruise WER87F. Can. MS Rep. Fish. Aquat. Sci. In Press.
Miller, D.J. and J. Schmidke. 1955. Report on the distribution and abundance of Pacific herring (*Clupea pallasi*) along the coast of central and southern California. Cal if. Fish. Game 42: 163-187.
Morita, S. 1985. History of the herring fishery and review of artificial propagation techniques for herring in Japan. Can. J. Fish. Aquat. Sci. 42 (Supply.1): 222-229.
Reid, G.M. 1971. Age composition, weight, length and sex of herring, *Clupea Pallasi*, used for reduction in Alaska, 1929-66. NATO. Mar. Fish. Serv. Specs Rep. No. 634: 25p.
Richardson, S.L. 1981. Spawning biomass and early life history of northern anchovy, *Engraulis mordax*, in the northern sub population off Oregon and Washington. Fishery Bulletin 78: 855-876.
Rounsefell, G. 1930. Contribution to the biology of the Pacific herring *Clupea pallasi*, and the condition of the fishery in Alaska. Bull. U.S. our. Fish. 45: 227-320.
Schweigert, J.F. and C.W. Haegele. 1988. Herring stock estimates from diving surveys of spawn for Georgia and John stone Straits in 1985. Can. MS Rep. Fish. Aquat. Sci. 1972: 63
Pratt, J.D. 1981. Status of the Pacific herring *Clupea harengus Pallasi* resource in California, 1972 to 1980. Fish. Bull. Cal if. Dept. Fish. Game 171: 1-107.
Taylor, F.H.C. 1964. Life history and present status of British Columbia herring stocks. Fish. Reg. Bud. Canada. Bull. 143. 81 p.

·Ware, D.M. and G.A. McFarland. 1987. Relative impact of Pacific hake, sable fish and Pacific cod on west coast of Vancouver Island herring stocks. Bull. Into N. Pac. Fish. Comma 47: 67-77.

D.E. Hay
Department of Fisheries and Oceans
Biological Sciences Branch
Pacific Biological Station
Nanaimo B.C. V9R 5K6

14 THE INFLUENCE OF OCEAN SURFACE TEMPERATURE VARIABILITY ON SAURY SHOAL DISPLACEMENT

R.S. Goldman, V.I. Ilyichev and L.P. Smirnykh

Abstract

The paper deals with problems of fishery oceanology in which the understanding of a fishery situation is based on the combined analysis of fishery data and space- and time-averaged data on the distribution of ocean surface temperature. It is supposed and illustrated that the determination of quantative relationships between the ocean surface temperature variability and the migration behaviour of fish schools enables forecasting probable fishing grounds.

The information support for this investigation was files of isothermal charts of surface temperatures averaged over three and ten days for the NW Pacific, daily sea water temperature measurements made by fishing and search vessels, data on daily output of the entire fishery fleet, archives of fishery and hydrological retrospective information.

The data was processed by personal computers. Data files on computer were created. Techniques and algorithms for analyzing fishery and hydrological information in an interactive mode were developed, with quantative characteristics being evaluated and fishery situations displayed graphically.

Procedures for drawing hydrological plots on specified scales from routine temperature measurements were developed.

Such approach allowed to relate the quantative variability of hydrological fields of the ocean surface to the migration of saira schoals during the actual fishery.

Gathering detail information on the actual fishing ground, along with developing algorithms and programs for analyzing the information on personal computers, allows to build an information model of the fishing ground. The model can be used in the future for solving scientific and applied problems of fishery oceanography which aim is to evaluate the parameters of the relationship between environment parameter variation and fish schoals migration. For fishery industry, the model can be used for the developing of processes of assimilating and using various types of information of managing and forecasting fishery.

14.1 Introduction

The present level of the study of intensive ocean fishery regions is related to sampling and analysis of information recorded regularly. These are mainly fishing-ship reports on activity results and ocean surface water temperatures. Methods of determining relationships between the fishery and hydrological situations were considered [1].

The present study continues these investigations using personal computers as a technical means of handling algorithms of relation determination.

Surface temperature charts [2] and reports of fishing and searching ships were used in experimental computations. It is supposed and shown that obtained quantitative relations between ocean surface temperature variation and fish cluster displacement allow to predict economically significant fishery situations.

Brief characteristics of basic fishery situations are presented [3] which could be related to ocean surface temperature variability and other environmental parameters. The appearance of new technical means of data sampling and processing has led to

V. I. Ilyichev and V. V. Anikiev (eds.), Oceanic and Anthropogenic Controls of Life in the Pacific Ocean, 193–201.

a higher qualitative level of study and detailed analysis of individual fishery fragments.
Let us consider the the system of the ocean surface temperature sampling and processing.

During a commercial fishing expedition the fishing and research ships measure water temperatures. Several times a day this information is transmitted through radio channels as measurement site coordinates and temperature values. A generalized hydrological plane-table is constructed using this data to determine the hydrological regime in the fishery region. Data is introduced into a personal computer and presented in the form shown in Figure 14.1a.
During data recording on the computer carrier, the possibility of measurement correction is provided. Before construction of the hydrological plane-table, the report set is formed, which is a sample for the given coordinates of the region in which the measurements were made and for a given time interval (Figure 14.1b).
In order to construct a plane-table the region and elementary square size are predetermined. For each elementary square the average, maximal and minimal temperature values and number of measurements are calculated. In Figure 14.1c a fragment of hydrological plane-table is shown for measurement period 26.09.87 - - 29.09.87 in region 46°30'N, 154°30'E; 47°N, 155°E (elementary square size is 10'x10').
The possibility of the simplest comparative analysis of two or more hydrological plane-tables is provided with result output to graphic terminals. In Figure 14.1d, a fragment is shown of a comparison of two plane-tables for 01.- 03.10.87 and 26.09.87 - 29.09.87 in region 46°N, 154°E; 46°0'N, 155°E where the elementary square size is 10'x20'. Average temperatures are also shown.
The great possibilities of the personal computer allow us to carry out a visual analysis of plane-tables in real time, to trace the variability of ocean surface temperatures in a fishery region in a multivariant comparison regime.
Practical experience in commercial fishery has shown that situations occur which can not be explained by temperature variability within the fishery region only. In [3], for instance, examples of temperature characteristics are given which were measured in squares located at a distance of some hundred miles from the area under study but which determined the fishery situation at a given time. For this, ten-day charts of surface temperatures in the NW Pacific were used. A more detailed analysis of ocean surface temperature is based on the use of charts of another spatial-temporal scale. In the personal computer, a chart archive of three-, four-, five- and ten-day averaging was formed for coding mesh 30'x 30' and 1°X 1° in region 20°-39° N, 46°-50° N and from 140° E to 160°-180° E. The charts are introduced into the personal computer as a temperature field (temperature values are given for squares of coding mesh). Due to big volume of data, archives are kept on magnetic tapes and the working file is organized on magnetic disk for operative use. Examples showing the use of surface temperature charts will be presented below.

Figure 14.1: Fragments of water temperature data processing.

```
-----------------------------------------------
! point  date coordinates horiz.,m temperature !
-----------------------------------------------
  1  06.08.87 4414N 14715E  0       10.4
  2  06.08.87 4406N 14715E  0       12.2
  3  06 08.87 4407N 14716E  0       12.6
  4  06.08.87 4409N 14707E  0       12.5
-----------------------------------------------
                      a)
-----------------------------------------------
catalog of bathometric data
-----------------------------------------------
set No.13     Date of construction    20.09.87
-----------------------------------------------
Coordi-  Longitude Latitude
nates       146E       43N        29 -
            147E       44N    measurement number
-----------------------------------------------
day interval 11-13
month 07,year 1987,horizon 0,m
-----------------------------------------------
                      b)
-------------------         -------------------
charts bathimetric          Comparative chart for
points for 26-29.09.87        bathimetric points
horizon 0,m; paramet.:
average value               No.1 01.10-03.10.1987
point number
min value,max value         No.2 26.09-29.09.1987
------------------47°
:      : 10.2 : 10.2        ---------------46°50'
:      :  6   :  9          : 8.9:      :
:      : 10.0 :  9.0        :     : 8.1: 9.3
:      : 10.4 : 11.4        ---------------- 40'
------------------50'       : 8.8: 9.4: 9.0
: 8.9 : 10.8 : 11.0         : 8.3: 8.2: 9.5
:  5  :   4  :  4           ---------------- 30'
: 8.1 : 10.4 : 10.8         :     : 9.8: 9.4
: 9.7 : 11.2 : 11.2         :     :     : 9.6
------------------46°40'   ---------------46°20'
30    40      50   157     30    40    50    151

         c)                         d)
```

14.2 Organization of the fishery information archive and its processing

Several times during the day fishing ships transmit by radio reports in which the ship's name, its affiliation, coordinates, catches and type of fishing are indicated. To process this information, program complexes were developed for the personal computer allowing the introduction of data into the computer and accumulation on the carriers and as well as performing of the analysis.

Saucy catching is done at night, therefore, the cycle "evening-morning" was taken and divided into four parts by ship's bells.

Figure 14.2: Fragments of fisher data archive.
c) S migration regions, 20.08-30.09 1987.
d) fishery regions 04,09,07.09; saucy at 47°-49°N, sardina at 43°-45°N.
e) fishery regions shift during 4.09-7.09.

No.	company	ship name	coordinates lat.	long.	catch (cent.)	note
1	FFC	sych	4414N	14715E	100	4 cath.
2	FFC	sala	4406N	14715E	020	
3	PFC	aniv	4407N	14716E	000	search.
4	PFCU	olin	4409N	14707E	060	light.

a)

catalog of fishery
27 days

No.	Date	Fish	Reports
1	1.10	saury	050
2	2.10	saury	065
3	3.10	saury	048
	. . .		
4	27.10	saury	075

b)

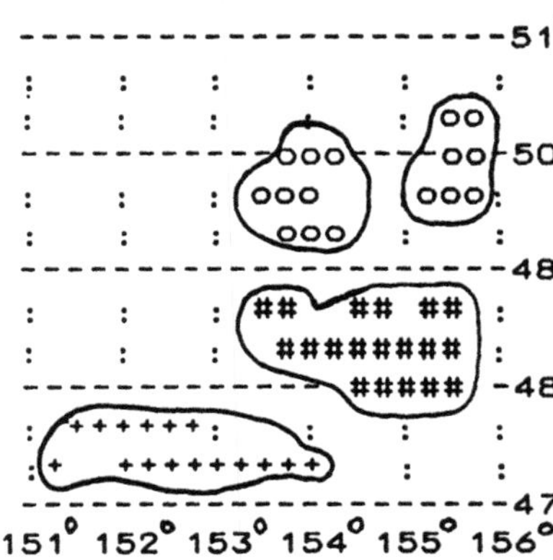

c)

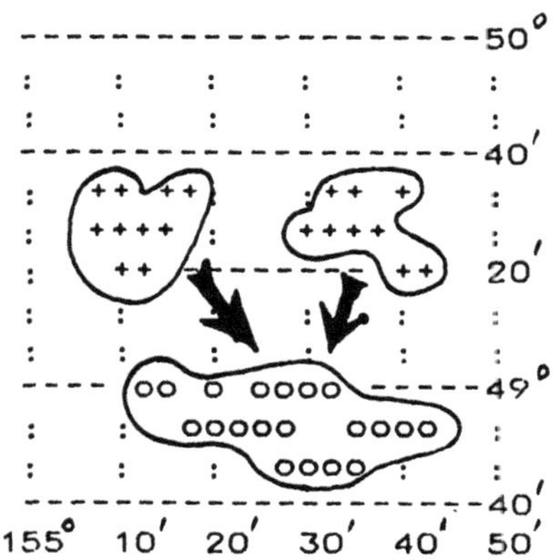

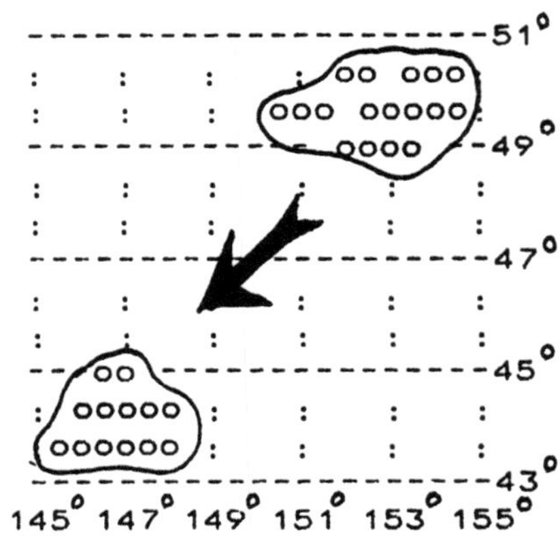

e) d)

In Figure 14.2a, an information fragment is shown for fishery on 18th August 1987. To obtain different generalized characteristics, the possibility of forming generalized files of ten-day periods and monthly averaging is provided and obtained results are output to graphic terminals. In Figure 14.2b, a catalogue fragment of generalized data for one month is presented.
Fishery report processing involves fishery plane-table construction and calculation of some quantitative characteristics. The plane-table construction is performed by means of computer graphics. In Figure 14.2b and 2c, examples of fishery plane tables are shown. The possibility for simultaneous output of one, two or four charts to the terminal is provided. The scale is automatically selected to mark all ships on the plane-table. Any symbol may be used to mark the ship on the plane-table. A list of fishery days could be given in the regime of graphical dialogue for which the construction of plane-tables is nesessary. This approach allows us to obtain migration paths of shoals for a given period. In order to analyze the fishery region in more details, any region of fishery plane-table could be presented on the chart separately. In Figure 14.2e, the fishery region and movements of ships during the fishery cycle are shown in enlarged form.
For ships shown on the plane-table some quantitative characteristics are computed: fishery region, total catch, effort number. The possibilities of fishery characteristic processing will be presented in more detail for saucy catching analysis in 1987.

14.3 Analysis of saucy catching by personal computer

In the search for fishery regions, practical recommendations for the fishery fleet were based on the following activities. Ship report processes plots on the regions surveyed by ships day and night showing whether catches occurred, the same way the fishery regions are generalized and their characteristics are computed. Coordinates of fishery regions are compared with those for proceding days and regional shift and variation of its characteristics are determined. Simultaneously, environmental parameters' variability was computed. Based on charts of ship water temperature measurements and facsimile charts, a comparison between current temperature and the preceding one was made. Simultaneously with analysis of operative information, fishery situations were determined, which were similar to current ones or were very different in respect to retrospective data. An evaluation was made of how water temperature influenced fishery region dynamics in general. As a result of comparison between retrospective and current hydrological information, an assumption was made on the possible nature of fishery variation. This assumption of a possible fishery region shift in the next 2-3 days was verified by the searching ship and also by neighbouring fishing ships.
The principally new aspect of this work lies in the fact that all recommendations were based on quantitative computations of environmental and fishery variability. Let us consider some fragments of fishery analysis.

1.1. Initial period of saucy catching

Comparison of ten-day charts for the NW Pacific showed that temperature off the S Kuriles is supposed to be at a level of late fishery beginning in July, as in 1979, 1983. Measurements of surface temperature end object observation were performed at night. To select observation regions we orientated on the years with late fishery beginning when the first approaches were to the E of 147°. These measurements have confirmed the prediction of low temperatures in the traditional fishery region. In Figure 14.3a, the chart is presented for the region in 10-14 July. Temperature values in squares 4400-4420, 14800-14820 in 1969 and 1978 were 10.6 and 11.4, i.e. 2°-3° higher than in 1987. A few small saucy were observed at night. For solving the question of search continuation in the S Kuriles area or of expansion of the search area, comparative computation were performed using ten-day facsimile charts for the second ten-day period of July. In Figure 14.3b, comparison of surface temperature data of 1987 with other years is shown for the S Kuriles area. Saucy fisheries in 1971 and 1974 began in mid July, and fishery during the second half of July was made mainly to the S and SE of Shikotan Island. Temperature in 1987 was 2°-3° lower than in 1971, 1974. Still higher temperature difference was observed in comparison to 1968 and 1978 when fisheries occured in early July. Therefore, the conclusion was made that fishery off Shikotan Island was unlikely during the third ten-day period of July. Analysis for the second ten-day of July revealed that low temperatures, characteristic for late fishery beginning, as for instance in 1981 and 1983, occur off the S Kuriles area. Therefore, comparative analysis was performed for other regions of area under study. In Figure 14.2e, comparison data is shown for those years when fishery regions were located to the E of 150°, particularly in 1986. The year of 1987 was found to be characterized by positive temperature anomalies as a whole in regions 4500-5000, 15200-15800. The assumption was made, which was subsequently confirmed, that the onset of fishery was possible in late July in the Free Strait.

1.2. Fishery regions of S migration

In the third ten-day period of August, the water temperature dropped in fishery regions.
Figure 14.4a shows a comparison of the charts for August 25-27 and August 21-23. At latitude of 45° to the E of 157°, a temperature drop of 1-2° is observed. In Figure 14.4b and 14.4c, fishery regions are presented for August 24-27. S migration occurs in two directions. A portion of saucy shoals shifted to the E and fishery regions were in areas 5000-5020, 15640-15700. Another portion of shoals shifted to the SW. The shoal bifurcation is clearly seen in Figure 14.5d.
From the beginning of S migration a significant variability of fishery regions occurs continually. The problem of timely prediction of new regions becomes more and more urgent, moreover, it is important not only to fix the temperature variations effected on saucy displacement but to evaluate cluster concentration. Shoals are observed in a fairly large area but catches differed in regions significantly. Sometimes, two or more groups of ships were only 10-15 miles distant from one another, however, catches in one group were 1.5-2.0 times larger than in other groups. This caused ship transitions from one region to other, which allowed indirect evaluation of fish shoal concentration.

These small fragments of the general picture of analysis of practical saucy catches show the possibilities of personal computer use for solving similar problems of fishery oceanography.
The accumulation of detailed information on saucy catches in 1987 together with algorithms and programs of its analysis on the personal computer, allowed us to obtain the fishery model. In the future, this informative model could be used for solving scientific and practical problems not only for saucy catches but other fisheries.

Figure 14.3: Examples of hydrological chart comparison in determination of the onset of fishing.
c) chart of temperature difference 1987-(1980.1981,1986,1979). Horizon 0,m. Fragment of comparison of years with late fishery, fishery region is located at 153°-155°.

```
-------------------          ----------- -----------
!Charts for bathimetr.       temperature temperature
!points from 10-14.07        difference  difference
!horizon 0,m                  87-(71,74)  87-(68,78)
!parameters:                 horizon,0,m horizon,0,m
!average value
-----------------30'  147°---------- -----------
:    :7.0:7.4:6.6:        :-1.0:-1.0: :-4.0:-2.0:
-----------------20'      :-3.0:-1.5: :-6.5:-2.5:
:6.1:7.4:7.1:6.6:     148°----------- -----------
-----------------10'      :-2.0:-2.5: :-3.5:-2.0:
:8.5:    :7.5:6.9:        :-3.0:-2.0: :-6.0:-3.0:
-----------------44°  149°----------- -----------
:8.8:7.6:7.6:    :        :-1.0:-2.0: :-3.0:-2.0:
-----------------50'      :-2.0:-2.5: :-6.0:-3.5:
:    :8.0:7.8:8.0:    150°----------- -----------
-----------------40'      43°      44° 43°      44°
20' 40' 148° 20' 40'
        a)                             b)
```

```
45°       46°       47°       48°       49°       50°
-----------------------------------------------------152°
:  0.0  :  1.5  :  1.0  :         :         :
:  0.0  :  1.5  :  1.0  :         :         :
:  3.0  :  3.0  :  2.0  :         :         :
:  0.5  :  1.5  :  0.5  :         :         :
:  1.0  :  0.5  : -0.5  :         :         :
-----------------------------------------------------153°
:  0.0  :  0.0  :  1.0  :  0.5    :         :
:  0.0  :  0.5  :  0.0  :  0.0    :         :
:  2.5  :  2.0  :  1.5  :  1.5    :         :
:  0.0  :  1.0  :  0.5  :  0.0    :         :
:  0.5  :  0.5  : -0.5  : -0.5    :         :
-----------------------------------------------------154°
:  0.8  :  0.5  : -0.5  :  1.0    :  2.0    :
:  0.0  :  0.0  : -0.5  :  0.0    :  1.0    :
:  2.0  :  1.0  : -0.5  :  1.0    :  0.0    :
:  0.0  :  0.0  :  0.0  :  0.5    :         :
:  0.0  :  0.0  : -1.0  :  0.0    :  0.0    :
-----------------------------------------------------155°

                    c)
```

Figure 14.4: Fragments of analysis of S migration shoals.
b) catches: 24.08 - #; 25.08 - +; 26.08 - o
c) fishery region shift to the SW.
d) fragments of analysis of S migration shoals.

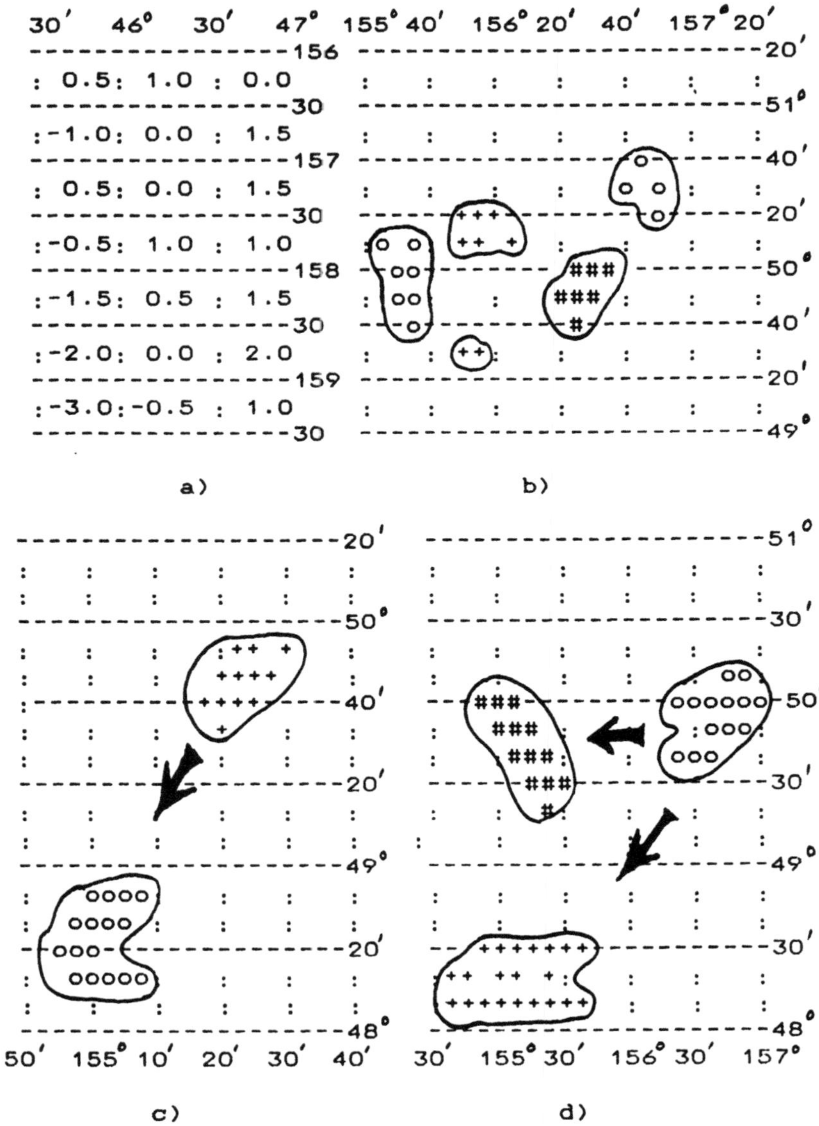

References

1. A.A.Voronov, R.S.Goldman, V.I.Ilyichev. Algorithmic methods in prediction problems of fishery regions in the ocean. Automation and telemechanics, No.5, 1985, pp.146-157,(in Russian).

2. L.P.Smirnykh. Bank of ten-day charts of surface temperatures in the Northwest Pacific. Meteorology and Hydrology, No.2, 1987, pp.63-69,(in Russian).

3. R.S.Goldman, L.P.Smirnykh, G.I.Yurasov. Algorithmic technique of identifying discriminative ocean areas in fisheries oceanography problems. GeoJournal, 16.1, 1988.

R.S.Goldman and V.I.Ilyichev
Pacific Institute of Oceanology
USSR Academy of Sciences
Far Eastern Branch
Vladivostok
USSR

PART IV
EXPLOITATION OF OCEANIC RESOURCES AND ESTIMATION OF ANTHROPOGENIC INFLUENCE ON MARINE ECOSYSTEMS

15 TWO THEORIES IN THE EXPLOITATION OF CHEMICAL RESOURCES OF SEA WATER

Zhang Zhengbin and Liu Liansheng

15.1 Introduction

With continental resources recovered and consumed, chemical resources of sea water become more and more important. For a long time the extaction of various salts from sea water is refered to major elements in sea water. The principle of the extraction is the phase law, for example, producing salt from sea water with solar energy, carnallite ($KCl.MgCl_2.6H_2O$) from bittern by concentration, sodium sulfate and magnesium sulfate by cooling. Another method of salt extraction is based on addition of chemicals, for example, precipitation of magnesium hydroxide, extraction of bromine from sea water or brine with oxidation. On the other hand, for extraction of minor, or trace, elements from sea water only concentration could be used. Separation and extraction of minor elements from sea water after appearance of chemical sorption approaches was based on the theory of stepwise ion/coordination particle exchange and the composite model for the ion/coordination particle exchange kinetics at the liquid-solid interface in sea water. With this background research on uranium extraction from sea water has been developed in the 1960's by more than 10 countries such as P.R. China, USSR, Japan, Great Britain and others. So far highly effective extractor of uranium from sea water with concentration factor in the range of 0.5 to million has been successfully produced and a plant with production of 1000 tons per year will be built by the end of this century. Meanwhile the extraction of potassium iodine, bromine, strontium, lithium and rubidium and other elements from sea water has also been investigated with this sorption method and many fundamental research achievements have been obtained.

This article deals with two new theories in the exploitation of chemical resources of sea water: (1) Theory of interfacial stepwise ion/coordination particle exchange; (2) The composite model for the ion/coordination particle exchange kinetics at the interface between liquid and solid phases in sea water.

15.2 Theory of Interfacial Stepwise Ion/coordination Particle Exchange

1. General classification and interpretation of solid-liquid sorption isotherm in sea water reaction isotherms of the interaction between trace metals and suspended particulate matter in sea water were studied with or without added organic matter. Seventy systems have been studied, including minor, or trace, metals [U(VI), Cu(II), Zn(II), Pb(II), Cd(II), Cr(III), Cr(VI)], solid particulate matter (hydrous titanium oxide, amorphous hydrous ferric oxide, goethite, hematite, δ-MnO_2, γ-MnOOH, manganite, illite, kaolinite, montmorillonite) and organic matter (HA, FA and amino acids). Approximately one hundred isotherms, including stepwise isotherm and ion exchange-precipitation isotherm, were observed to occur in the above systems, the result of some experiments that show new types of isotherm in sea water medium have not been reported before.

V. I. Ilyichev and V. V. Anikiev (eds.), Oceanic and Anthropogenic Controls of Life in the Pacific Ocean, 205–223.

Figure 15.1: Classification of solid-liquid sorption isotherm in seawater.

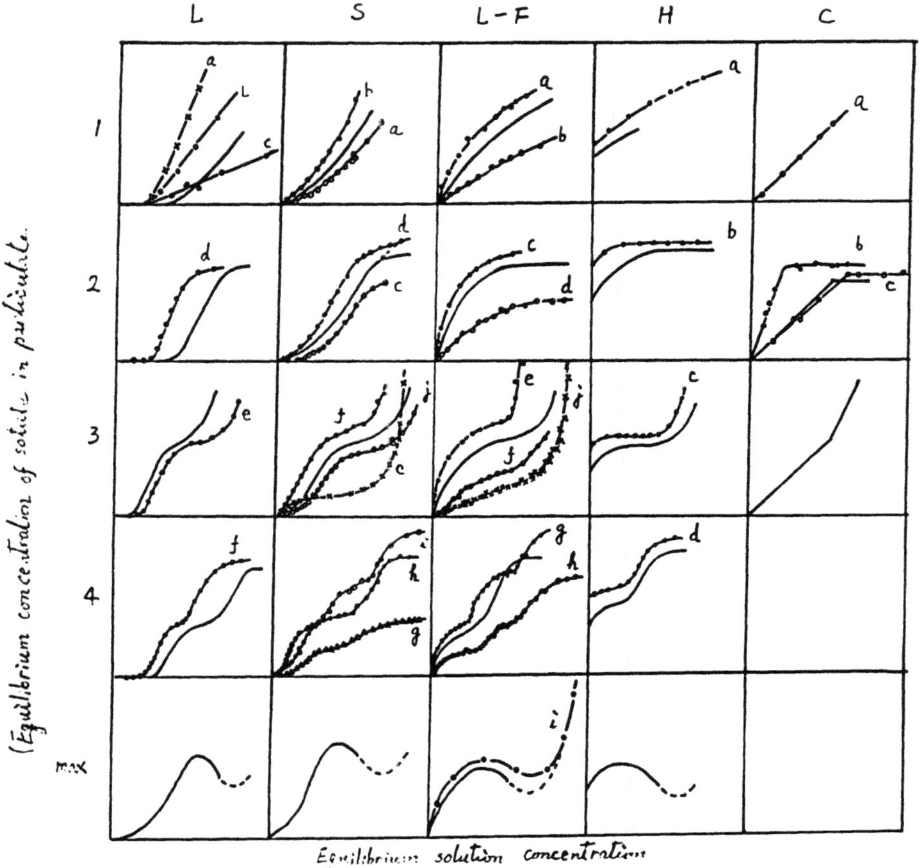

(1) Classification

It is possible to classify almost all of those isotherms into the pattern shown in Figure 15.1, the Classsification system of SLISSW (Solid-Liquid Interfacial Sorption in Sea Water). This divides them all, by their initial slope, into five main classes, named for convenience the L (Low affinity), S (Sigmoid shaped), L-F (Langmuir-Freundlich type), H (High affinity) and C (Constant partition). As the concentration of solution increases the curve develops in a certain way shown as subgroups 1,2,3,4 which could be extended if necessary to 5, and so on, if additional inflections are recorded. The subgroup "Max" is allocated to those curves which show a maximum. A number of experimental examples of these isotherms have been tabulated in Table 15.1.

Table 15.1: Examples of ion/coordination particle stepwise exchange isotherm of the five classes.

Class	Curve	System	Scales (box limits)	
			Horizontal	Vertical
S-Curve	a	Cr(III)-hydrous titanium oxide in seawater (pH = 4.5, hydrous titanium oxide 0.1000 g)	0–60 ppm $(Cr)_{eq}$	0–0.6 (μg/mg)
	b	Cu(II)-hydrous titanium oxide (seawater) (pH = 4.75; HTO 0.1000 g)	0–200 ppm	0–3 (μg/mg)
	c	Cd(II)-Goethite in seawater (pH = 6.50, Goethite 0.1000 g)	0–25 ppm	0–400 (μg/mg)
	d	Zn(II)-γ-MnOOH in seawater (pH = 4.3, γ-MnOOH 0.1000 g)	0–28 ppm	0–500 (μg/mg)
	e	same as c	0–40 ppm	0–550 (μg/mg)
	f	same as d	0–60 ppm	0–1000 (μg/mg)
	g	Cd(II)-Hematite in seawater (pH = 8.30, Hematite 0.1000 g)	0–60 ppm	0–800 (μg/mg)
	h	U(VI)-hydrous titanium oxide in seawater (pH = 3.50, HTO 0.1000 g)	0–3 ppm	0–150 (μg/mg)
	i	same as f	0–80 ppm	0–1600 (μg/mg)
	j	Pb(II)-Montn. in seawater (pH = 8.10, M 0.1000 g)	0–2 ppm	0–4 (μg/mg)
F-L-Curve	a	Cu(II)-Amorphous hydrous ferric oxide (pH = 5.00, AHFO 0.1000 g)	0–30 ppm	0–400 (μg/mg)
	b	Cu(II)-Goethite in seawater (pH = 5.00, Goethite 0.1000 g)	0–25 ppm	0–400 (μg/mg)
	c	Cu(II)-δ-MnO_2 in seawater (pH = 2.20, δ-MnO_2 0.1000 g)	0–110 ppm	0–2300 (μg/mg)
	d	same as a	0–42 ppm	0–100 (μg/mg)
	e	same as c	0–140 ppm	0–2400 (μg/mg)
	f	Zn(II)-δ-MnO_2 in seawater (pH = 4.30, δ-MnO_2 0.1000 g)	0–90 ppm	0–1600 (μg/mg)
	g	same as c	0–400 ppm	0–4000 (μg/mg)
	h	Cu(II)-γ-MnOOH in seawater (pH = 3.5, γ-MnOOH 0.1000 g)	0–400 ppm	0–4000 (μg/mg)
	i	0.1 ppm Lysine-Cu-Illite in seawater (pH = 8.00, illite 0.1000 g)	0–1.0 ppm	0–1.0 (μg/mg)
	j	Cu(II)-Illite in seawater (pH = 8.10, illite 0.1000 g)	0–1.2 ppm	0–2.0 (μg/mg)
L-Curve	a	Activated Silicon-$Al(OH)_3$ (pH = 8.2 $\sim$ 8.4, 13 Cl°/$_{oo}$)	0–16 ppm	0–2.5 ($mgSiO_2$/mgAL)
	b	same as a (pH = 8.2-8.4, 6.5 Cl°/$_{oo}$)	0–16 ppm	0–2.5 ($mgSiO_2$/mgAL)
	c	same as a (pH = 7.2, 6.5 Cl°/$_{oo}$)	0–30 ppm	0–2.5 ($mgSiO_2$/mgAL)
	d	U(VI)-hydrous titanium oxide in SW (pH = 4.00, 10°C, HTO 0.1000 g)	0–1.5 ppm	0–80 (μg/mg)
	e	same as d	0–2.2 ppm	0–110 (μg/mg)
	f	same as d	0–3.0 ppm	0–150 (μg/mg)
H-Curve	a	same as F-L-a, but pH = 8.10	same as F-L-a	
	b	same as F-L-c, but pH = 8.10	same as F-L-c	
	c	same as F-L-f, but pH = 8.10	same as F-L-f	
	d	same as F-L-g, but pH = 8.10	same as F-L-g	
C-Curve	a	U(VI)-hydrous titanium oxide in SW (natural condition, Keen's data)	0–3 ppm	0–8 (μg/mg)
	b	Cu(II)-Bentonite Clay in SW (pH = 8.00)	0–50 ($10^{-7}\frac{moles}{L}$)	0–120 ($10^{-6}\frac{moles}{L}$)
	c	Cd(II)-Bentonite Clay in SW (pH = 8.00)	0–50 ($10^{-7}\frac{moles}{L}$)	0–120 ($10^{-6}\frac{moles}{L}$)

(2) Qualititative interpretations [1-4]

A simplified qualitative theoretical interpretation of the factors that determine the difference in shape of these isotherms have been developed below.

(A). The factors that determine the difference in five classes L, S, L-F, H and C are as follows:

(a) Suspension pH. A large number of studies in our laboratory has shown the importance of pH as a factor controlling the shape of isotherm. For cation exchange, the transformation sequence in shape of isotherm of the five classes is; L → S → C → L-F → H with the suspension pH increasing, while the other conditions remain constant.

(b) The adsorbate (minor metals) - substrate (suspended particals) affinity. The affinity of these metals with suspended particals falls in the following sequence: H > L-F > S > L type isotherm, C type isotherm represents conditions in which the adsorbate-substrate affinity remains constant as the concentration of adsorbate increases (i.e. the number of adsorption sites remains constant as the concentration of adsorbate increases).

(c) The ratio of (amount of adsorbate/amount of adsorbent). When the ratio is relatively low, the L-type and S-type isotherms should be produced; when the ratio is relatively high, the H-type and F-L type isotherms should appear.

(d) Lateral interaction. The Langmuir equation can be further reduced to a straight line at very low concentration; however, highly precise data indicate that the isotherm may sometimes still be curved at low concentration (an exponential relationship). This phenomenon is perhaps caused by the lateral forces existing among the molecules or radical groups on the surface layer. As a result, we have [5]

$$\frac{\theta}{(1-\theta)} = [K_0 \exp(-\varepsilon_i \theta / kT \ ] \tag{1}$$

From the nature of Eq. (1) it can be seen that the lateral interaction made the isotherm in the low concentration region have S type or L type.

(e) Effect of activity coefficient, effect of temperature, etc.

(B). The factors that determine the differences in subgroups 1,2,3,4 .. "Max" are as follows:

(a) Concentration of adsorbate. As the concentration of adsorbate is increasing, the shape of isotherm transfomrs from subgroups 1, 2, 3, .. to "Max".

(b) We suggest that the cause for the stepwise isotherms is that there are more than one type of active sites on the surface of solid. Loanathen and Burau (1973) suggested that Mn and Fe in hydrous manganese and ferric oxide, for example, may exchange with metal ions (or complex ions) in solution, in addition to hydroxy groups on the surface of particulates. Schindler (1981) suggested that the OH groups on the surface may also be associated with 1, 2, or 3 Mn or Fe in hydrous manganese or ferric oxide. We also suggest that in addition to OH group on the surface, Na, K and Mg ions in the lattice of clays may also compete with surface sites for metal ions in solution.

(c) Formation of precipitation. We suggest that the interaction of Cu(II), Pb(II) with kaolinite, illite and montmorillonite in sea water, is ion exchange reaction at lower concentration range, whereas precipitation dominates at higher concentration range. These isotherms are similar to S-3 and L-3 isotherms. The ion-exchange-precipitation isotherms are generally represented by modified BET equation, modified Toth equation and de Boer-Zwikker equation:

$$\frac{\kappa\left(\frac{c}{c_0}\right)}{\bar{n}\left(1-\kappa\frac{c}{c_0}\right)} = \frac{1}{\bar{n}_{max}b} + \frac{b-c}{\bar{n}_{max}b}\kappa\left(\frac{c}{c_0}\right) \qquad \text{(BET)}$$

$$\theta = \frac{\bar{n}_1}{\bar{n}_{max}} = \frac{\kappa\left(\frac{c}{c_0}\right)^i}{1+\kappa\left(\frac{c}{c_0}\right)^i-\kappa_s\left(\frac{c}{c_0}\right)^j} \qquad \text{(Tóth)}$$

$$\log\left[RT\log\frac{c_0}{c}\right] = \log\frac{\varepsilon_0}{2.303} - \frac{K\left[\frac{\frac{x}{m}}{\left(\frac{x}{m}\right)_{max}}\right]}{2.303}$$

(d) Effect of organic matter. Subgroup "Max" isotherms has a close relationship with this effect.

2. Theory of interfacial stepwise ion/coordination particle exchange and quantitative of the general classification.

(1) Theory of interfacial stepwise ion/coordination particle exchange [1,4,6].

(A) Basic hypothesis: There are many ligands in sea water, such as OH^-, Cl^-, HCO_3^-, CO_3^{2-}, F^- etc. Therefore, beside hydration [forming $M(H_2O)$], M may also form $M(H_2O)_{m-x}(OH)_{x(n-x)}^{+}$ and $M(H_2O)_{m-y}L_{y(n-y)}$ (L = Cl^-, HCO_3^-, CO_3^{2-}, F^- etc.). In sea water they exist in a state of equilibrium. These hydrated ions, hydrolyzates and various complexes are distributed between sea water and the interface of solid particles. Such solid-liquid distribution may be considered as taking place in the mode of stepwise ion/coordination particle exchange equilibrium [1], as shown by the schematic diagram of Figure 15.2.

Figure 15.2: Ion/coordination particle exchange phenomena at solid particle-water interface.

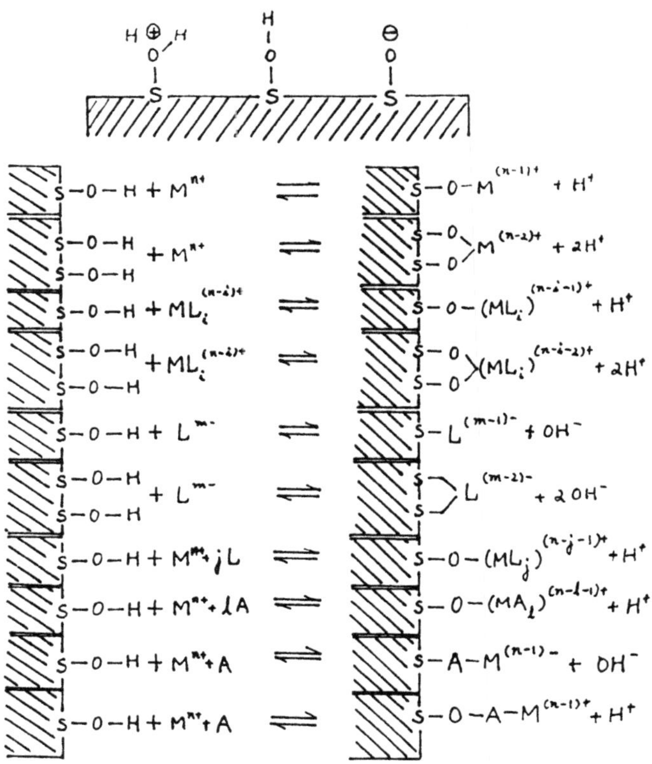

The stepwise ion/coordination particle exchange at solid-liquid may also be represented as:

$$nS\text{-}OH + iM + jL + MA \rightleftharpoons (SO)_nM_iL_jA_m + nH$$

Where $N = 1, 2, \ldots n$; $m = 0, 1, 2, \ldots$; $i = 1, 2, \ldots I$; $j = 1, 2, \ldots J$.

(B) General isotherm equation

If we regard the solid particles distributed in solutions as subgroups and regard sea water as medium, then according to basic hypothesis the minor element M therein (complex ions M_mL_i, M_mA_j etc. are omitted for simplicity; the same herinafter) can be distributed to subgroups as follows each phase cell being able to contain i number of M, N being the highest ligand number. According to statistics, the distribution probability of the variable system of this kind of particle number is:

$$\omega\ (i,\varepsilon)\ =\ C(\frac{i\mu - \varepsilon_i}{\kappa T}) \tag{2}$$

where C is the normalization coefficient;

μ is the chemical potential, related to temperature and the concentration of minor element M in solution;

when in balance the μ of the phases is the same;

ε_i is the energy of the subgroups in holding i number of minor element M;

where i is taken as equal, the ε_i of the subgroup is approximately equal; when $i=0$, $\varepsilon_i=0$.

κ and T is Boltzman constant and absolute temperature, respectively.

The condition for normalization is:

$$\sum_{i=0}^{N} \omega\ (i,\varepsilon)\ =\ 1 \tag{3}$$

or

$$C\{1+\sum_{i=1}^{N} \exp\ (\frac{i\mu - \varepsilon_i}{kT})\}\ =\ 1 \tag{4}$$

In each cell at the state of equilibrium the average surface ion exchange number θ distributed to R is:

$$\theta\ =\ \sum_{i=1}^{N} i\omega\ (i,\varepsilon)\ =\ \sum_{i=1}^{N} iC \exp\ (\frac{i\mu - \varepsilon_i}{kT})\} \tag{5}$$

Where θ is also the ratio of sorption (i.e. θ = quantity of sorption ν saturation quantity of sorption) or their average surface complexing number by the theory of complexation.

Substituting Eq. (4) into Eq.(5) we get:

$$\theta\ =\ \frac{\sum_{i=1}^{N} i \exp\ (\frac{i\mu - \varepsilon_i}{kT})}{1+\sum_{i=1}^{N} (\frac{i\mu - \varepsilon_i}{kT})} \tag{6}$$

According to chemical thermodynamics we have:

$$\mu = \mu^0 + kT\ln\left(\frac{\mu}{kT}\right) \tag{7}$$

Setting standard state $a^0_M = 1$, let $\mu = \mu' - \mu^0$, we get:

$$a_M = a^0_M \exp\left(\frac{\mu}{kT}\right) \tag{8}$$

Again we let:

$$K_i = K^0_i \exp\left(\frac{-\varepsilon_i}{Kt}\right) \tag{9}$$

substituting Eqs. (8) and (9) into Eq. (6), we get the surface stepwise ion exchange equilibrium of solid particle as follows:

$$\theta = \frac{\sum_{i=1}^{N} iK_i a^i_M}{1+\sum_{i=1}^{N} K_i a^i_M} \tag{10}$$

(C) Quantitative treatment of the general classification

The combination of modified Tóth equation and stepwise isotherm equation can produce a generalized equation:

(11)

where M and j are the empirical constants. Eq. (11) can describe all patterns of isotherms that we observed in sea water as shown in Figure 15.1. On the basis of Eq. (2), the general interpretation and quantitative treatment were shown in Table 2.

When $K_i a^j_M = 0$, i = i' = 1.2; from Eq. (11) we get

$$\theta = \frac{K_1 a_M + 2K_2 a^2_M}{1+K_1 a_M + K_2 a^2_M} \tag{12}$$

Table 15.2: A general interpretation of solid/liquid sorption isotherm in seawater.

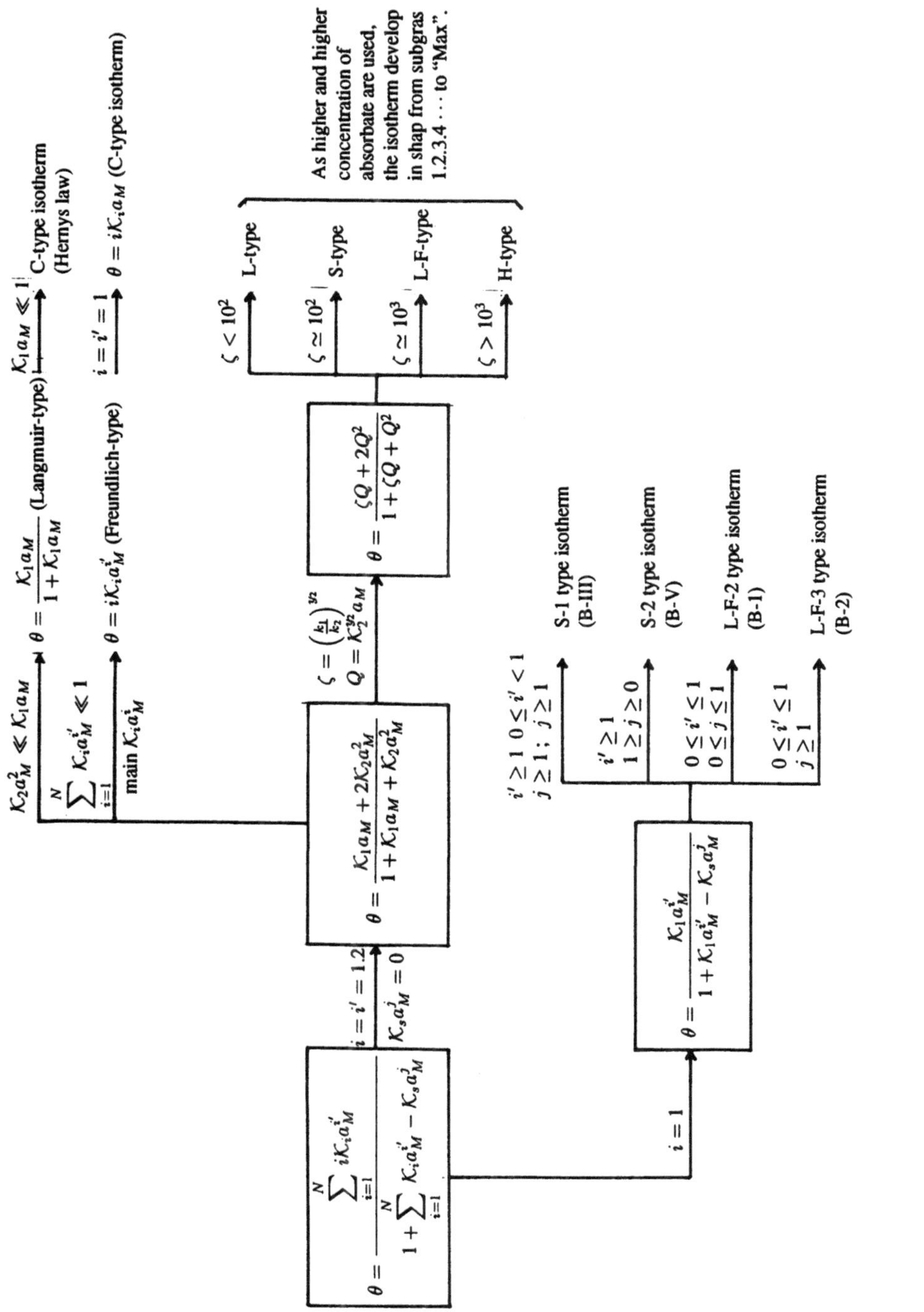

Now we define

$$\zeta = \left(\frac{K_1}{K_2^{1/2}}\right) , \quad Q = K_2^{1/2} a_M$$

and rearrange Eq.(3) to

$$\theta = \frac{\zeta Q + 2Q^2}{1 + \zeta Q + Q^2}$$

A plot of θ vs. log Q at various

$$\zeta = \left(\frac{K_1}{K_2^{1/2}}\right)$$

ratios is given in Figure 15.3.
Figure 15.3 agrees with all patterns of isotherm as shown in Figure 15.1 (H, F-L, S, L, and C-types).

Figure 15.3: The normal curve obtained with different values of (K_1/K_2).

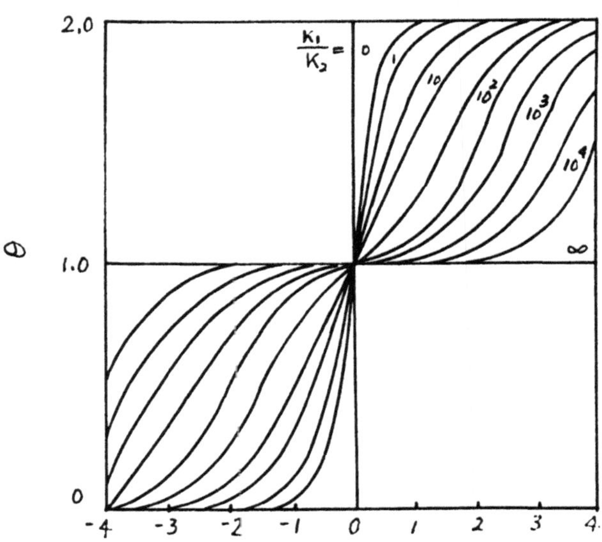

15.3 The Composite model for the ion/coordination particle exchange kinetics at the interface between liquid and solid in sea water

1. Introduction

Sea water is a moderately concentrated electrolyte solution system ($I=0.7$) with multi phases and multi compontents. The investigation on the transfer of ions among several phases has become an essential task in marine chemistry. The ion exchange between the heavy metal ions and solid particles (e.g. hydrous metal oxide, clay minerals) is one of the most important ways of such transfer processes. There are many works done on their equilibria properties. But untill now, little systematical work has been done in the aspect of kinetics. As most processes have not been attained equilibria in ocean, the investigation of the ion exchange kinetics between liquid and solid phase in sea water is very important for marine geochemistry, environmental chemistry and the sea water resource chemistry.

Boyd et al. (1974) showed that the rate determining step of the exchange reaction is diffusion, which was, furthermore, divided into liquid film diffusion and particle diffusion. The equation for the liquid film control process is

$$F = 1-\exp(-Rt) \tag{13}$$

where F is the fractional attainment of equilibrium, R is the liquid film diffusion constant. The particle diffusion equation is

$$F = 1-\frac{6}{\pi^2}\sum_{n=1}^{\alpha} \frac{1}{n^2} \exp(-Bt) \tag{14}$$

where B is the particle diffusion constant.

In the recent two decades, accomplishment has been achieved in the two aspects. Firstly, many combined film and particle diffusion controlled kinetics models have been proposed [2,9]. Secondly, the boundary progressive phenomenon has been observed.

For the ion exchangers with weak acid functional groups, since the products are site-binding ones, an obvious boundary between the exchanged part and the unexchanged core exists, which is towards the interior. Helfferich (1965) was one of the first to predict such possibility. Then Miller, Bolto and Warner (1970), Selim (1973), Native (1975), and Dana (1974) proved the existance of moving boundary phenomenon. Zhang et al. (1979) showed that the boundary progressive phenomenon also exists in the ion exchange reactions between heavy metal ions and the inorganic particles in sea water. But they have different explanation about the moving boundary. Native emphasised the concept of shell progressive diffusion. Dana proposed a composite mechanism of film diffusion and shell diffusion.

Generally, we can express the ion exchange reaction in sea water as follows:

$$RB + A \rightarrow RA + B \quad (15)$$

where RB or R^{n-} is the exchangable group, A is heavy metal ions. B could be H^+ or other cations such as Na^+, K^+ and Ca^{2+}.

According to the properties of sea water system, Zhang did not consider the exchange as a quasi-homogeneous spherical particle. Most solid particles in sea water consist of small crystal particles (hydrous metal oxides, clay minerals), so that, they are exchangers which are order in small region and disorder in large region. Zhang pointed out that ions can diffuse to the unexchanged groups through the porous solution in the exchanger. As a result, he proposed a liquid film progressive model for the ion exchange reaction in sea water. This model considers the porous solution as the extend of the out liquid film. Furthermore they have the same property and could be treated with the same mathematical method. The resulting equation is

$$[1-3(1-F)^{2/3}+2(1-F)] = \frac{6DC_{AO}}{C_{SO}R^2}t \quad (16)$$

which seems similar to that of Native's from mathematically point.

2. The Composite model of liquid film diffusion control and progressive liquid layer diffusion control.

According to our new experimental result we think it is better to treat the outer liquid film and the internal porous solution in differnt ways. We define the porous liquid as equivalent liquid layer which consists of a part of porous solution, which contains ions A and is connected with RA. This liquid layer has boundaries with the unexchanged core and the internal porous solution which contains RB, and with the outer liquid film respectively. The three regions divided by the two boundaries have obviously different physical-chemical properties. For example, there could be some mobile co-ion (e.g. Cl) in the liquid layer for there is less Donna repulsive force caused by free R groups. Furthermore, this liquid is a more structured one than that of the outer liquid film.

Refered to the ideas mention above, we developed a composite model of liquid film diffusion control and progressive liquid layer diffusion control for the ion exchange reaction between heavy metal ions and solid particles in sea water. The main principle of our model is as follows (refer to Figure 15.4):

(1) The exchanger is a multi-porous spherical particle which is order in small region and disorder in a large region.

(2) The heavy metal ions A and the reactive groups R form site-binding results.

Figure 15.4: Scheme for the composite model of liquid film diffusion control and progressive liquid layer diffusion control.

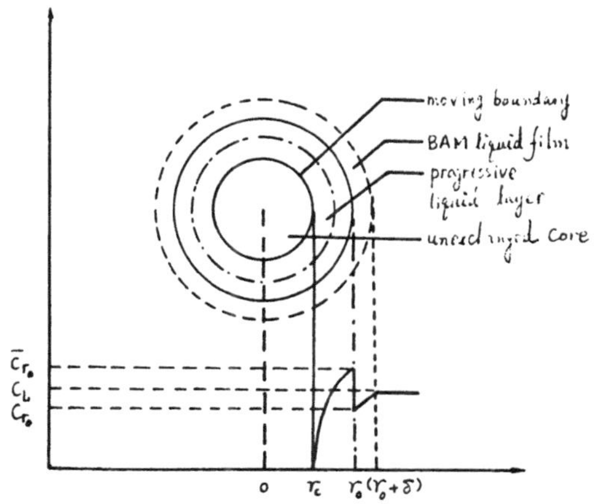

(3) The diffusion of ion A inside the particle through the exchanger atrix has two ways.
First, it diffuses directly through the exchanger matrix. Second, it diffuses through the porous liquid. The velocity of the latter is much faster than the former, so that the later is the main way.

(4) The ion exchange reaction of A with the reactive group R is controlled by the diffusion both in the liquid film and in the progressive liquid layer. At low concentration (for example, in sea water) especially during the initial stage the exchange process is mainly controlled by the liquid film diffusion.

(5) There is an obvious boundary between the exchange part (RA) and unexchanged core (R or RB), which is moving towards the core.

(6) The liquid film is considered as a plane and the ions diffusion through it is treated as a steady state.

(7) Assumed that the diffusion in the liquid layer is also steady state.

(8) The concentration of heavy metal ions in sea water maintains a certain value. According to the above model assumptions, we have the diffusion equation in the liquid layer as follows (Fick's second law):

$$\overline{D}\left(\frac{\partial^2 \overline{C}}{\partial r^2} + \frac{2}{r}\frac{\partial \overline{C}}{\partial r}\right) = \frac{\partial \overline{C}}{\partial r} = 0 \qquad (17)$$

where the second equaity is made from the assumption (7),
"-" indicates that the quatity is connected with the solid particle.

Due to point (6) and (8) boundary condition in the surface of exchange particle is:

$$\overline{D}(\frac{\partial \overline{C}}{\partial r})_{r=r_0} = k_f(C_L - \overline{C_{r_0}}/k) \tag{18}$$

Where $k_f = D/\delta$, D is diffusion constant of A in the liquid.

From point (2), the moving boundary should obey the following conditions:

$$\overline{C_{r_c}} = 0 \tag{19}$$

and

$$\overline{D}(\frac{\partial \overline{C}}{\partial r})_{r_c} = \overline{C}_{SO}\frac{\partial r}{\partial t} \tag{20}$$

Where C_{SO} is the concentration of R. The condition between the liquid film and the bulk solution is:

$$C_{r_0+\delta} = C_L \tag{21}$$

The initial condition is:

$$C_{L(t=0)} = C_{LO} \tag{22}$$

C_{LO} is the concentration of A in the bulk solution before the reaction. Furthermore according to point (3), there should be:

$$C_{L(t)} = C_{LO} \tag{23}$$

Equation (17) could be solved under the condition of (6) to (11). At last we get:

$$\tau \equiv \frac{t}{t_{eg}} = \frac{1}{1+\phi}\{-\ln(1-F) + \phi[1-3(1-F)^{2/3} + \alpha(1-F)]\} \tag{24}$$

where τ is a parameter which measures the relative importance of liquid layer, diffusion layer and liquid film diffusion in controlling the exchange process.

$$\phi \equiv \frac{1}{2}(\frac{D}{\overline{D}})(\frac{r_0}{\delta})(\frac{C}{\overline{C}}) \tag{25}$$

Equation (24) has two limiting cases. First, when $\phi \to 0$ due to the definitions, the liquid film diffusion velocity is much slower than that of the liquid layer. Eq. (24)

is reduced to

$$\tau = -\ln(1-F) \tag{26}$$

This is BAM's liquid film control equation. Second, when $\phi \to \infty$, the liquid film diffusion velocity is much faster than that of liquid layer diffusion. In Eq. (24) the second term > > the first term, so that we have

$$\tau = 1-3(1-F)^{2/3}+\alpha(1-F) \tag{27}$$

This is Zhang's liquid film progressive equation.
From what we have discussed above, we know that BAM's liquid film diffusion equation and Zhang's model are two limiting cases of our new composite model. We have calculated the τ - F relationship (F=0.005 to 0.995, step length 0.005) in different ϕ values. Then ϕ can be a fitting parameter and determined from the experimental values during the fitting process.

3. Experimental Proof and Discussion

We investigated the kinetics of ion exchange reaction between Cu(II) and FeOOH and manganite.
The experiments were carried out in a spherical glass flask with four necks. The pH of the system determined by pHS-2 pH meter (made in China). HCl or NaOH were added to maintain the pH values constants (within 0.05 unit).
Samples were taken as a function of time to follow the reaction. About 20 ml of suspension was withdrawn into a syringe and then filtered into a vacant tube through 0.45 μm membrane filter (the time required for phase separation was about 10 seconds).
We determined the orders and rate constants of the reaction by means of differentiation (using four different initial concentrations, 10^{-5} M). They are first order reactions. The values of their activation energy are 5,8, 4,9 and < 1 kcal.mole^{-1} respectively, which is in the region of liquid film diffusion control.
Let us use BAM's equation to examine our results. We could not get a straight line by either Eq. (13) (Figures 15.6 and 15.7 when ϕ=0) or Eq. (14) (Figure 15.5). When using Zhang's equation we got a straight line with the initial points dropping down (Figures 15.6 and 15.7, when ϕ=20). As we know, the combination of heavy metal ions with hydrous metal oxides produces site-binding products, there might exist boundary progressive phenomenon. And also due to our research on the initial reactions we know that the reactions are controlled by the liquid film diffusion initially. Thus, we proposed the above composite model to treat our experimental results. As Figures 15.3 and 15.4 show, when ϕ changes, the line shape changes. When =8.5 and 8 for Cu(II)/α-FeOOH, Cu(II)/Am-FeOOH and Cu(II)/Manganite (Figures 15.6 to 15.10), we get straight lines passing through the origins, respectively. The theoretical model conforms with the experimental results quit well.

Figure 15.5: Relationship between Bt and t (Cu(II)-α-FeOOH, pH=7.0, 25°C).

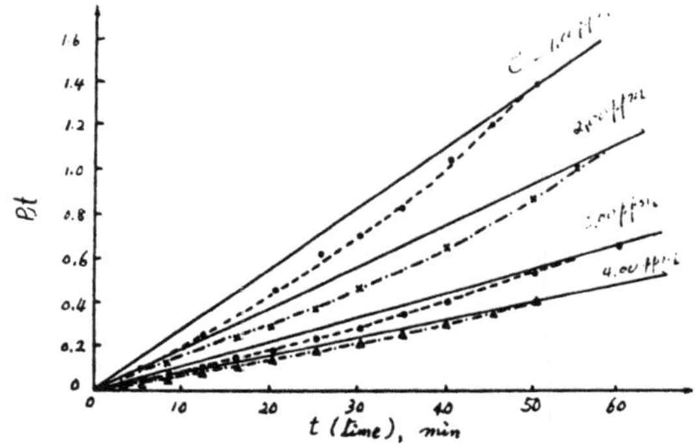

Figure 15.6: Relationship between $1/1+\phi\{-\ln(1\text{-}F)+\phi[1\text{-}3(1\text{-}F)^{2/3}+\alpha(1\text{-}F)]\}$ and t (Cu(II)-α-FeOOH, C°=2.00 ppm, pH+7.3, 25°C).

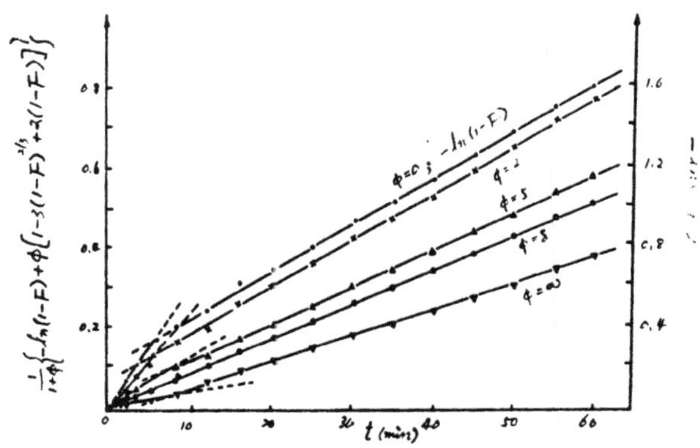

Figure 15.7: Relationship between $1/1+\phi\{-\ln(1\text{-}F)+\phi[1\text{-}3(1\text{-}F)^{2/3}+\alpha(1\text{-}F)]\}$ and t (Cu(II)-am-FeOOH, C°=1.00 ppm, pH=7.3, 25°C).

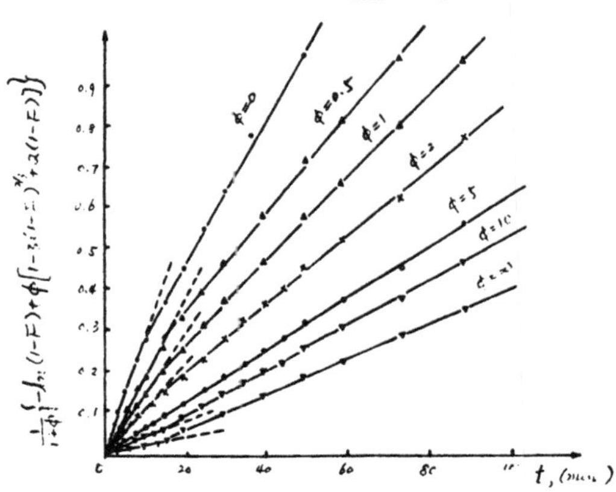

Figure 15.8: Relationship between $1/1+\phi\{-\ln(1\text{-}F)+\phi[1\text{-}3(1\text{-}F)^{2/3}+\alpha(1\text{-}F)]\}$ and t ($Cu(II)$-α-FeOOH, $C^{\circ}=2.00$ ppm, pH = 7.3, 25°C).

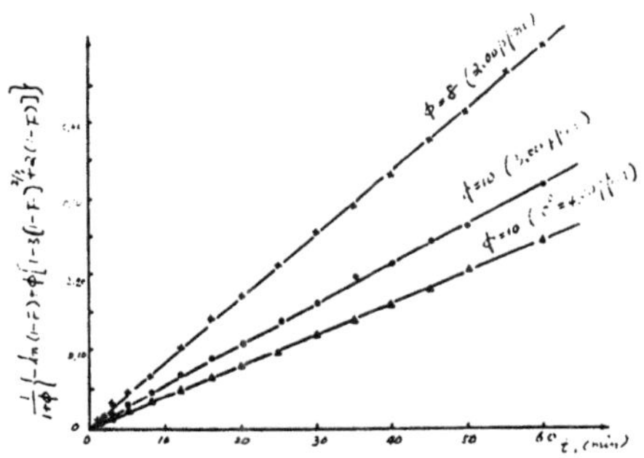

Figure 15.9: Relationship between $1/1+\phi\{-\ln(1\text{-}F)+\phi[1\text{-}3(1\text{-}F)^{2/3}+\alpha(1\text{-}F)]\}$ and t ($Cu(II)$-am-FeOOH, $C^{\circ}=1.00$ ppm, pH = 7.3, 25°C).

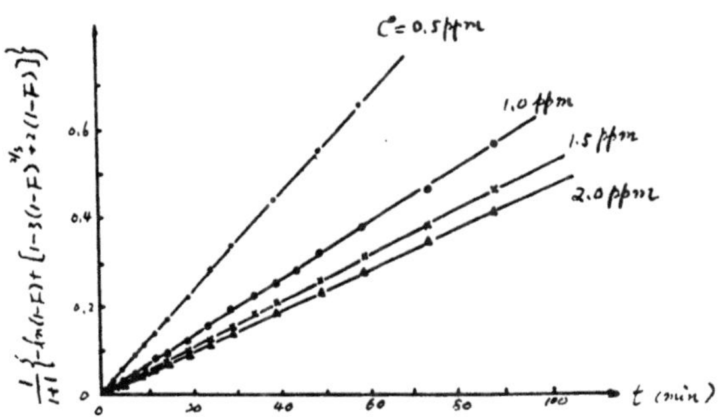

Figure 15.10: Relationship between $1/1+\phi\{-\ln(1\text{-}F)+\phi[1\text{-}3(1\text{-}F)^{2/3}+\alpha(1\text{-}F)]\}$ and t ($Cu(II)$-Magnetite system, $C^{\circ}=2.00$ ppm, pH = 4.00, 25°C).

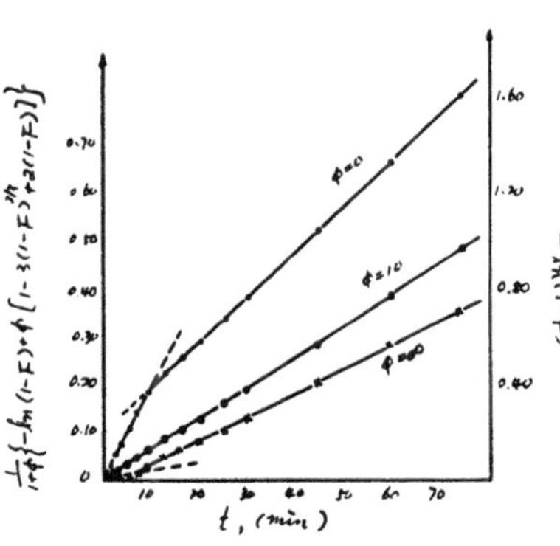

In Eq.(14) we are mostly interested most in paramaters. Calculation shows, that when $\phi << 1, \tau$ changes with ϕ greatly, when $\phi > 10$ τ changes slowly and there is almost no difference between $\phi = 100$ and $\phi = \infty$. For our experimental results, ϕ takes 5 to 10. This means in most cases that the reaction was controlled by the diffusion of ions Cu(II) in the liquid layer. For real ocean system (with much lower concentration, 10^{-8} M) ϕ might be small value, that is to say the whole process might be controlled by liquid film diffusion mainly. Besides, there are some evidens that ϕ may change with pH.

According to Eq.(24), the slope of the straight line is the equilibrium time. For system Cu(II)/α-FeOOH when C°(Cu) = 2.00, 3.00 and 4.00 ppm, t(eq) are 120, 190 and 230 min. respectively. For Cu/Am-FeOOH, t(eq) = 90, 160, 190 and 220 min for C°(Cu) = 0.5, 1.00, 1.0 and 2.00 respectively. Those are almost the same was what is measured in experiment.

From what have been mentioned above, it is clear that our concept on boundary progressive diffusion is different from that of Native and Dana, and it is more clear and detailed than that of Zhang's liquid film progressive model. Here we must emphasize what is different from general ideas. Generally it is thought that ions must exchange with almost every exchangable group and form site-binding results, which forms a moving boundary. This is true for the case of organic ion exchange resin with uniform exchangable groups. But in sea water system, the component of the solid particles is very complex, their exchangable groups are different not only in energy but also in species. So that the progressive phenomenon happens for high reactive groups. Furthermore, investigation shows, in certain conditions when the reaction proceeds to a certain extent, the determined step will change from diffusion into the exchange reaction. In this way we get a kind of stepwise kinetic curves.

The above discussion shows, that a two step diffusion process in the ion exchange reactions between Cu(II) and hydrous metal oxides exists. Initially the reaction is controlled by the diffusion in the outer liquid film, then it is controlled by the diffusion in the interior liquid layer which is moving towards the unreacted core. The whole process could be described by the composite model.

References

[1] Zhang Zhengbin, Liu Liansheng, 1985, Theory of interfacial stepwise ion/coordination particle exchange and its applications, China Ocean Press, Beijing , p. 356 (in English).

[2] Giles, C.H., Smith, D. and Nuitdon, A., 1974, A general treatment and classification of the solute adsorption isotherm. I.Theoretical, J.Colloi Interface Sci., 47: 755-765.

[3] Giles, C.H., Dsilva, A.P. and Easton, I.A., 1974, A general trdatment and classification of the solute adsorption isotherm II. Experimental interpretation. J.Colloid Interface Sci., 47. 766-778.

[4] Zhang Zhengbin, Liu Liansheng, 1989, Marine Physical Chemistry, Science Press, Beijing (in Chinese).

[5] Zhang Zhengbin, et al., 1984, Marine Chemistry Vol. 1 and Vol. 2, Shanghai Sci. Tech.Press, pp 770 (in Chinese).

[6] Zhang Zhengbin and Liu Liansheng, 1985, Interfacial stepwise ion/coordination particle exchange isotherm and isotherm equation on suspended marine particle, Chines J. Oceanol. Limnol. 4(1), 81-99 (in

English).
[7] Boyd, G.B., Adamson, A.W. and Myers, L.S., 1947. Kinetics of ionic exchange adsorption processes, J.Amer.Chem.Soc., 69: 2836-2848.
[8] Span, J., 1970, Self-diffusion of iodide ions into ion exchange resin Dowes 1*8, J.Chem. Pgys., 52: 3097-3100.
[9] Huang, T.C., Li, K,.Y., 1973, Ion exchange kinetics for calcium radio tracer in batch system, Ins.Eng. Chem.Fundam., 12:50-55.
[10] Welfferich, F., 1965, Ion exchange kinetics V., J.Phys.Chem., 69; 1178-1187.
[11] Liu Liansheng, Zhang Zhengbin, Cai Weijun and Zhang Hao, 1986, A kinetical study of the inorganic ion exchange of copper-goethite system in seawater, Oceanol.Limnol.Sinica, 17(5): 436-448 (in Chinese).

Zhang Zhengbin and Liu Liansheng
Ocean University of Qingdao
Qingdao
P.R. China

16 POSSIBLE ECOLOGICAL CONSEQUENCES OF FERROMANGANESE NODULE MINING FROM THE PACIFIC OCEAN FLOOR

A.V.Tkalin

16.1 Introduction

Ferromanganese nodule (FMN) mining from the ocean floor is associateded with many difficulties: technological, law, ecological and others. The present study attempts to evaluate the possible ecological consequences of FMN mining from the Pacific Ocean floor on the basis of literature data. To evaluate the effect of FMN mining on physical, chemical and biological processes occurring in the marine environment it is necessary to know their discharge volume, composition, and physico-chemical properties. This data was obtained by the USA scientists during the experimental works for FMN mining in 1978 /1/. The data on chemical composition of pore and bottom water for the study area is also available /2, 3, 4/.
Extrapolation for the case of commercial mining was made by authors of the works cited considering the proposed mining volumes and type of the systems used. The discharge of waste solid products containing bottom sediments and nodule debris is evaluated to be 1.4 10^4 gps (grams per second) at the surface and 5.5 10^5 gps near the bottom. The liquid phase of the surface discharge will be presented by the near-bottom water of 34.7‰ salinity and 7°C temperature. The total density of the discharge material is 1.06 g cm 3. The estimates were made for the daily volume of Fe mining of 5000 t (dry weight) and the waste solid products content about 15% in the material lifted (by volume). Table 16.1 presents these estimations together with the average data obtained for the experimental FMN mining /1/.
The estimates made by a different method confirmed the obtained results. The following parameters are recommended to be used for practical calculations: the discharge rate of waste solid products at the surface is 2.8 x 104 gps and near the bottom is 9.6x10^5 gps.
To reveal the possible consequences of FMN mining it is necessary to have the information about physico-chemical properties of the discharge material. The bottom sediment density in the area of proposed FMN mining is 1.2-1.3 g cm with particles of 2-4 μm in size predominating here. However, the nodule debris of different sizes is contained in the discharge material together with the bottom sediments. The deposition rate of solid particles in the discharge material is 0.06 cm s^{-1} which is consistent with the equivalent particle diameter of 25 μm. This may indicate the flocculation and agglomeration of particles in the discharge material /1/.

The Fe and Mn contents (in suspended form) could be used as chemical tracers of the discharge material behaviour as they are sufficiently high in the waste solid products. The Fe and My contents are 5-6 and 0.9-18% in the discharge material respectively. For further estimates let us consider the dimensions of zones which are subject to FMN mining in the surface and near-bottom ocean layers. Typical values of parameters needed for computations are: an average ocean layer depth 53 m, surface current speed 25 cm s^{-1}, bottom current speed 4 cm s^{-1}, vertical eddy diffusion coefficient in the surface layer 100 cm s^{-1} horizontal eddy diffusion coefficient in the surface layer 105 cm^2 s^{-1} and that in the near-bottom layer 104 cm^2 s^{-1} /1,5,6/. As mentioned above, the discharge volume of waste solid products is estimated to be 2.8 10^4 gs^{-1} (at the surface) and the particle deposition rate 0.06 cm s^{-1}.

V. I. Ilyichev and V. V. Anikiev (eds.), Oceanic and Anthropogenic Controls of Life in the Pacific Ocean, 225–234.

Table 16.1: Characteristics of surface discharges for experimental (1) and commercial (2) mining of nodules.

Characteristics	1	2
Discharge rate, l s^{-1}	120	270
Waste solid product content, g l^{-1}	8.5	54
Waste solid product discharge, g s^{-1}	1.0 10^3	1.4 10^4
Discharge temperature, °C	7	7
Total discharge density, g cm^{-3}	1.032	1.057
Nodules sampled, t d^{-1}	345	5000
Waste solid product content in lifted material, % of volume	2.8	15

For the steady-state regime of the work of nonmoving mining ship, the longitudinal (along the flow direction) dimensions of zones of increased suspension content are 60 km at the surface, 100 km at 50 m depth and 125 km at depth of 100 m /5/. The zone dimensions are shown by 5 μg l^{-1} contour of excess concentration (the background content of suspension in the given area is about 30-40 ng l^{-1}). The maximal suspension concentration observed at the surface near the discharge site is approximately 10 mg l^{-1}. At a distance from the mining ship, the suspension content decreases and is 100 mkg l^{-1} and 10 mg l^{-1} at a distance of 20 km and 45 km respectively /5/. At a depth of 100 m, the suspension content does not exceed 300 mgk l^{-1} in the plume. The zone width varies slightly with a depth and is approximately 5 km on the average. Therefore, the area of 500 km^2 is the subject to the Fe mining process. For alternate ship movement (passes about 30 km), the plume acquires the zigzag form. In this case the dimensions of zone having increased suspension content are retained. Time of suspension content decrease to the background level in moving water mass is 80-100 hours.
Interesting to note that for the above environmental parameters and the discharge material the layer of density jump does not considerably affect the suspension deposition /5,7,8/.
The turbulent mixing intensity appears to be sufficient for equating the concentration of the fine-dispersed suspension phase (at the deposition rate less than 10^2 cm s^{-1}) throughout the upper quasi-homogeneous layer. The particles of greater deposition rate (10^2 - 10^{-1} cm s^{-1}) do not overcome considerable obstacles as they pass through the layer of the density jump.
In the near-bottom layer, the longitudinal dimensions of zone of increased suspension content (Contour of 5 ng l^{-1}) are about 160 km and lateral ones are mainly determined by pass length of mining ship (30 km) /6/. The area of the zone

is approximately 5000 km^2 and the suspension deposition time is 1-2 months. Near the mining mechanism, the suspension content exceeds 1 mg l^{-1} while it is less than 100 $\mu g\ l^{-1}$ at a distance of 30-40 km /6/.
Based on the general theory and literature data analysis /8,9/ the basic negative consequences accompanying FMN mining could be determined for marine environments. As to the surface ocean layer, the main affecting factors will be supply of suspended matter (containing organic carbon, trace elements, biogenic elements, etc.) as well as dissolved forms of trace elements and biogenic materials. Relative division of the effects is: physical, chemical and biologic (in marine environments these effects will be exhibited together).

Physical effects:
- bigger suspended material (SM) content will cause the sea water turbidity increase (a transparency decrease);
- transparency decrease will affect the absorption of solar shortwave radiation and heat distribution in the surface ocean layer;
- a discharge of colder near-bottom water can have some effect on the sea surface thermal regimes
- SM content increase will impact the process of sea water evaporation.

Chemical effects:
- increased content of suspended and dissolved forms of trace elements, biogenic matter, organic carbon;
- variations of basic chemical composition (dissolved gases, elements of carbonate system, salinity) of surface waters can occur under the impact of SM discharge and colder near-bottom water,for air-lifted system saturated by gases;
- SM and related chemical compounds accumulation can occur above the layer of density jump;
- some chemical effects are related to the biological consequences of FMN mining.

Biological effects:
- a direct impact of the discharge material on larvae and spawn of marine organisms, zooplankton and fish behaviour which inhabit the given area;
- decreasing of the primary production of phytoplankton due to solar radiation attenuation caused by increased turbidity;
- increased content of suspended and dissolved trace elements (and toxic heavy metals (HM) can have a toxic effect on phyto- and zooplankton, that would cause variations of plankton biomass, quantitative and species composition;
- for HM accumulation in the food chain is possible.
Also, the a possibility for the opposite biologic effects exists:
- input of biogenic matter together with the discharge material will cause a growth of primary phytoplankton production;
- input of suspended matter enriched by organic and biogenic elements will cause the bacteria plankton growth. This could in turn cause an increased consumption of oxygen dissolved in the sea water;
- bacteria growth would serve as an additional food source for plankton organisms.

Negative consequences of FMN mining for the bottom layer could be considered according to a similar scheme. In this case, mechanical destruction of the upper layer

of bottom sediments caused by collector activity and perturbation of sedimentation process in its clove proximity should be considered as basic physical effects. Basic chemical effects are: increased contents of suspended and dissolved HM forms, biogenic elements, organic carbon and variations of gaseous composition related to the biochemical oxidation processes. Biological effects of different directions should be dwelled upon: benthos destruction on the path of the collector and burying of organisms in its close proximity, change of organism habitat conditions and their death due to violation of the trophic processes in the intensive impact zone which would especially affect the sestonphages predominating in the area under study, increasing of the bacteria biomass, additional food supply for hydrobionts (when bottom sediments are soiled and activity of microorganisms increases) at the periphery of the impact zone.
In the following sections some chemico-biological effects of FMN mining on marine environments will be considered in more details.

16.2 Content of trace elements

The behaviour of trace elements including toxic heavy metals (HM) in the system "sea water-bottom sediments - suspended matter" is fairly difficult to be described theoretically. In view of this, processes of the trace element transfer between the above phases are usually modelled in laboratory conditions for controlled environmental parameters. These studies are carried out not only for the FMN problem but for investigations of the behaviour of waste products buried in the sea and of polluted grounds extracted during dredging operations, when trace elements are investigated in the system "river-sea". Let us consider the results for modelling of HM transfer process from the solid phase to the sea water.
Canadian scientists /10/ studied a transfer of dissolved HM from the shaft buried at the sea bottom to sea water. Simulation was carried out in 25-litre reservoir with constant aeration and periodic agitation. The shaft additions (in pulp form) were 1,10,100 and 1000 mg l^{-1}, samples were taken in 1,3,8,27,80,243 and 728 hours. The dissolved HM forms were determined. HM concentration in waste solid products was (mg kg 1) Fe (2-8) x 10^4, Cu 700, Mn 650, Zn 80, Ni 20, Pb 20, Cd 3. The experiments showed that dissolved HM, contained in the liquid phase of the discharge which was 60% by weight, did not contribute to the total metal mass in the sea water. The HM transfer from solid phase occurs through different mechanisms (ion exchange, oxidation, electrochemical reactions). On addition of discharge materials of 10 mg l^{-1} concentration (which corresponds to maximal SM concentration for the waste products of discharge for FMN mining) no substantial variations of Mn, Ni, Cd and Zn concentrations were observed. Fe, Cu and Pb concentrations increased almost twice (Fe up to 2.8, Cd 0.2, Pb 0.02 $\mu g\ l^{-1}$). Maximal concentrations were observed for Fe and Cu in 3 hours and for Pb in 24 hours. Later on the concentration decrease up to initial or lower values occurred. According to the authors, one of the mechanisms of HM recovery from the solution is co-deposition with the Fe hydroxide. Fe, Cu and Pb transfer to sea water was 1,3 and 5% of their content in waste solid products respectively. For the rest of the metals, even in the case of discharge material addition (1000 mg l^{-1} concentration), a transfer to sea water did not exceed 2%. It should be noted that the above Cu and Pb concentrations are less than threshold ones under which the toxic impact on the hydrobionts is exhibited /11/.

British researchers /12/ studied the dissolved HM transfer from the polluted bottom sediments to sea water in some experiments. The simulations were carried out in the 14-litre tank with constant aeration and mixing with temperature, salinity, pH and O_2 content monitoring. The addition of bottom sediments was 500 mg l^{-1} (which is an order of magnitude higher than SM concentrations observed for FMN mining), samples were taken hourly over first six hours and twice per day during the following three days. HM content in sediments was for Fe (3.0-3.2) 10^4 mg kg^{-1}, for Mn 595-945 mg kg^{-1}, for Zn 790-835 mg kg^{-1}, for Cu 69-84 mg kg^{-1}. Cu and Zn concentrations decreased quickly after the bottom sediment addition, Cu from 6 to 1 mg l^{-1} over 3 hours, Zn from 40 to 10 μg l^{-1} per day. On the contrary, Fe and Mn concentrations sharply increased: Fe from 10 to 1000 μg l^{-1} over 3 hours, Mn from 5 to 75 μg l^{-1} over 10 hours. After that, Fe and Mn contents gradually decreased down to initial level. Fe and Mn transfer to sea water was 1 and 21% respectively. As a result of performed experiments, it was established that boiling of the polluted bottom sediments with oxygen saturated sea water did not cause a substantial growth of HM concentrations in dissolved phase. On the contrary, bottom sediments in this case could be considered as HM accumulator.
Data from investigations of mercury transfer from the polluted bottom sediments to the sea water obtained by american scientists could also be considered. Mercury content in the ground varies from 0.1 to 1.0 mg kg^{-1}. After ground addition (25% of weight), concentration of dissolved mercury did not exceed 1 μg l^{-1}, which corresponds to its concentration level in the three studied regions of the USA coasts zone. Laboratory experimental data was confirmed by field observations in the regions of bottom dredging and ground discharge /13/
The Pb and Cd behaviour in the "sea-water-bottom sediments" system /14/ was investigated in one more series of experiments. Modelling was repeated three times over 5 weeks in the 20-litre reservoirs with constant mixing. The additions of polluted bottom sediments were 1, 10 and 100 mg l^{-1} and dissolved HM were determined. Cd content in the ground was 3 mg kg^{-1} and Pb concentration was not presented. The organic carbon content was 5.6%. The concentration of dissolved Cd for SM concentration of 1 mg l^{-1} did not change with the time. For SM concentration of 10 mg l^{-1} Cd content decreased from 80 to 40 ng l^{-1} over the first hour and after 48 hours it increased and approached 120 ng l^{-1} at the end of the experiment. For SM content of 100 mg l^{-1} Cd concentration decreased almost down to zero and increased up to 500 ng l^{-1} at the end of the experiment. The author of the cited paper attributes Cd concentration decrease to related organic impact and subsequent resorption to the oxidation and decay of humid complexes. The dissolved Cd concentrations found in the experiment are sufficiently less than toxic ones /11/.
For SM concentration of 100 mg l^{-1} the dissolved Pb concentration increases monotonously throughout the experiment (from 10 to 200-300 ng l^{-1}). For HM concentrations of 1 and 10 mg l^{-1} the Pb content growth changes into some decrease after 50-100 hours of experiment. Maximal Pb concentrations were 100 and 200 ng l^{-1} respectively which is sufficiently less than the toxic concentrations. From analysis of the laboratory experimental data it follows that the sufficient growth of dissolved HM content would not occur during the first day after the discharge for SM concentrations observed in close proximity from the discharge site. On the contrary, their concentrations may decrease due to sorption /15, 16/. It should be also assumed that substantial dilution of the discharge material would occur over a day (SM content would decrease by some orders of magnitude /17/). The influence of

additional HM input from porous and near-bottom waters could also be neglected because of a substantial mixing with the surface water (by 2-3 orders of magnitude). The data obtained during the experimental FMN mining confirms the conclusions made. The content of dissolved HM decreases down to background level at the distance of 1 km from the mining ship /5, 8, 18/.
The behaviour of dissolved HM depends mainly on the behaviour of suspension. If SM concentration, HM content in the bottom sediments and background concentration of dissolved HM are known, one could approximately determine the areal of the discharge material expansion in the surface layer and in the underlaying horizons. The computations show that for suspension concentration less than 100 $\mu g\ l^{-1}$ (i.e. at a distance of 20 km) the contents of suspended forms of Cu, Ni and Pb do not exceed their concentrations in dissolved forms at the background level. For Co, Zn and Mn this condition is met for suspension concentration less than 10 $\mu g\ l^{-1}$ (at a distance of 50 km).
In the near-bottom layer, increasing the dissolved and suspended trace elements concentrations would be more substantial. It is difficult to predict their toxic impact on the bottom population since the bottom organisms are able to accumulate toxic materials, which, in turn, can cause sterilization, death of adult organisms or death of their posterity. And quantitative scarcity of deep sea bottom fauna and low-mobile way of life do not favour the fast occupation of the bottom areas which suffered from FMN mining.

16.3 Biogenic elements and dissolved oxygen

As shown above, concentrations of nitrates and silicon in porous waters slightly exceed their contents in the near-bottom ocean layers. Phosphate content in the silt waters is less than in the near-bottom waters and concentrations of nitrites and ammonium nitrogen in porous waters are negligible. Thus, the deep water discharge will have a basic effect on the chemical composition of the surface ocean watery. The silicon and nitrate contents in the near-bottom waters exceed their surface concentrations by 15 times and phosphate concentration by 10 times /1-4, 8/.
Let us evaluate this effect quantitatively. The collector moves in 1-2 km^2 area during a day /1/. Assume that the mining ship moves also in area of 1 km^2, For the rate of the near-bottom water discharge of 270 $l\ s^{-1}$, its volume would be around 25000 m^3 over a days for the jump layer depth of 50 m, the water volume in upper mixing layer, where the discharge material arrives, is $5\ 10^7\ m^3$. Thus, the near-bottom waters would be diluted by 2000 times. For this relation between discharged and surface waters, the concentration of biogenic elements does not substantially change. Similarly, the influence of the near-bottom water discharge could be evaluated by the content of dissolved oxygen and other substances in the surface ocean layer. The conclusions made are confirmed by the results of the field observations /8,18/. According to this data, the dilution factor is 10^3-10^4 /8,15,18,19/.
It is interesting to estimate the possible changes of O_2 content in the oxygen minimum layer due to acceleration of the processes of biochemical oxidation of organic matter (in the surface layer, which contacts atmosphere, similar effects would be less exhibited). For the organic carbon content of 0.1% in bottom sediments and concentration of the discharge material of 100 $ng\ l^{-1}$ at the depth more than 100 m, the organic carbon addition would be 0.1 $\mu g\ l^{-1}$. The background content of suspended organic carbon in the Pacific subsurface waters is 10-100 $\mu g\ l^{-1}$ and that

of dissolved one is by an order of magnitude higher /20/. Consequently, it is the growth of bacteria biomass (and biochemical consumption of oxygen) due to SM input rather than additional input of organic carbon from bottom sediments that is of a critical value. It is difficult to obtain the quantitative estimate of biochemical consumption of oxygen change. In the full-scale experiments, some increase of bacteria biomass was observed due to FMN mining. It should be noted, however, that non-oxic conditions in the oxygen minimum layer are unlikely /8/.

In the near-bottom ocean layer increase of concentration of ammonium nitrogen (for the transfer from porous waters) and microorganisms activity due to great amount of HM should be expected. However, the oxygen saturation in the deep-sea waters is fairly high, therefore, appreciable variations of O_2 content are unlikely. As noted earlier, small additions of organic carbon from bottom sediments do not cause substantial changes of biochemical oxygen consumption.

16.4 Phytoplankton

As was mentioned above, the concentrations of dissolved HM decrease to background level at a distance less than 1 km from the mining ship. For SM concentration less than 1 mg l^{-1}, the content of suspended HM does not exceed their threshold toxic concentrations /11/ (estimates were made for Pb, Cu and Zn). The content variations of biogenic elements in the upper ocean layer under effect of FMN mining are also negligible. Consequently, increase of water turbidity and, hence, attenuation of solar radiation have a main effect on phytoplankton.

The photosynthesis intensity in the euphoric layer decreases due to turbidity that increased proportionally to SM content. Decreasing of the primary production is 10% at a distance of 50-60 km from the mining ship, 50% at a distance of 15-20 km and 70% or more in close proximity to the ship /18/. The zone area subjected to 50% decrease of the photosynthesis intensity is around 30 km^2 and duration of action is 1 day. Therefore, the large-scale and long-term variations of phytoplankton productivity due to FMN mining are unlikely in the Pacific region under consideration.

It should be also noted that natural variations of subaquatic illumination in the ocean are 60-70% of the average values that are comparable with effect of SM content growth /8/. A conclusion is confirmed on the negligible effect of toxic metals and biogenic matter supplied with the discharge material. The experiments indicated that phytoplankton species composition and primary production did not change over three days even on addition of the discharge material of 10 mg l^{-1} concentration /8,18/.

16.5 Zooplankton and fish

The toxic action of the discharge material of 0.1-1.0 mg l^{-1} concentration on zooplankton was investigated in laboratory experiments. In experiment lasting for 1-2 days, no statistically significant variations of zooplankton death-rate were observed /8/. Nevertheless, additions of the discharge material caused some increase of zooplankton death-rate in comparison with control. The full-scale observations did not permit to reveal direct impact of FMN mining on zooplankton death-rate either /8/.

On the other hand, zooplankton faecal pellet rate increased 3-4 times with the

discharge material concentrations of O.1-1.0 mg l^{-1}. And the withdrawal rate of the suspension (with trace elements absorbed) filtered by zooplankton, will also increase. The rate of pellet deposition saturated by HM was 20-220 m day^{-1} /8/.
The surface catching made during experimental FMN mining showed that for HM concentrations less than 1 mg l^{-1} the biomass decrease and chemical composition variations were not observed; the data processing for species composition was not completed /8/. The data allows us to conclude with sufficient probability that the negative influence of FMN mining on zooplankton is transient and small-scale.
The Pacific region under study is the habitat area for tinny and other fish species. The waste products discharge for FMN mining can have a direct impact on adult individuals,spawn and larvae. The adult individuals according to literature data, have easily withstand SM concentrations more than 10 mg l^{-1}. Laboratory experiments with the discharge material (1-10 mg l^{-1}) did not reveal negative effects on tinny /8/. Zones of increased SM concentration can both attract and scare away the adult fish. Since the impact of FMN mining on phyto- and zooplankton has a local and transient nature there are no reasons for large-scale variations of fish biomass. The action of FMN mining process on the spawn and larvae is still uncertain. In close proximity to the mining ship, they will submerge and perish under the impact of the discharge material.

16.6 Benthos

Benthos is a subject to direct and indirect impact during FMN mining. It is completely destroyed on the path of collector (operating width is around 20 m). At 25 km distance from the collector the depth of newly formed layer of sediments decreases from to 20 to 1 cm. Decreasing the layer thickness down to 1 mm is observed at a distance of 100-200 m and down to 0.1 mm at a distance of 600 m /6,8/. With a considerable thickness of sediment layer, benthos death will be caused both by burial of organisms themselves and their food. For collector speed of 1 m and burial zone width of 400 m, the area about 35 km^2 will be covered by sediment layer of 1 mm over a day. Over a year (a time of ship operation is 300 days), this area will be 10000 km^2. For simultaneous operation of 20 mining ships, the total area of direct impact on benthos over a year is 1.5% of the area of possible FMN mining zone (5°-20°N, 110°-180°W), 13000000 km^2. Similar estimates of the area (0.6-3.0%) were obtained in other papers /9,15/.
The increased SM content in the near-bottom water layer will have an additional impact on benthos. Positive or negative effects, depending on SM concentration, should be expected. In the zone of intensive impact (this zone dimensions can be determined only experimentally), the animal diet will considerably change which will have an unfavourable effect on the sestonphages. A positive effect is possible at the periphery of the impact zone where the feed conditions should be improved at the expense of increased amount of suspended particles.

16.7 Conclusion

From the analysis of the available data presented in the previous sections, it follows that basic negative factor for Fe mining is a flow of great volume of suspended matter (SM) both to the surface ocean and near-bottom layers. In the surface layer this causes a decrease of the sea water transparency which, in turn, has a negative

effect on phytoplankton productivity. However, this effect should be considered as a transient and small-scale (in space). Considerable change of trace elements, biogenic matter and dissolved oxygen concentrations in the surface layer is unlikely. The discharge material will have a direct impact on the fish spawn and larvae and zooplankton organisms which are at the collector site but this effect is also local.
FMN mining has a considerable negative effect on the bottom population. In addition to direct impact of collector, the effect of organisms and their food burial by rolled bottom sediments and destruction of lifetime processes of filtrating and detritus eating organisms are observed. This effect is inevitable due to the present-day technology of FMN mining. The positive factor is the efficiency of Boston area coverage other than 100% for FMN mining. This permits the recolonization of perturbed bottom areas by benthos organisms.
nor the surface layer, the basic measures aimed at decreasing the negative consequences for marine environment is to reduce SM discharge or the areal of their extension and time of their contact with water mass. This can be achieved by some methods. First, one can increase the efficiency of mined nodule separation from bottom sediments. However, for tested collector samples, the efficiency was already 96-98%. Evidently, further efficiency increase will be related to specific technological difficulties. Second, the separation of bottom sediments from discharged water aboard the mining ship is possible. For experimental mining, the centrifugal separators, settling tanks, etc. have been already used. Consequently, further improvement of suspension separation is also connected with considerable complication of the treatment technology of mined material.
Decreasing of the areal of extension and contact time of SM with environmental water mass could be achieved by increasing the deposition rate of discharge material. It is possible if the mean size of particles is increased through chemical reagents (coagulators or flocculants). In the case of considerable consumption of discharged water (around 300 l s^{-1}), an experimental determination of optimal volumes of reagent additions (on the one hand, as minimal as possible and, on the other hand, sufficient for effective impact) will be of importance. This approach permits to avoid the necessity for the discharge material pumping in depth and minimizes all possible consequences of FMN mining for the surface ocean layer. As mentioned above, it is unlikely to avoid perturbations in the near-bottom layer with available technology of mining.

References

1. Ozturgut E. et al. Estimated discharge characteristics of a commercial nodule mining operation. Marine Mining, 1981, vol.3, No. 1/2, p.1-17.
2. Klinkhammer G. et al. Metal diagenesis in toxic marine sediments. Earth and Planetary Science Letters, 1982, vol.61, No.2, p.211-219 .
3. Jahnke R. et al. Pore waters on the central Pacific Ocean: nutrient results. Earth and Planetary Science Letters, 1982, vol.61, No.2, p.233-256.
4. Emerson S., Bender M. Carbon fluxes at the sediment-water interface of the deep sea: calcium carbonate preservation. Journal of Marine Research, 1981, vol.39, No.1, p.139-162.
5 . Lavelle J. MY., Ozturgut E. Dispersion of deep-sea mining particulate and their effect on light in ocean surface layers. Marine Mining, 1981, vol.3, No.1/2, p.185-212.
6. Lavelle J.,V. et al. Dispersal and resedimentation of the benthic plume from deep-

sea mining operations: a model with calibration. Marine Mining, 1981, vol.3, No.1/2, p.59-93.
7. Ozturgut E., Lavelle J.W. The influence of the pycnocline on the oceanic settling of manganese nodule mining waste. - Marine Environmental Research, 1984, vol. 12, No.2, p. 127-142.
8. Ozturgut E. et al. Impact of manganese nodule mining on the environment: results from pilot-scale minning tests in the North Equatorial Pacific. In: Marine Environmental Pollution. Vol.2. Dumping and Mining ,Ed. by R.A. Geyser. Amsterdam, Elsevier, 1981, p.437-474.
9. Brown E.D. Pollution from seabed mining: legal safeguards. Environmental Policy and Law, 1933, vol.10, No.4, p.122-134.
10. Hoff J.T. et al. Heavy metal release from mine tailings into sea water - a laboratory study. Marine Pollution Bulletin, 1932, vol.13, No.8, p.283-286.
11. Patin S.A. Influence of pollution on the biological resources and World Ocean productivity. Pishthevaya promyshlennost, 1979, pp.304.(in Russian)
12. Hurst J.M., Aston S.R. Behaviour of copper, zinc, iron and manganese during experimental resuspension and reoxidation of polluted anoxic sediments. Estuarine, Coastal and Shelf Science, 1983, vol.16, No.5, p.549-558.
13. Grimwood C. Effects of dredging on adjacent waters. J. Environ. Eng. Div./Proc. Amer. Soc. Civ. Eng., 1983, vol.1O9, No.1, p.47-67.
14. Stukas V.J. Release of cadmium and lead from False Creek sediments: a laboratory study. Canadian Contractor Report of Hydrography and Ocean Sciences, 1984, No.17, p.4-10.
15. Morgan C.L. Environmental impacts of deep-sea mining - the importance of manganese. In: Marine Environmental Pollution. Vol.2, pumping and Mining/Ed. by R.A. Geyser. Amsterdam, Elsevier, 1981, p.415-436.
16. Gideiri Y.B.A. Impacts of mining on central Red Sea environments. Deep-Sea Research, 1984, vol.31, No.6-8, p.823-828.
17. Scientific Aspects of Pollution Arising from the Exploration and Exploitation of the Seabed. GESAMP Reports and Studies, 1977, No.7, New York, United Nations, 37 pp.
18. Chad A.T., Anderson G.C. Environmental investigation of the effects of deep-sea mining on marine phytoplankton and primary productivity in the tropical eastern North Pacific Ocean. Marine Mining, 1981, vol.3, No.1/2, p.121-149.
19. Ichiye T., Carnes M. Sediment dispersion and other environmental impacts of deep-ocean mining in the eastern tropical Pacific ocean. In: Marine environmental pollution. Vol.2. Dumping and Mining, Ed. by R.A. Geyser. Amsterdam, Elsevier, 1981, p.475-517.
20. Pacific Ocean , Ed by: Leontjev O.K., Moscow, Mysl, 1982, p.183-189.

A.V. Tkalin
Far Eastern Regional Hydrometeorological Institute
Vladivostok
USSR

17 THE EXPLOITATION AND UTILIZATION OF MARINE LIVING RESOURCES

Wu Baoling

Abstract

In this paper, the exploitation, utilization and enhancement of marine living resources in the world, especially in China, have been bescribed and discussed. The main contents of the present paper is summarized as follows:

1. The term "marine living resources" is used here in a broad sense and includes all the living bodies (microbes, algae, seagrass, mangrove and animals) in the marine environment (including estuaries). We should discuss them from the point of view of ecosystems.

2. China is a country rich in marine living resources. Nearly ten thousand species of marine organisms are found from the coastal zone and the shallow water of the continental shelf of China.

3. The amount of photosynthetic production in waters of the Pohai Sea, the Yellow Sea, the East China Sea and the South China Sea is about 3.5×10^8 tons of carbon/year. According to the theoretical estimates from data on primary production the total annual production of fish biomass is about 16 million tons.

4. Marine living resources are exploited and utilized as food, raw meterials for industries, feed, fertilizers, new energy resources and ornaments. Marine pharmacognosy which plays an important role in Chinese traditional medicine has attracted attention of the public in China and other countries.

5. Though overfishing in some areas in the ocean took place, the future of world production of marine fisheries will be bright.

6. Mariculture is greatly developed in China. At present, over 40 species of marine algae and animals are cultivated, the annual yield have increased to more than 50×10^4 tons.

7. Protective management is very important in maintaining the high, abundant biomass of marine living resources.

The exploitation and utilization of marine living resources is an important component in the development of the ocean usage and have received close attention of the various coastal states of the world.

17.1 The Marine Living Resources

The living resources of the ocean refer to all the living organisms living in the ocean (together with river mouths), including microorganisms, lower and higher plants, invertebrates and vertebrates. Although at present mankind utilizes only a small part of the marine biologic resources (as fishery resources), yet the various marine organisms are interrelated, constituting an inseperable component of marine

V. I. Ilyichev and V. V. Anikiev (eds.), Oceanic and Anthropogenic Controls of Life in the Pacific Ocean, 235–240.

ecological system. Therefore studying the marine biological resources should be done from the viewpoint of ecosystems.

China borders the Bohai Sea, the Huanghai Sea (the Yellow Sea), the East China Sea and the South China Sea, with a coastline 18,000 km long and rich offshore biologic resources. According to incomplete statistics, at present the main species living in the coast of China include: about 100 species of dinoflagellates, about 300 species of diatoms, 300 species of benthic algae, 60 species of sea grass and mangrove plants, 1000 odd species of foraminifers and radiarians (including ancient species), 40 odd species of ciliates, 100 odd species of sponges, 800 species of coelenterates, about 500 species of moss animals, nearly 800 species of polychaetes, 2000 odd species of molluscs, nearly 2000 species of crustaceans, about 30 species of cheatognaths, 100 odd species of echinoderms, 6 species of phoronids, 40 odd species of protchordates, about 1500 species of marine fishes (among which are 200 odd species in the Huanghai and Bohai Sea), 400 odd species in the East China Sea and nearly 900 species in the South China Sea) and, in addition, several tens of sea beasts and sea snakes. As Chinese Sea areas extend across the temperates, subtropical and tropical zones, species compositions in different sea areas differ significantly.

According to the estimations of J. Bunt (1981), the primary productivity in the W Pacific waters is roughly as follows: phytoplankton $6.6\text{-}9.3 \times 10^{15}$ gC/a, coral reefs 0.25×10^{15} gC/a, mangrove 0.13×10^{15} gC/a, macroalgae $< 0.06 \times 10^{15}$ gC/a, salt marshes $< 0.01 \times 10^{15}$ gC/a. In addition, sea grass and benthic microalgae are not estimated due to lack of appropriate data. In all, the primary productivity of the W Pacific areas is approximately 6.6×10^{15} gC/a. But some researches consider this value as low, possibly it will increase to 43.5×10^{15} gC/a.

In recent years, Chinese scientists have carried out investigations on the primary productivities of the Bohai Sea, the Huanghai Sea and the East China Sea areas, using chlorophyll and ^{14}C methods. From some of the acquired results, it can be seen that high value appears in the upwelling areas of the coast of Zhejiang (27°30'-30°30' N, W of 124°00' E), with an average primary productivity 1.25 gC/m^2.d (Ning Xiu-ren et al., 1985). The second high value appears in the S of the Huanghai Sea and in the N of the East China Sea (28°45'-35°N, 122°30'-128°30' E) with an annual average of 180.2 gC/m^2 (Chai Xinyu et al. 1986) and in some coastal waters, for instance, the annual mean primary production for the front sea areas of Qingdao is 157 gC/m^2 (Guo Yujie et al. 1984) and in the Bohai Sea and some open areas of the Huanghai Sea the primary productivity is repsectively 90 g C/m.a and 65 gC/m^2.a (Lu Peiding et al., 1984, 1981). The primary productivity of the waters of the South China Sea is the lowest. According to the determination of the local waters of the S of the South China Sea by A. Lusinsap et al (1974), the annual mean was only 40 gC/m^2. On the basis of the above mentioned data and using the nutrient dynamical method for rough estimation, the annual phytoplankton production (fresh weight) in the four coastal waters of China is about 7×10^9 t (about 3.5×10^{14} gC) in total. According to the theoretical calculations, the annual production of top level organisms (including benthic, mesopelagic and pelagic fishes, shrimps, crabs and squids) may reach 1.6×10^7 t and an estimated maximum sustainable yield (MSY) is 5×10^6 t.

17.2 Exploitation and Utilization

At present, the utilization of marine living resources by mankind can be summerized as follows:

1. As food

This is the main use. At present several thousand species of seaweeds and marine animals have been known to be edible. Abalones scallops, sea cucumbers and shark's fins are also considered as nourishing health-building foods.
Of the total global catch fishes amount to 85-90%, invertebrates 8-12%, and seaweeds 2-4%. Among the hundreds species of the main marine fishes, herring occupy the first place, reaching $1.37 x 10^7$ t in 1980; next comes the cod, with a yield over $1.0 x 10^6$ t. The species each with annual yield exceeding $1.0 x 10^7$ t include mackerel, sardine, anchory, scad, tuna and flounder. Among the invertebrates, those species with highest yield include oysters, mussels, scallops, clams euphausiids and penaeids.

China's marine fishing industry (not including Taiwan) is also developing rapidly with an annual yield starting from $53.6 x 10^4$ t in 1950, topping $3.0 x^6$ t in 1974 and reaching $3.62 x 10^6$ t in 1983 which has increased 6.5 times than the yield in 1950, amounting to 66% of the national yield of the production of fishery products, of which 90% came from offshore. In 1983 the species of marine fishing were: Trichiurus $4.93 x 10^5$ t, Cantherines modestus (Gunther) $2.66 x 10^5$ t, Decapterus maruadse (T. & S.) $1.77 x 10^5$, Pneumatophoreus japonicus (Houttuyn) $1.07 x 10^5$ t, Scomberomorus $6.1 x 10^4$ t, Pseudosciaena crocea (Richardson) $5.9 x 10^4$, Stromateoides argenteus (Euphrasen) $5.7 x 10^4$, Loigo and Sepiella $5.0 x 10^4$ t, Pseudosciaena polyactis Bleeker $3.1 x 10^4$ t, Muraenesox cinereus (Forskal) $2.3 x 10^4$ t, Clupea pallasi Cuvier et Valenciennes $2.3 x 10^4$, Penaeus orientalis (Bennett) $1.7 x 10^4$ t, Ilisha elongata (Bennett) $1.3 x 10^4$ t and Trachurus japonicus (T. et S.) $4.0 x 10^3$ t.

2. As drugs

In the coast of China, 51 species of seaweeds, 3-5 species of mangrove plants, 34 species of coelenterates, 138 species of marine shellfishes, nearly a hundred species of crustaceans, 10 species of echinoderms and a hundred odd species of fishes, snakes and sea beasts have been found for use as drugs.

In recent years, the extraction of anticancer drugs from marine organisms has received great attention and some positive achievements have been acquired. For instance, Paolin III extracted from molluscs has been found to have antitumor activity and mercenene has been verified to be a highly effective anticancer substance. Moreover, according to the experiments by Japanese scientists, addition of a small amount of Laminaria japonica Aresch into the feed of experimental mice can markedly reduce the incidence of cancers.

3. As raw materials for industries

Agar which is extracted from Gelidium and Gracilaria is used as a clusture medium for microorganisms, in food and medicine. Carrageenan is extracted from Eucheuma and widely used in food and light industries (e.g. tooth paste, soft sweets etc.). At present the global annual yield of Gelidium is 5-6x10^4 t, but only 1000 t are produced in China, with its supply falling short of demand. In fishes and sea beasts almost every part of their bodies is a treasure and they are also important raw meterials for industry. Fish scales can be made into fish scale glue, which is an important raw material for making motion-picture film, fish skin can be cooked to make glue for use as adhesives in timber processing, shark's skin and hide of sea beasts can be used to make leather, fish oil and whale blubber can be used to make soap and lubricating oil and in leather tanning and fish livers are used to extract liver oils.

Reed is produced in some river mouth areas along the coast of N China. Using 1 ton of reed to make paper can save 2 m^3 of timber. It is calculated that the tidal flats of the coast of S China have about 1.3x10^5 mu of mangrove, which has much economic benefit, and possess the effects of protecting the coasts and against tides, enhancing fishery production in river mouths and bays, stabilizing sediments, and extending newly formed land in beaches. Coral reefs can be used as raw materials for making lime and as building materials.

4. As pesiticides, energy resources, food and fertilizers

The most successful example of utilizing marine organisms as pesticides is the extraction of nereistoxin from the nereid (Lumbriconereis heteropoda). This toxin has good pesticidal property but is harmless to human beings and domestic animals. It may be called a pesticide without environmental pollution. At present great attention has been paid to using macroalgae, such as the giant kelp (Macroceptis pyrifera) to produce marsh gas. Seaweeds and animals have been widely used as feeds and fertilizers in the coastal areas of China. Besides, some marine organisms are utilized as precious ornaments and good handicraft articles.

17.3 Present Status of Fishery Resources

In the thirty-odd years Chinese marine fishery scientific workers have been doing research on the resources of Trichiurus, Pseudoscianena crocea, Pseudosciaena polyaetis, Stromateoides argenteus and Penaeus orcintalis and have obtained quite good results.

The present status of the world fisheries is: the global marine fishing catch has been pacing up and down for ten years, and at least 25 species of fish resources which are of the highest economic value have also been seriously depleted. China's marine fishing yield also increased very slowly after topping the yield of 3.0x10^6 t since 1974 and the resources of some economic fishes have also been seriously depleted.

Fishery resources are renewable resources. It depends mainly on 3 factors: whether the development of fishery is successful or not, the biology of the fishery species, environmental factors (biotic and abiotic) and fishing. The activity of these factors is often joint. With regard to the situation in China, the depletion of some economically important fishes is mainly the result of ineffective management of fishery production and overfishing.

In the light of the depletion of some of the economically important fish resources in the coast of China, no doubt that it is necessary to conserve and renew the resources by adopting measures for limitation, reduction and fishing ban within a definite period of time, but it is difficult to meet the increasing demand with each passing day for the requirement for fish products. We consider that marine propagation should be actively developed besides adopting the above-mentioned measures, that is, we should combine cultivation with fishing, fully utilize space and material base of the ocean, make great efforts to increase the recruitment of the kind of fisheries,to raise the growth rates and reduce natural mortality so that fishery resouces may be propagated an utilized rationally.

In the recent ten-odd years, cultivation has developed quite rapidly. According to statistics, mariculture amounted to 75% of the total production. As far as regions are concerned, developments show considerable disequilibrium, and cultivation is most developed in Asia. With regard to the yield of cultivated species, fishes amount to 37.1%, molluscs 36.7%, seaweeds 25.4% and crustaceans 0.8%. As fishes are taken as the dominant factor in freshwater cultivation, at present the highest yield in mariculture is still attributed to molluscs.

China was the first country in the world which developed aquaculture. Since liberation mariculture has developed greatly. Starting from 1950s when kelps were transplanted southwards and laver norhtwards, and up to 1960s when Penaeus orientalis was succussfully cultivated, wide fields have been opened up for mariculture. The species cultivated have reached the number of several tens and the area reached $2.8x10^6$ mu. Artificial fish shelter construction and artificial release experiments have been also developed. It is estimated that $20x10^8$ mu of beaches are still suitable for cultivation in the coast of China and the area of sea waters for cultivation is equal to or exceeds that of farmland.

From now on great attention should be paid to the utilization and transformation of the whole bays or local waters.

We should implement the 'Marine Environmental Protection Law of the People's Republic of China" and the "Regulations regarding the Propagation and Conservation of Aquatic Resources", prevent pollution of coastal waters and strengthen the conservation and management of living organisms and aquatic resources. Many examples have proved that strengthening management is an effective measure for conservation of marine living resources. Some countries that have paid attention to the conservation of living resources in coral reefs have achieved good results.

References

[1] Cushing, D.H. 1981. Fisheres Biology, The University of Wisconsin Press, pp.1-245.

[2] Mann, K.H. 1982. Ecology of Coastal Waters, University of California Press. Berkeley, pp.232-259.

[3] Nienhuis, P.H. 1981. Marine Organic Chemistry, Ed. by E.K. Duursma and R. Dawson, Elsevier Scientific Publishing Company, Amsterdam, pp. 31-69

[4] Roje, R. et al. ed. 1978. Upwelling Ecosystems, Springer Verlang, p.:3-300.
[5] Rounsefell, G.A. 1975. Ecology, Utilization and Management of Marine Fisheries, The C.V. Mosby Company, pp.1-392.
[6] Ruddle, K. 1981. Southeast Asian Seas, Frontiers for Development, Eds.: Chia, L.S. et al. Singapore, Mcgraw-Hill, pp. 136-176.
[7] Ruddle, K. 1982. Man, Land and sea: Coastal Zone Resource Use and Management in Asia and the Pacific. Eds.: Coysa, C. et al. The Agricultural Development Council, pp. 15-27.
[8] Ryther, J.H. 1969. Science, 166, pp.72-76.
[9] IXth International Seaweed Symposium, 19-25 June, 1983. Qindao, China, Abstracts, pp.1-306.
[10] Proceeding of the Fifth International Coral Reef Congress 1985. Tahiti, 27 May-1 June, Vol. 2: Abstracts, pp.1-424.

Wu Baoling
The First Institute of Oceanography
P.R.China

18 PHYSICAL, CHEMICAL AND BACTERIAL DESTRUCTION OF OIL ON THE SHELF OF OKHOTSK AND JAPAN SEAS

V.V. Anikiev, M.N. Mansurov and G.N Moiseyevsky

18.1 Definition of the problem and choice of the study objects

During the recent decade Soviet geologists and geophysicists have completed prospecting for oil and gas deposits in the marginal seas of the Pacific Ocean. Large reserves of hydrocarbon matter have been found, in particular, on the Okhotsk Sea shelf. The industrial development of deposits is to be started in the immediate future. It is also expected that by the end of this century similar deposits will be opened in the Bering Sea shelf.
The industrial development of deposits may result in contamination of marine environment due to a constant influx of small quantities of oil as a result of output and transportation. Accidental spills are also possible.
Since the industrial output of oil and gas is planned in areas which until present are spawning grounds for salmonids, herring and pollock, the more strict demands must be put on organizational and technological measures aimed at minimizing the possible damage to ecological systems.
Taking all this into account, we performed a preventive research in ecological rating of possible oil pollution of some regions of the NE shelf of Sakhalin Island.
For another study object we choose the Amursky Bay (Sea of Japan) which is ice-covered during two to three months a year and experiences oil hydrocarbon (OH) pollution as a result of sewage water discharge from ships and industrial enterprises of the city of Vladivostok. Additional sources of oil pollution are river and storm runoffs. Essentially, the Amursky Bay has already attained the ecological situation which is still awaiting the shelf of the Okhotsk Sea when stable anthropogenic sources of oil pollution are put into operation.
Comparison of the results of research conducted on both objects makes enables to use the information obtained for selection of a criterium for estimation of negative consequences of man's impact on marine ecosystems with highly variable natural factors.
The recent concept of ecological rating of sea environment is based on the idea of assimilative capacity A_i implying the influx of such amount of pollutant i to a certain water volume as can be processed at the expense of dynamic, physico-chemical and biological processes and have no negative consequences for water ecosystem (Izrael, Tsyban, 1983; Izrael et al.,1988).

$$A = K_i \frac{V}{\tau_i} C_i \qquad (1)$$

where: V is the volume of the ecosystem;
C_i, the average pollutant concentration;
τ_i the residence time of pollutant in sea ecosystem.

Here K_i is storage ratio which mirrors the possible negative effect of pollutants when C attains the critical value.
All mentioned above reveals a strong demand for complex studies aimed at the

V. I. Ilyichev and V. V. Anikiev (eds.), Oceanic and Anthropogenic Controls of Life in the Pacific Ocean, 241–255.

assessment of the impact of natural processes, both individual and in combination, on the fate of pollutants.

We have found earlier that the fate of OH as a surface film depends on the following combination of natural processes which can be divided by the time basis (Anikiev, 1987):

1. evaporation and photochemical oxidation, several hours;
2. emulsification and bacterial destruction, several days;
3. destruction of emulsified fraction due to physico-chemical and bacterial oxidation, months and longer. Possibly, to hundreds of years, the composition of the bulk OH remains constant inside oil clumps.

The rate of these processes depends on a set of environmental conditions (water temperature, sea roughness, intensity and spectral characteristics of sunlight, availability of biogenic elements, amount and species composition of microorganisms etc.) and on physico-chemical characteristics of oil which depend on the ratio between light and heavy fractions.

The effect of photochemical processes and bacterial destruction on physico-chemical properties of oil film we studied in laboratory and field experiments (Anikiev et al., 1988). However, the using this data to forecast the behaviour of the oil film is difficult as it is impossible to consider the whole set of local conditions.

As the dependence of destruction rate of OH in natural conditions is complex, we performed semi-field experiments on evaluation of variations in chemical composition and physicochemical properties of Sakhalin oil in the sea medium with the effect of weather conditions and bacterial processes, typical of the NE coast of Sakhalin.

At present, the topic of common concern is the admissible extent of risk in the use of large-scale technologies, e.g. sea drilling platforms (Larichev,1987). Measures providing the absolute safety render these technologies unprofitable and, at the same time they cannot eliminate accidental spills completely.

We believe that a criterium of risk may be the extent of damage to water medium as a result of accidental spill of pollutants under hardly probable circumstances caused by human errors (Larichev, 1987). Analysis of damages to sea pipelines used for transportation of oil and gas in the Gulf of Mexico and the North Sea has shown that 47 to 49% of damages were caused by sea roughness, due to lithodynamic factors, whereas unidentified causes made 7 to 12% (Mansurov, 1986). It is supposed that the number of spills in the Okhotsk Sea will be bigger due to ice and seismicity. This presupposes the need to carry out a research in the behaviour of oil films of different thickness forming on the water surface as a result of accidental spills. It was taken into account in our experiments.

When oil is constantly supplied to the sea in small quantities, an association of microorganisms consuming OH is formed. In this case the efficiency of OH-oxidizing bacteria is determined not only by water temperature but also by their relations with other components of the ecosystem (Gulina et al.,1988). For this reason the main objective of studies in the Amursky Bay was to evaluate the role of bacterial processes in the self-purification from oil with the help of the following:

- elucidating the nature of relation between the number of OH-oxidizing bacteria to particular elements of the Amursky Bay ecosystem.
- finding out the possibility to use the number of bacteria as an index of pollution of a sea region with the constant presence of oil in water.

18.2 Simulating of physico-chemical and bacterial oxidation of oil films at the laboratory and in the field

Laboratory experiments were conducted using an installation consisting of a source of light, a filter, a thermostat and reactor systems to measure the film tension and devices to measure the rate of water evaporation (Figure 18.1). During the experiment the light intensity of two lamps, each 1.5 kW, with spectral characteristics close to of the sunlight due to filters, was controlled by a photodiode and within the wavelength range 320 to 450 nm was equal to $(0.97 \pm 0.10) * 10^{16}$ quant/cm^2*s. Thermostat-provided reactors contained 50 ml of distilled, or sea water and on its surface (27 cm^2 area) 3 ml oil was irradiated during an hour. To evaluate the effect of dark oxidation, one of the reactors was shaded during the experiment.

Figure 18.1: Experimental installation for the study of photochemical oxidation of organic films.
1, light source; 2, filter; 2, film tension-meter; 4, photodiode; 5, reactor; 6, screen; 7, thermostat; 8, plate; 9, vibrodamper; 10, water vapor absorber.

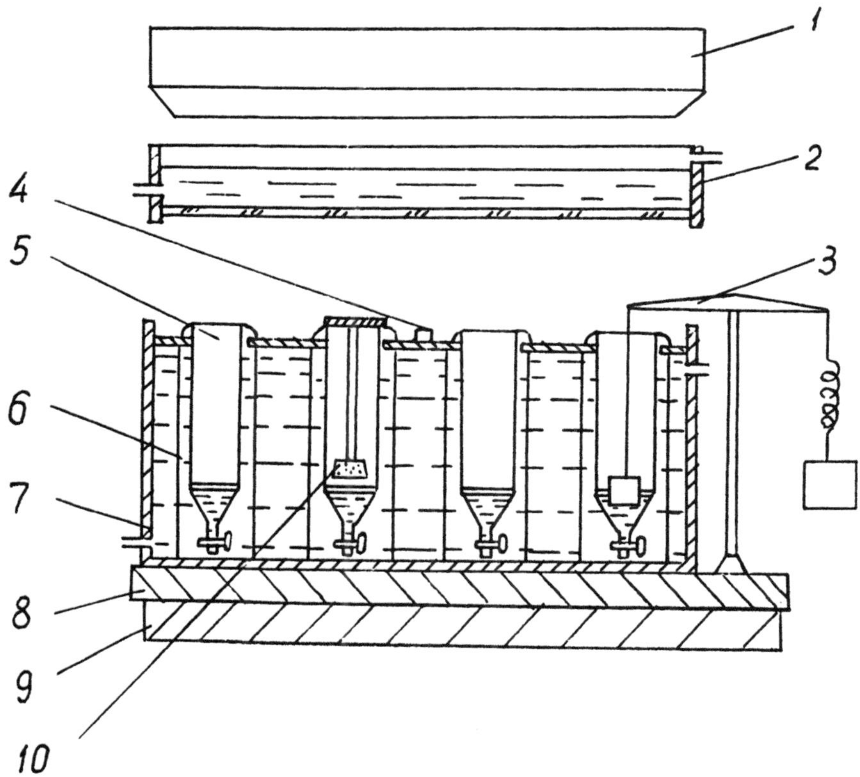

In certain time intervals the organic layer was removed from the next-in-turn reactor in which the concentration of peroxides was determined by iodometric method, as they serve as all purpose indicator of organic compound oxidation.

In our experiments we used oil from the Odoptu-Morye deposit, which had the following physico-chemical characteristics: density 0.84 g/cm^3, viscosity 2.72 cs (20°), resins, 15%, silicagel resins 5.16%, paraffins 1.61%, asphaltenes 0.63%, sulphur 0.15%. The sea water for experiments was sampled at Sakhalin Coast in the area of deposits and had salinity 32.33 ‰, pH 7.88, and the number of oil-oxidizing bacteria $4.5*10^4$ cell/l.

Ions and some heavy metal compounds are known to be able to act as either catalysts or inhibitors. So in different experiments peroxide concentration could change. Figure 18.2 shows the rate of formation of peroxide concentrations (C) to be 1.8 and 2.3 times less in the sea water sampled near Sakhalin and in the Amursky Bay (Sea of Japan) respectively, than in distilled water. This is due to higher content of heavy metals. Thus increased concentration of heavy metals inhibits photochemical oil oxidation.

Figure 18.2: Variations in peroxide concentrations in oil films at different temperatures.
a, sea water; b, distilled water.

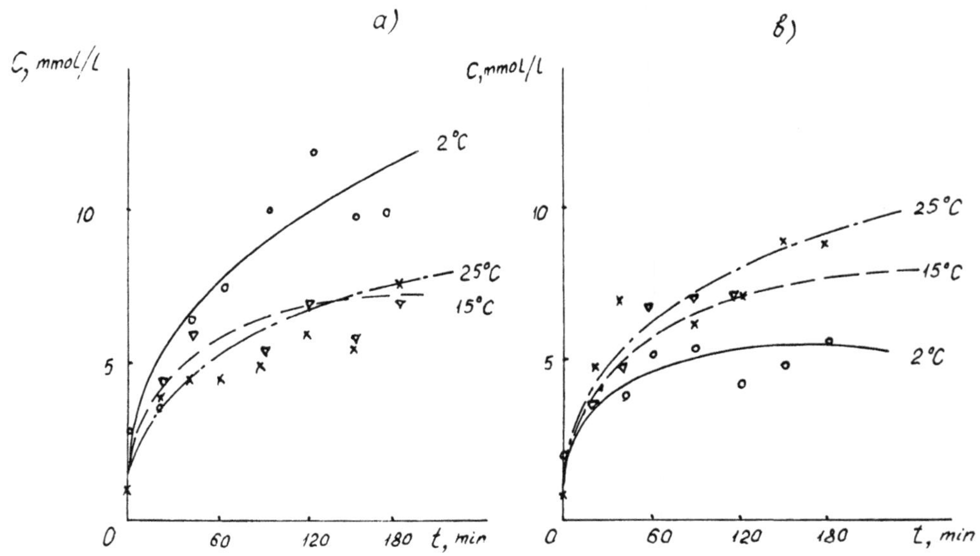

Physico-chemical properties of sea water seem to induce the reverse relation of oil oxidation to temperatures unlike in distilled water Figure 18.3). Thus laboratory simulation showed the rate of photochemical oil destruction on the sea water surface to depend on its composition including trace elements.

Figure 18.3: Variations in peroxide concentration (C) in oil film at 15°C during photo-oxidation over the distilled water /o/, seawater of Amrusky Bay /Δ/ and on Okhotsk shelf /∇/.

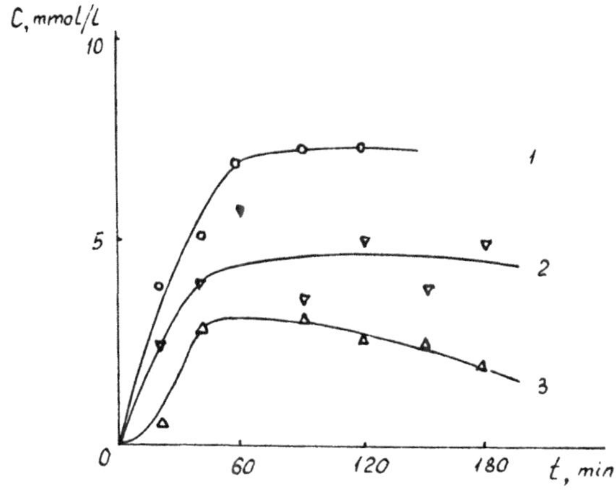

Figure 18.4: An experimental installation for the study of oxidation rate of oil films.
1, cell; 2, reactor; 3, thermostat; 4, anemometer; 5, actinometer.

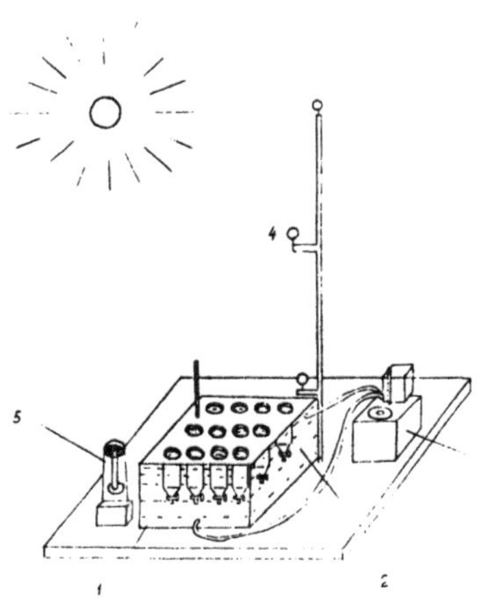

To estimate the effect of the environmental conditions at the NE coast of Sakhalin on the rate of oil oxidation, we carried out semi-field experiments using a special installation (Figure 18.4). It enabled us to measure wind velocity at three levels, the intensity of sunlight and also to maintain 5-litre reactors filled with sea water at 5°C. In certain time intervals we removed oil film from one of the reactors for subsequent determination of the normal and aromatic OH, resins and asphaltenes.

As seen from Figure 18.5, the relative contribution of light OH (C_9-C_9) decreased abruptly to zero during the initial 4 hours perhaps due to evaporation, promoting at first an increase in the relative contribution of the group C_{10}-C_{12} and with its subsequent decrease later. Relative content of heavy OH (C_{13}-C_{25}) increased due to their lower volatility and higher resistance to destructive processes (Figure 18.6).

Figure 18.5: Variations in the ratio of normal hydrocarbons /C_i/ to total normal hydrocarbons C_7-C_{25} /C_Σ/ depending on time of exposure in experiments in the area of Odoptu deposit.
1, time, hours.

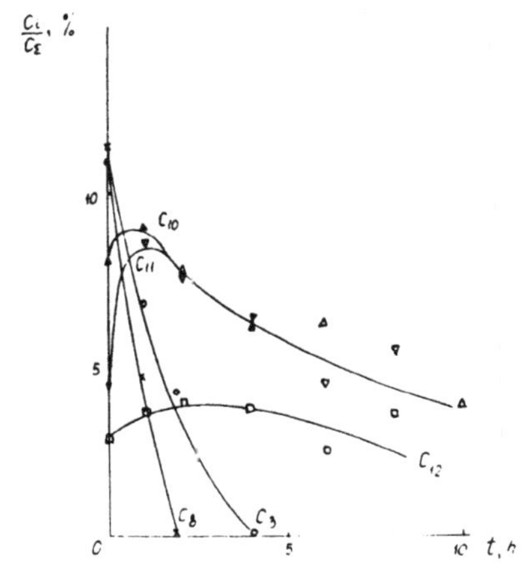

Generally in the daylight with decreasing content of normal OH, the peroxide concentration increased with a simultaneous rise of resin and asphaltene content in the film (Figure 18.7). A direct relation between the intensity of daylight and photo chemical oil destruction in the film is evident from the comparison of Figure 18.7 and 8, when on a sunny day the peroxide concentration increased by 1.5 times as compared with the data of experiment with the variable cloudiness. During a month-long exposure of oil in autumn, when air temperature dropped from 11.4 to 1.3° and wind velocity increased from 3.3 to 10 m/s, the following laws of changes in group composition were observed (Figure 18.9):
- the combined effect of evaporation, dissolution and photochemical oxidation during the initial 24 hour period results in decline in the content of mobile aromatic OH with a simultaneous rise in concentrations of resins and asphaltenes;
- the maximum asphaltene content is reached on the 5th day of exposure, apparently at the expense of destruction of light aromatic OH and resins, attaining constant levels during this period.

To elucidate the role of physico-chemical and microbiological processes in oil destruction, experiments were conducted in systems with inhibited and non-inhibited biological activity. We found that at the average water temperature 2°C, bacteria affected, but little, the destruction of aromatic OH, resins and asphalthenes in coastal waters of Sakhalin when the exposure lasted for 20 days.

Figure 18.6: Variations in the ratio of normal hydrocarbons /C_i/ C_{13}-C_{25} to total normal hydrocarbons /C_{Σ}/ C_7-C_{25} depending on time in experiments in the area of Odoptu deposit.
1, time, hours.

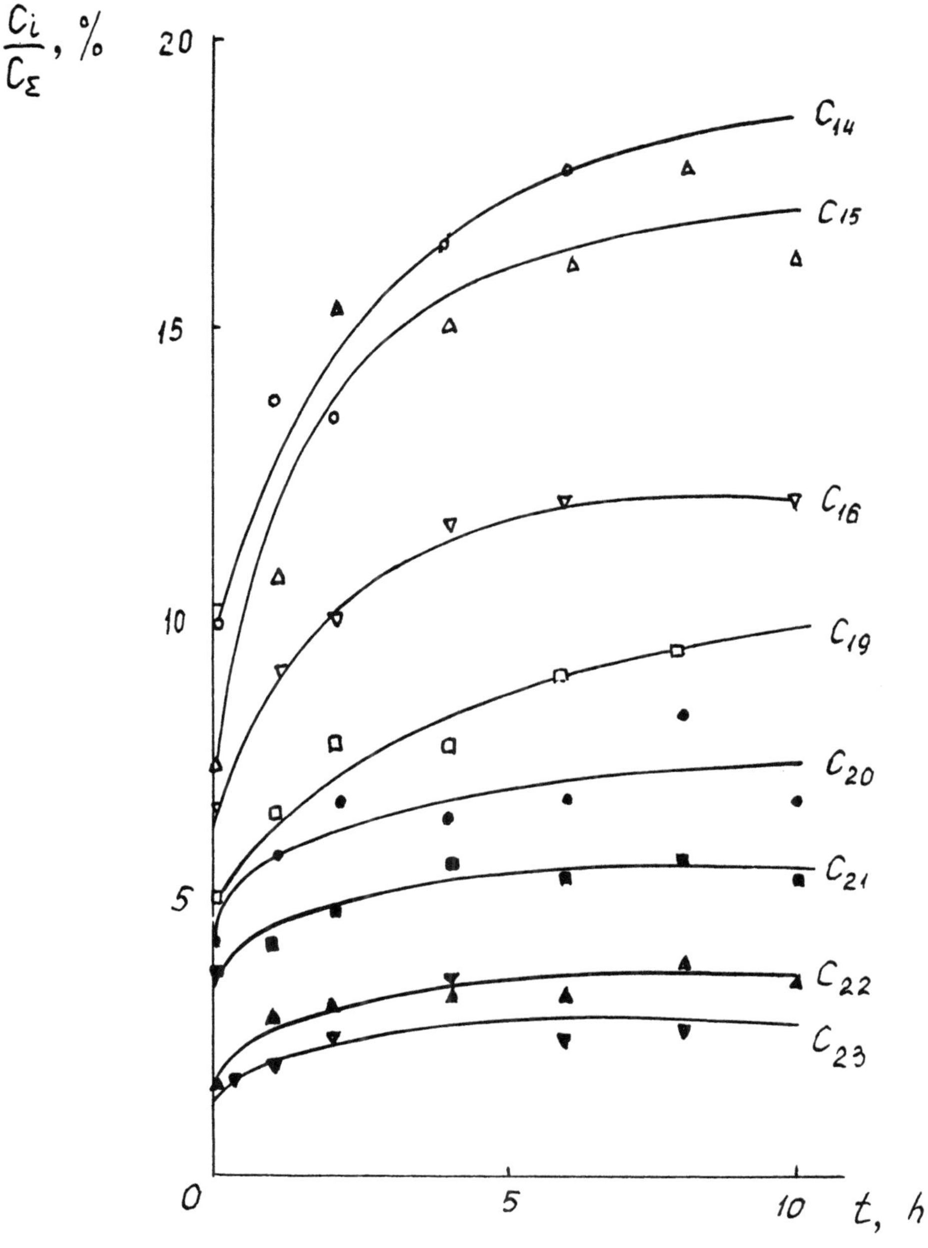

Figure 18.7: Variations in peroxide concentration /C_{PO}, •/, total normal OH/ SUM i=7 to 25 H-C_i, ∇/, resin content/Δ/, asphaltene content /■/ at sunlight intensity /I_0,x/ in the Odoptu oil films depending on duration of experiment
1, time of the day, hours.

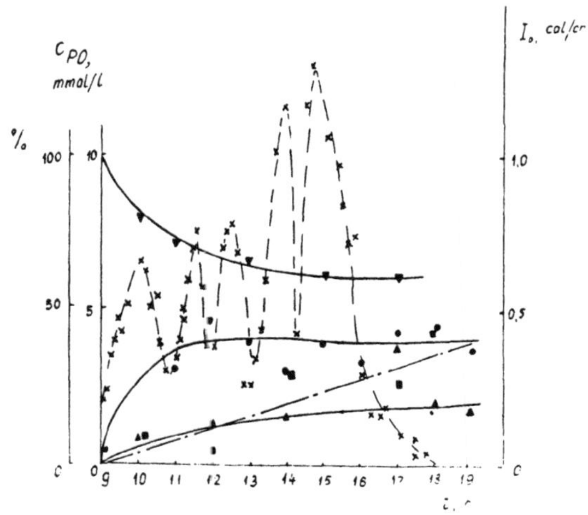

Figure 18.8: Variations in sunlight intensity /I_0,x/ and peroxide concentration /C_{PO} •/ depending on time of the day.

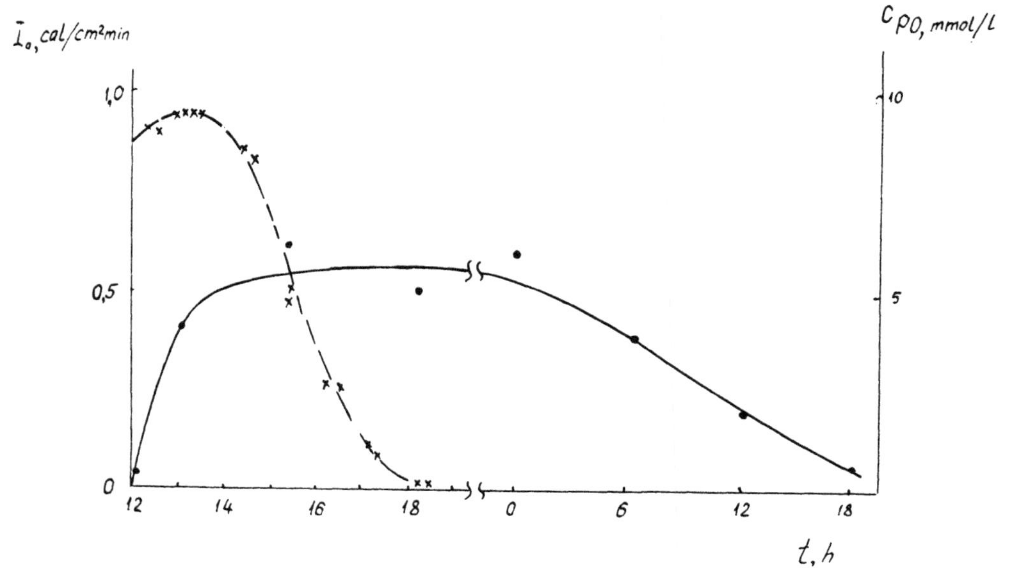

Figure 18.9: Relative variations in concentration of oil hydrocarbons from Odoptu-Morye deposit while exposed on the Sakhalin coast.
C_τ/C_0 the ratio of i-hydrocarbon concentration at time τ to the initial one. of 1, • aromatic OH; 2, ⊙ resins; 3, × asphaltenes.

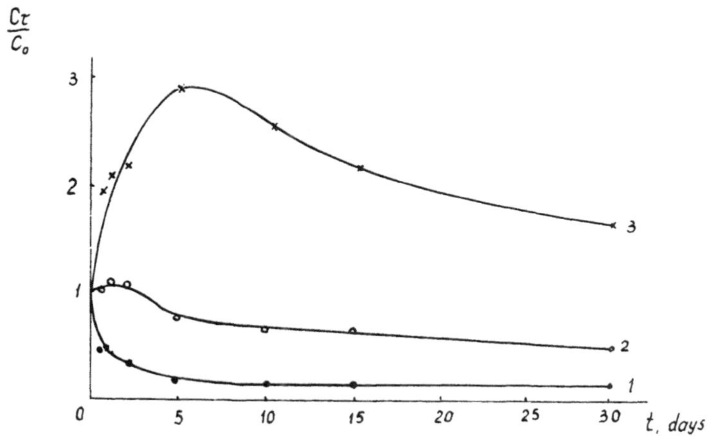

It is known at the same time that at low temperatures high rates of bacterial destruction of OH can be observed, since for OH-oxidizing bacteria of Arctic waters the range of the optimum temperatures may be shifted towards lower values (Sakhina, Shcherbakov, 1987). Indeed, in our laboratory experiments with a fresh oil we found that bacterial destruction was about 10 times more intensive due to physico-chemical oxidation in the dark. After temperature dropped from 15° to 5°C, the half-life of oil affected by bacteria increased 1.4 times and the rate of physico-chemical oxidations decreased 9 times.

Thus, we may come to a conclusion that in different periods the fate of oil is determined by the following dominant processes in definite time intervals after the spillage:

1. evaporation, photochemical oxidation - first hours;
2. dissolution, emulsification, photo chemical oxidation, physico-chemical oxidation, bacterial destruction - the initial 24 hours;
3. bacterial destruction - 1 to 5 days;
4. physico-chemical oxidation - 5 to 20 days;
5. a combination of environmental factors - months and years. It should be noticed that a drop in water temperature results in increase in the contribution of bacteria into oil destruction with respect to abiotic oxidation in the dark.

18.3 Estimation of the effect of bacterial association on oil pollution level

The assimilative capacity of the coastal Sakhalin water with respect to oil due to bacterial processes was estimated on the basis of the experimental value of the total amount of oil-oxidizing bacteria $4.5*10^4$ cell/l and the rate of destruction of OH by one bacterial cell (Zobell,1964). We found that the given number corresponded to the intensity of oil destruction, $5.4*10^4$ mg/m^3*day which indicates big potential of self-purification of Sakhalin water from oil.
By the order of magnitude, the total number of OH-oxidizing bacteria in the coastal water of Sakhalin, determined in the period of biological spring, is comparable with the average number of these bacteria in the Amursky Bay in the spring-autumn period, equal to $5.3*10^{-4}$ cell/l. In this case the maximum number of OH-oxidizing bacterial plankton in the Bay exceeded the average nearly 6 times, which, using the above-mentioned approach to estimation of potential bacterial oil oxidation, enabled us to evaluate the intensity of this process in the Bay at $3*10^{-2}$mg/m^3
A more detailed analysis will permit a differentiated approach to the estimation of self-purification of separate sea areas. For example, the quantitative distribution of OH-oxidizing bacterial plankton in the Amursky Bay at the summer/autumn boundary permits isolation, on one hand, of the area of increased amount of plankton ($2.5*10^5$ cell/l) in the central Bay, and, on the other hand, of areas with less plankton in the mixing zone of fresh water of the Razdolnaya river and sea ($1.0*10^4$cell/l) and in the open S Bay ($1*10^3$ cell/l). Generally for the entire Bay the bacterial destruction of oil, corresponding to the amounts of OH-oxidizing bacteria, varies within $1*10^{-4}$ mg/ml* day to $3*10^{-2}$ mg/m^3*day.
For the Razdolnaya estuary, the comparison of the velocity and the volume of the OH influx from the river to the Bay with the intensity of bacterial oil destruction in a zone with less amount of OH-oxidizing bacteria enables us to suggest the accumulation of OH in the given deposits. This assumption is based on the results of the analysis of quantitative distribution of OH-oxidizing bacterial benthos, which was not detected at the point of the minimum number of OH-oxidizing bacterial plankton (Table 18.1).

The data on quantitative distribution of OH-oxidizing bacteria will enable, at first approximation, to estimate the potential bacterial oil oxidation of the area. Further steps to obtain more precise values may be directed towards the account of interrelation between the number of the given bacteria with the elements of the ecosystem studied.
We have analyzed the data of a three-year-long series of observations over the ecosystem of the Amursky Bay. Data sets for factor analysis using the main component method were formed from the results of observations over the dynamics of 15 hydrochemical, 14 hydrobiological, and 6 meteorological variables (Table 18.2).

Table 18.1: The result of quantitative investigation of bacterial destruction within water mixing zone of Razdolnaya River and Amursky Bay.

Nos of station	S,%		TBN, 10^3 cells.ml^{-1}		10^3 cells.l^{-1}	HBN, cells.g^{-1}	cells.g^{-1} of DM***
	SWL*	NWL*	UWC	SM	UWC	SM	SLS
1	0.00	0.00	820	660	45	100	1300
2	0.00	11.46	600	660	45	100	-
3	27.20	32.00	340	260	30	50	-
4	29.96	32.33	260	250	10	20	0
5	31.70	32.54	400	110 0	20	50	-
6	31.97	32.65	420	160 0	50	160	6200

* surface water layer
** near-bottom water layer
*** deep mass Also, we used the data on 4 spatial and 4 time variables. All variables had been pre-standardized. For interpretation we isolated variables with a meaningful factor load at 99% confidence level.

A set of parameters, related to the number of OH-oxidizing bacteria (N) from the results of processing three one-year data sets (Table 18.2) may be divided into three groups:
Group 1 includes variables which connection with N showed up only once in some particular year. These are mainly Si, PO_4, Po, S, Ω. This interrelation with N may be considered as an episodic one.
Group 2 includes variables connected with N differently in different years. In this case one may speak of an ambiguous relation of the Bay ecosystem elements to N.
Group 3, the most interesting one, includes variables of stable long-term correlation with N. First of all this is fCos with a yearly period as well as quantitative indices of planktonic phytocenosis: N_{ph}, B_{ph} ,Chl, A and P. This group features a stable relation to N.
The division of N-related variables into groups and isolation of elements having

stable correlation with N, enabled us to define a set of constituents of bacterial oil oxidation model in the system examined. Thus, the quantitative distribution of OH-oxidizing bacteria in waters of the Amursky Bay has a stable correlation with parameters, characterizing the primary production which may indicate the paramount importance of biogenic autochthonous organic substrates in the diet of these bacteria. Also, the analysis of data for a three-year-long period of observations over the ecosystem of the Bay with the average OH concentration of 36 $\mu g/l^{-1}$ in water column, did not reveal any interconnection with oil concentration in the water. Hence, the parameter N cannot be an indicator of oil pollution in waters with insignificant though constant OH content.

Thus, we may conclude that:

1. The intensity of oil destruction on the shelf of freezing seas depends on the combined effect of physico-chemical and bacterial processes. The relative role of each process is time-variable due to transformations in the content and the form of occurrence of oil hydrocarbons.

2. Spatial variability of bacterial oil destruction intensity is defined by the interval ranging from $1.0*10^{4}$ to $3*10^{-2}$ mg/m^{3} day and reflects the heterogeneous distribution of OH-oxidizing bacterial plankton.

3. Spatial distribution of oil-oxidizing bacteria in coastal waters, subject to constant pollution, is closely related to the primary production.

Table 18.2: Variables employed in analysis.

Group	Conv. unit	Variables
Hydrochemical variables	T_W	Water temperature
	S	Salinity
	pH	Sea water pH
	O_2	Oxygen
	$Si_2O_3^{2-}$	Silicon as $Si_2O_3^{2-}$
	PO_4^{3-}	Phosphorus as PO_4^{3-}
	P	Total phosphorus
	N_3^-	Nitrogen as NO_3^-
	NO_2^-	Nitrogen as NO_2^-
	NH_4^+	Nitrogen as NH_4^+
	N_0	Total nitrogen
	Tr	Transparency
	OHC	Oil Hydrocarbons

	Ph	Phenols
	Dt	Detergents
Meteorological variables	U	Wind velocity
	F	Wind direction
	T_a	Air temperature
	H	Humidity
	Pr	Pressure
	Q	Total solar radiation
Hydrobiological variables	Nph	Total phytoplankton number
	Bph	Phytoplankton biomass
	Mph	Number of phytoplankton species
	Chl A	Chlorophyll A
	Chl B	Chlorophyll B
	Chl C	Chlorophyll C
	Nz	Total zooplankton number
	Bz	Zooplankton biomass
	Nm	Total number of microorganisms
	Ns	Saprotrophic bacterias' number
	N	HC-oxidizing bacterias' number
	P	Primary production
	D	Destruction
	P/D	Production/destruction ratio
Spatial variables	λ	Longitude
	ϕ	Latitude
	Sd	Sea depth
	St	Stratum
Time variables	Y	Year
	C	Number of day in a year
	T	Time
	fSin	Year-period function
	fCos	Year-period function

Table 18.3: The elements of the Amursky Bay ecosystem correlated with hydrocarbon-oxidizing bacteriae number.

1984	1985	1986
-Dt	-	Dt
-Ph	-	-
Nph	-	Nph
Bph	-	Bph
-	$Si_2O_3^{2-}$	-
-	PO_4^{3-}	-
-	P_0	-
-	P	-
-	-Bz	-
-	Ns	-
Q	-Q	-
- f Cos	- f Cos	- f Cos
-	S	-
-	-	Chl A
-	-	- Nz
-	-	λ

References

Izrael, Yu.A., Tsyban, A.V., On assimilative capacity of the World Ocean. Doklady AN SSSR, v.272 No.3, p.702-705, 1983, (in Russian)

Izrael, Y.A., Tsyban, A.V., Venzel, M.V., Shigayev, V.V., Generalized model of assimilative capacity of a marine ecosystem. Doklady AN SSSR, v.298, No.2, pp.459-462, 1988.
(in Russian)

Anikiev, V.V., Short-period geochemical processes and ocean pollution. Moscow, Nauka, 193 p. 1987,(in Russian)

Anikiev, V.V., Mishukov, V.P., Moiseyevsky G.N., Tkalin, A.V., The effect of oil film on water evaporation and oxygen content in sea water. Geojournal, vol.16, No.1, 1988.
Larichev, O.I., Decision-making problem with account of risk and safety factors. Vestnik AN USSR, No.11, p.38-45,1987,(in Russian)
Mansurov, M.N., Ways to minimize environmental sea pollution during exploitation of subwater pipelines on the shelf of freezing seas. In: Environmental protection while searching prospecting and developing marine oil and gas deposits. VNIIMORGEO, Rigas,1986, (in Russian)
Gulina, G.N., Mishukov, V.F, Moiseyevsky, G.N., Yarosh, V.V., Factor analysis of oceanological fields of the coastal zone.
Doklady AN USSR. vol.298, No.6, p.1238-1242, 1988, (in Russian)
Sokhina, L.I., Shcherbakov, O.N., Propagation of bacterial plankton and processes of self-purification of water in the Bering Sea. In; Khimiya i biologiya morei. Ed. Simonov A.I. Gidrometeoizdat, p.191-200, 1987, (in Russian)
Zobell, C.E., The occurrence, effects and fate of oil polluting the sea. Adv.Waters Pollut.Res. vol.3,p.85-109,1964.

Anikiev, V.V.
Pacific Oceanological Institute
USSR Academy of Sciences
Far East Branch
Vladivostok
USSR

Mansurov, M.N.
SakhalinNIPIneftegaz
Mingazprom USSR
Okha
USSR

19 OIL AND GAS DEVELOPMENT AND ASSOCIATED ENVIRONMENTAL IMPACT STUDIES ON THE PACIFIC COAST OF THE UNITED STATES

James C. Kelley

Abstract

Oil was discovered in the United States in Titusville, Pennsylvania in 1863. In the 1880's, oil and gas development began on the West Coast and the first offshore well was drilled in the Summer land field, near Santa Barbara, in 1896 into the Monterey Formation. This activity has greatly intensified, particularly offshore in the last 20 years. The States of California, Oregon and Washington have jurisdiction from the shoreline to a distance of 3 nautical miles offshore. The Federal Government has jurisdiction from 3 to 200 nautical miles offshore in the U.S. Exclusive Economic Zone (EEZ). The U.S. Department of Interior, through the Bureau of Land Management and the Minerals Management Service, has been in the middle of an increasingly strident public debate, while trying to carry out an environmental studies program and management offshore leases in a timely manner. Those involved in the government, as advisors to the government and as contractors on the studies, have tried to develop an environmental assessment program which complements other research efforts along the U.S. Pacific Coast. Some good science has resulted, as well as better understanding between the participants: government, industry and academic science.

19.1 Introduction

Nearshore oil and gas development in the Pacific States of the U.S. (exclusive of Alaska and Hawaii) began in S California in 1896 (Figure 19.1). In the last 20 years offshore, or outer continental shelf (OCS) activity has greatly increased while activity in State waters has declined (Figure 19.2). At the present time OCS activity is confined to S and Central California, but the potential Pacific OCS area extends to Oregon and Washington (Figure 19.3). Oil and gas currently is produced from the Monterey Formation (Miocene) in California. A number of sedimentary basins along the coast are potential sources of oil and gas (Figure 19.4). The Pacific coast of the U.S. is a region of great natural beauty and is home to a large recreational industry. The region is also one of the largest and most productive coastal upwelling areas in the world and a large fishing industry, both commercial and recreational, depends upon this highly productive marine ecosystem. In addition, several marine mammals inhabit the region. Some, like the Northern elephant seal (*Mirounga angustirostris*) and the California sea otter (*Enhydra lutris*), breed there and California Greg whales (*Eschrichtius Robustus*) pass through the entire area on their annual migration from Alaska to the several lagoons in Baja California and Mexico when they breed. Other pinnipeds, notably the California sea lion (*Zalophus Californianus*) and the Stellar sea lion (*Eumetopias jubatus*), inhabit the waters of the California current system.

V. I. Ilyichev and V. V. Anikiev (eds.), Oceanic and Anthropogenic Controls of Life in the Pacific Ocean, 257–266.

Figure 19.1

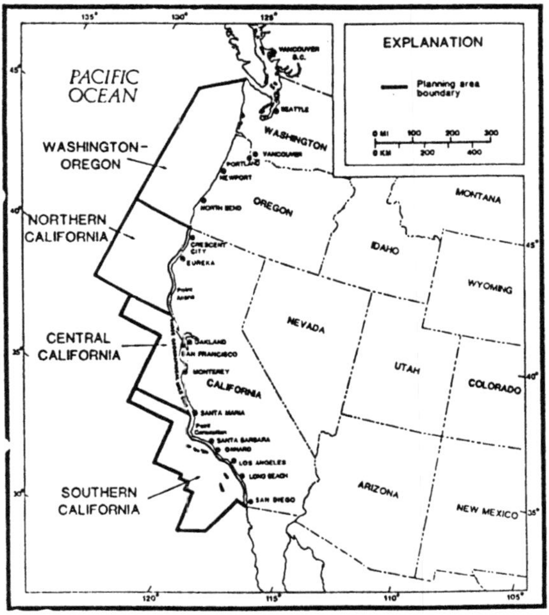

Figure 19.2

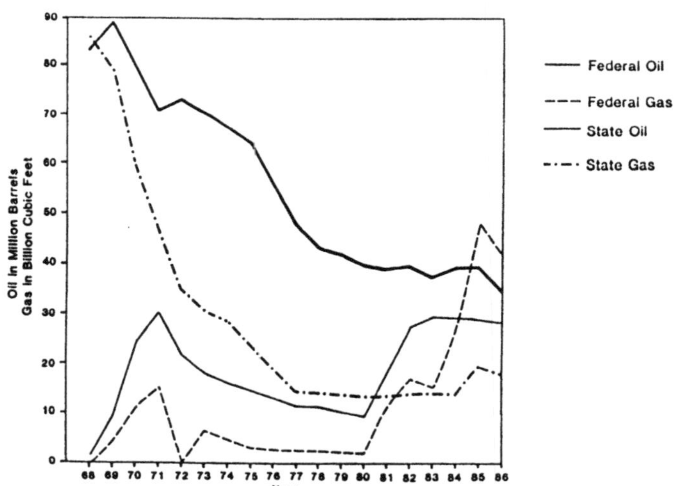

The U.S. government has afforded special protection for marine mammals under the Marine Mammals Protection Act of 1972. The requirements to protect marine mammals and other endangered species under the Endangered Species Act of 1973

Figure 19.3

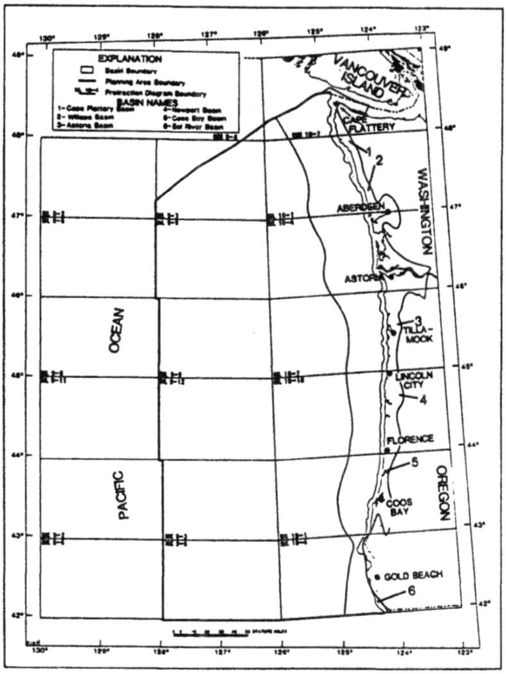

(The California sea otter is considered an endangered species: the other marine mammals on the Pacific coast are protected, but not considered endangered) have imposed constraints on the offshore oil and gas activity in the Pacific OCS. The oil and gas activities also conflict with the fishing industry operations. Finally, the Pacific States are home to a vocal and politically influential environmental protection movement which often objects to offshore drilling on both protectionist and aesthetic grounds.

The oil and gas industry in the Pacific OCS region produces $600 million of oil and $200 million of gas per year and has yielded $4 billion in bonuses and $1 billion in royalty revenues to the federal government, some of which is shared with the States. The fishing industry in the Pacific States produces $350 million wholesale fish value per year. The size recreational industry in the Pacific States is difficult to measure but may produce total revenues of over $3 billion per year.

Figure 19.4

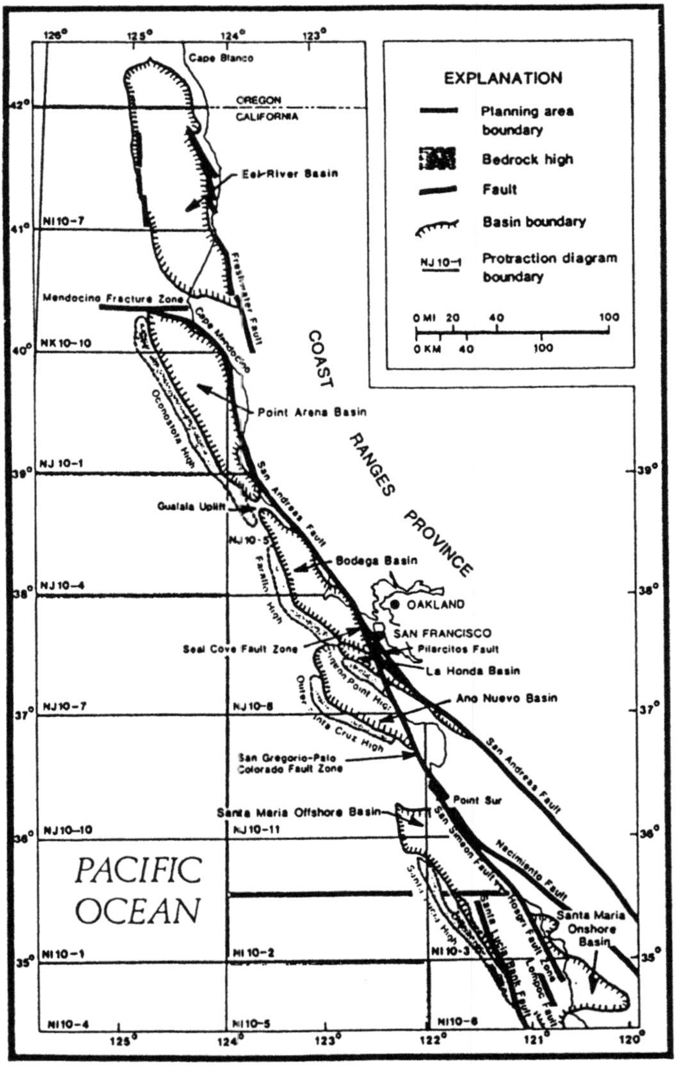

The federal government is required to resolve conflicts between these various uses of the OCS region and the task has been assigned to the Department of Interior. Originally the task was carried out by the Bureau of Land Management, but in 1982 the Minerals Management Service was formed to administer OCS minerals development, including oil and gas, as well as other strategic minerals such as polymetallic sulfides and phosphate and manganese deposits in the U.S. EEZ. The Department of Interior, through the Bureau of Land Management and the Minerals Management Service, has carried out an Environmental Studies Program since 1973 which is charged with contracting for and evaluating studies of the environmental, social and economic impacts of OCS oil and gas development. The program began by funding a series of baseline studies and for the past decade has supported research in the following categories: (1) Air Quality, (2) Physical Oceanography and Meteorology, (3) Biology and Ecology, (4) Endangered Species, (5) Fates and Effects of Oil Spills, and (6) Socioeconomic Impacts of OCS Development.

The Department of Interior has funded $67.5 million worth of research distributed as shown in Figure 19.5 since 1973.

Figure 19.5

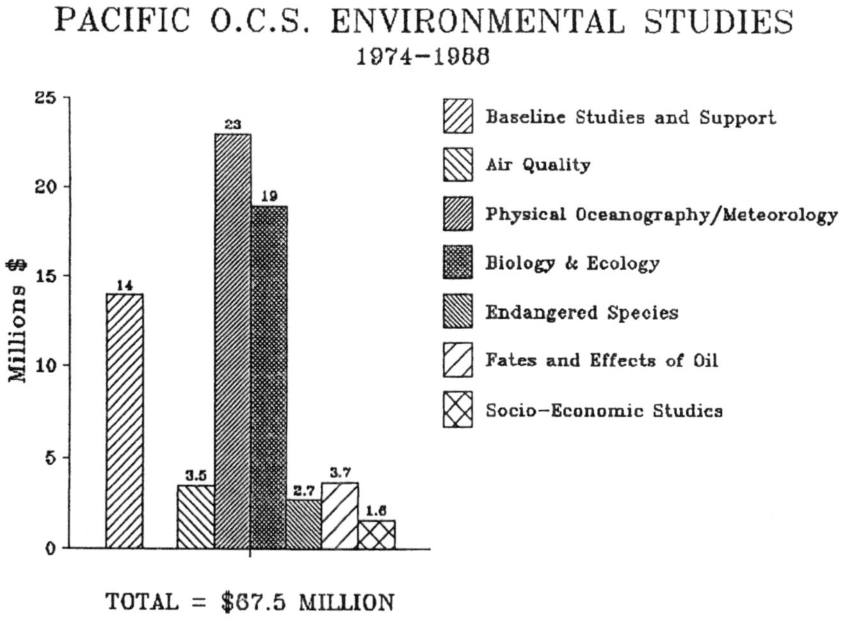

The Environmental Studies Program was designed to compliment the information previously collected by marine researchers, mainly those of the university community, the National Oceanic and Atmospheric Administration (NOAA) and the U.S. Navy. The major "blue-water" oceanographic institutions on the W coast of the U.S. were, until the last twenty years, Scripps Institution of Oceanography at La Jolla, California, Oregon State University at Newport, Oregon and the University of Washington at Seattle, Washington. The historical data base was distributed in a decidedly nonuniform fashion (Figure 19.6). Along the coasts of Washington and Oregon the data was mainly collected by the two regional universities and is densely distributed near shore, becoming gradually less dense offshore. In California the data distribution was dominated by a large grid designed by the California Cooperative Fisheries Investigation program (CalCoFI). The CalCoFI grid is fairly densely sampled S of Point Conception, but the most serious data gap was between Point Conception and the Oregon border. This area is covered by the Northern California and Central California Planning Areas of the MMS. Since 1978 studies on this region have been conducted under MMS auspices and the area has produced some rather remarkable results. Simultaneous with the increased MUMS attention has been the emergence of new academic and government research activity. The Moss Landing Marine Laboratories has developed as a major research institution. The U.S. Naval Postgraduate School and other academic institutions, such as Humboldt State University, the University of California at Santa Crux, the Bodega Bay Marine Biological Laboratory, the University of Southern California and the University of California at Santa Barbara, and the Hopkins Marine Station of Stanford University, all are involved in marine research in the region. A major research effort called the Coastal Transition Zone study is underway in Northern California. All this work has helped reveal the rather spectacular complexity of the flow regime and to fundamentally revise the picture of E boundary current flow in the Pacific.

The traditional picture of the California Current in the area N of Point Conception was that of simple equator ward flow along the coast, extending offshore several hundreds of kilometres. The only variation was thought to be a near shore pole ward current which appeared in the winter, called the Davidson Current. Actual measurement, as is usually the case, reveals a much different flow pattern. The most dynamic flow, which has been better understood with the aid of satellite imaging and drifters tracked by satellites and aircraft, is the offshore movement of features which have come to be known as "squirts" and "jets" (Figure 19.7). Two major features, which are temporarily persistent throughout the year, are attached to Cape Mendocino and Point Conception. The near shore regime is dominated seasonally by spring to summer wind-driven coastal upwelling, but the topographically induced upwelling at these to promontories produces these persistent upwelling jets and downstream eddies.

The traditional picture of the summer equator ward flow and the winter near shore pole ward flow of the Davidson current has been replaced by an upwelling model which applies in the Peru current as well, that of a pole ward undercurrent which surfaces nearshore. This surface expression can occur at any time of the year and was observed in Central California in July in 1985. Additional complexity is introduced in Monterey Bay when the Monterey submarine canyon (20 km wide, 100 km long and 2100 m deep) incises the shelf and steers upwelling to a site near Moss Landing. The cross-shelf flow in and out of Monterey Bay is targeted for intensive work in the future by the new Monterey Bay Aquarium Research Institute.

Figure 19.6

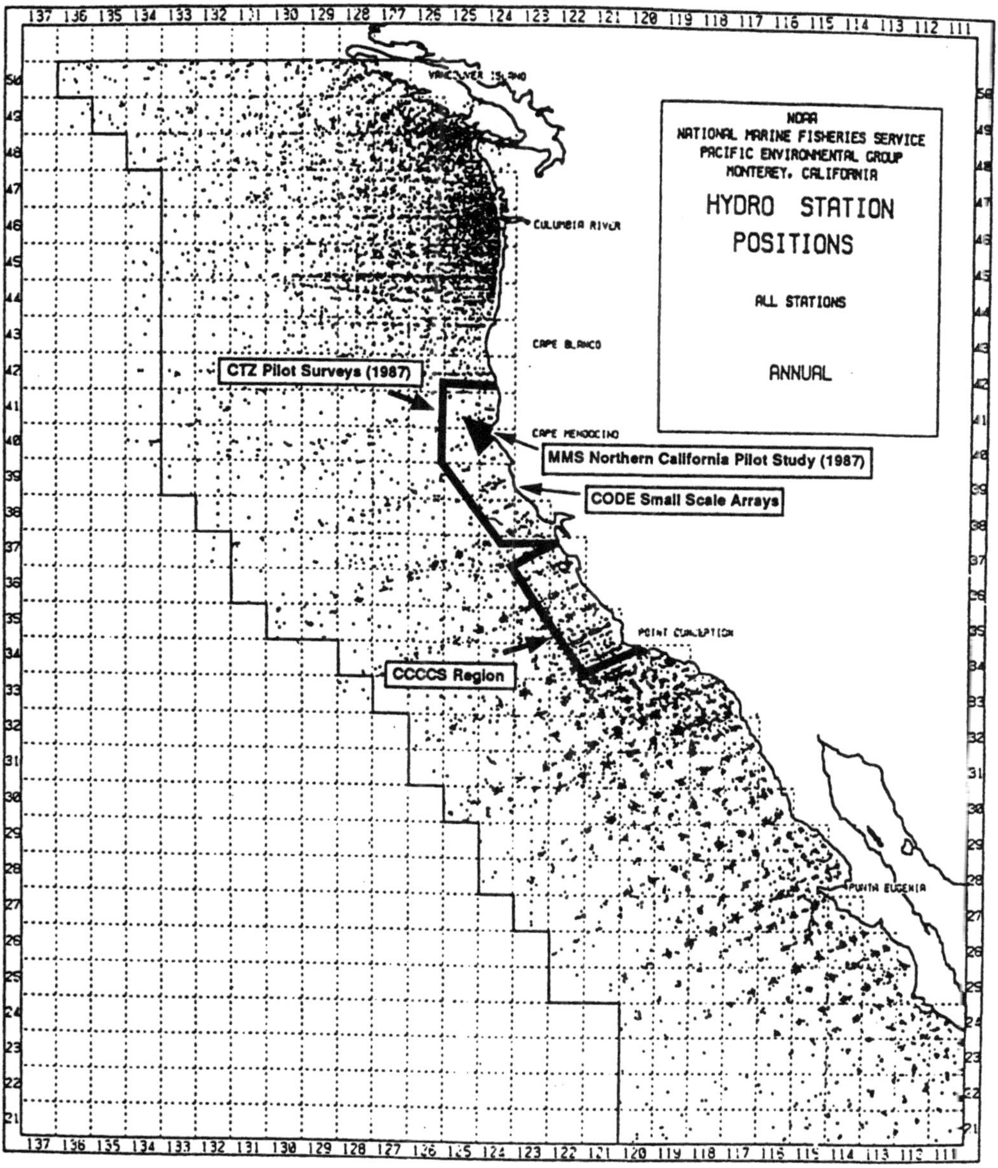
NOAA
NATIONAL MARINE FISHERIES SERVICE
PACIFIC ENVIRONMENTAL GROUP
MONTEREY, CALIFORNIA
HYDRO STATION
POSITIONS
ALL STATIONS
ANNUAL
VANCOUVER ISLAND
COLUMBIA RIVER
CAPE BLANCO
CAPE MENDOCINO
POINT CONCEPTION
PUNTA EUGENIA
CTZ Pilot Surveys (1987)
MMS Northern California Pilot Study (1987)
CODE Small Scale Arrays
CCCCS Region

Figure 19.7

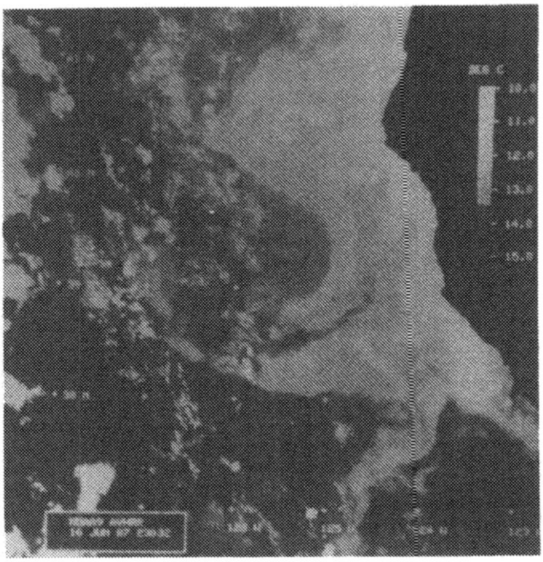

The development of offshore oil and gas reserves in the Pacific States has focused new attention on the level of understanding of the marine systems of the area. The federal government, through the MMS, has sought to develop the information base necessary to understand the system well enough to predict the effects of accidents associated with offshore drilling. The studies conducted with MUMS funding have complimented other work, funded by the National Science Foundation and NOAA. The previously severe data gap in the meteorology and physical oceanography is being filled and a new and much more sophisticated picture of the circulation in the area has emerged.

A number of important lessons have been learned.

(1) The government agency responsible for the Environmental Studies (in this case MUMS) needs to avail itself of the advice in the scientific community, as MUMS has increasingly done. In the long run this leads to better science conducted by the agency and less overlap with other work in the area. For example, MMS has responded to requests from the scientific community and a dense meteorological buoy network has been maintained for about a decade, something which the academic community would not have been able to support on its own.

(2) Government agency research, which is necessarily mission-oriented needs to have oversight by academic scientists to assure that high quality information is produced. The long range perspective of most academic scientists is sometimes frustrating in

short-term mission oriented studies, but can assure that data taken, even for short-term goals, has lasting value.

(3) No matter how carefully the government agency goes about its work, the strident political opposition, particularly that representing regional interests, will not be mollified. This is certainly not a problem peculiar to the United States.

Some elaboration of lesson 3 may provide an appropriate conclusion.

19.2 Conclusion

The existence of the MMS and the associated leasing schedule provides a focus for regional interests to express their views on a number of matters, some of which are unrelated to offshore oil and gas development. The rhetoric is often highly charged and emotions run high. It is not an environment conducive to objective scientific discussion and yet it is the environment in which the discussion takes place. To put the matter in perspective, the oil and gas revenues from the Pacific OCS are slightly less than $1 billion per year. The industry has a huge investment in the area. $5 billion has been paid to the federal and state governments as bonus payments and royalties. This is probably more than the total revenues which the industry has received thus far from the region. The fishing industry has revenues of about $350 million and the recreational industry has revenues of over $3 billion per year. The environmental protection interests, the preservation of endangered species, especially marine mammals, and the aesthetic and social concerns surrounding offshore drilling all are portrayed as competing in the political arena and are usually opposed to offshore drilling. Given this rather contentious setting, it is remarkable that so much progress has been made, not just in opening new leasing areas to drilling, but in understanding of the environment in which the drilling is being carried out.

The highly charged atmosphere surrounding offshore drilling in the Pacific States obscures the larger issue. Worldwide, the losses of oil to the ocean are, in the vast majority, losses from shipping oil -- tanker spills, and ballast discharges (Figure 19.8). In the 15 year period, 1969 to 1984, which includes the highly publicized Santa Barbara oil spill, the losses from all OCS operations in the U.S. equal only 1.5% of the worldwide losses from shipping. The Pacific OCS losses, including the Santa Barbara spill, are only 10% of the U.S. total, or 0.15% of the worldwide losses of oil to the ocean. The Pacific OCS is a highly eutrophic system, characterized by seasonal upwelling and persistent topographically enhanced upwelling centres. The Monterey Formation, the source of the currently produced OCS oil and gas, seeps naturally into the ocean in amounts probably considerably greater than OCS drilling losses, nearly all of which are the result of the single Santa Barbara spill in 1969.

Figure 19.8

OIL SPILLS ON THE OCEAN
FIFTEEN YEAR PERIOD 1969-1984

	(THOUSAND BBLS)
OIL SPILLS IN FEDERAL WATERS	348
PACIFIC OCS	36
WORLDWIDE TANKER LOSSES	21,837

OCS LOSSES = 1.56% OF TANKER LOSSES

(NATURAL SEEPAGE IN SANTA BARBARA CHANNEL IS BETWEEN 15-250 THOUSAND BARRELS/YEAR)

A tremendous weight of attention and money has been devoted to Environmental Studies related to offshore drilling which is responsible for less than 2% of the oil spilled on the ocean. Almost no attention is paid and little in the way of mitigating measures have been devoted to the cause of 98% of the problem. If we are to develop the resources of the Pacific Basin in an environmentally sound way, a critical problem to be addressed is the prevention of losses of petroleum from shipping accidents. This is a problem which must be addressed at the international level, since individual states can do little to regulate international tanker transport in their Exclusive Economic Zones.

James C. Kelley
San Francisco State University and California Academy of Sciences
San Francisco
California
U.S.A.

20 THE EFFECT OF ENVIRONMENTAL POLLUTION, HYDROCARBONS AND HEAVY METALS ON REPRODUCTION OF SEA URCHINS AND BIVALVES

P.M. Zhadan, M.A. Vashchenko, V.V. Malakhov, L.A. Medvedeva and R.V. Gareyeva

Abstract

The present work is devoted to two aspects of the effect of pollution on marine organisms. The first one is a search for biochemical and functional indices evidencing faults in the development of sex cells. The second one deals with the influence of pollutants on the early ontogenetic stages, such as fertilization, cleavage, and larval formation.

It has been shown in the present work that the prolonged effect of marine environmental pollution (field investigations) or of sublethal concentrations of hydrocarbons of diesel fuel oil (laboratory experiments) on sea urchins has resulted in the damage of gametogenesis which is seen in upset embryo genesis and larval development and, in some cases, in the inhibition of protein synthesis in sea urchin gonads and an increased ^{14}C-leucine and $^{45}Ca^{2+}$ transport through unfertilized egg membranes. Morphological abnormalities caused by heavy metals stress (copper, zinc and cadmium) have been analyzed at different stages of development of bivalve mollusc larvae. A specific effect of these metals have been revealed which shows up most vividly at the early veliger stage.

A concept of embryological monitoring has been formulated. It is based on the analysis of damages in gameto- and embryogenesis. Embryological monitoring will permit forecasting the quality of marine environment from the point of view of possible normal reproduction of marine organisms.

20.1 Introduction

The existing methods of hydrochemical control are not adequate for detecting all possible pollutants and products of their transformation, entering marine environment. In this relation it seems expedient to develop biological criteria of estimating negative effects of man-made pollution on ecological situation in coastal sea areas. The main objective in the development of such criteria is their applicability at the earliest stages of attaining ecologically significant levels of pollution. A plausible scheme of the effect of marine pollution on living organisms is given in Figure 20.1.

The earliest response of organism to pollution is changes in physico-chemical and biological processes within cells. If the impact trespasses the cell adaptability, the upsetting in their physiological and morphological characteristics occurs and, hence, the derangement of physiological functions of the organism as a whole. It results in a decrease in population rate, degradation and ruin of the whole communitie. However, changes at the populational level may be detected only some years after the pollution had attained ecologically significant levels. That is why in elaborating bioassay methods the strategy must be based on the search of such physico-chemical, biochemical, physiological and morphological indices that would constitute a basis for the early forecasting of possible changes at the populational level.

V. I. Ilyichev and V. V. Anikiev (eds.), Oceanic and Anthropogenic Controls of Life in the Pacific Ocean, 267–286.

Figure 20.1: A scheme of possible aftermaths of environmental pollution for living organisms.

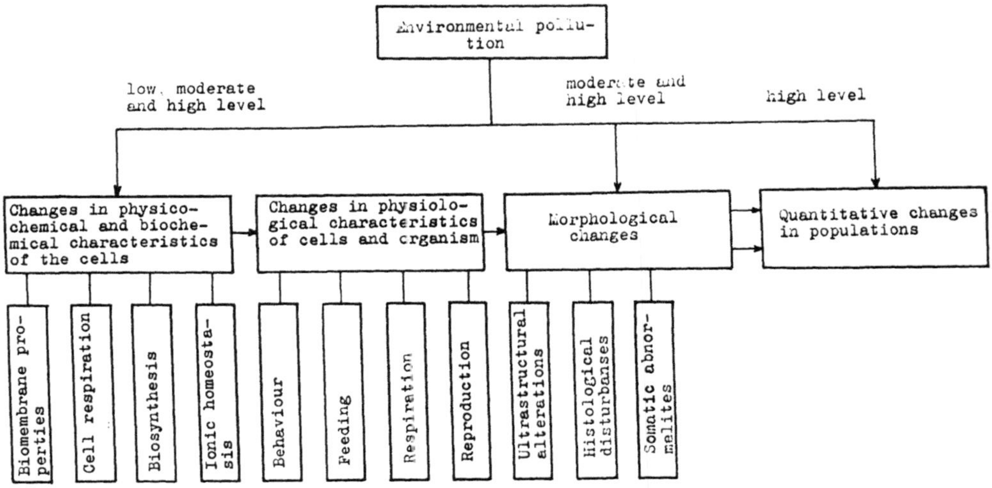

One of ways ensuring the implementation of this approach is the study of the effect of pollutants on processes responsible for reproduction. The functions of reproduction and development are known to be the most vulnerable for different damaging effects. Sea urchins and bivalve molluscs provide one of the most representative groups of sea animals and are frequently used in toxicological studies (Dixon,1983; Kobayashi, 1984). These animals provide a convenient object for the study of pollution effect on reproduction.

Two basic aspects of pollution effect on the reproduction of marine organisms have been considered in the prevent work. The first and the least studied aspect is to reveal alterations in biochemical and functional indices characterizing abnormal development of sex cells. As the criteria of gametes quality the offspring development, gonad morphology (field and laboratory observations) as well as protein biosynthesis and transport of ions and aminoacides through egg membranes (laboratory experiments) have been used.

The second aspect concerns the pollution influence upon the early stages of ontogeny, involving fertilization, cleavage and processes of larval formation which are the most sensitive to the affect of damaging factors. In sea urchins and bivalves these processes occur in the depth of water and are directly affected by toxic substances. We have studied morphological abnormalities in the development of bivalve larvae and revealed quantitative regularities defining the dependence of the embryo and larval mortality on the character of toxic agent and its concentration.

20.2 The effect of contaminated marine environment on gametogenesis in sea urchins

The characteristics of stations. To investigate the effect of pollution on the gametogenesis in sea urchins, the animals were sample at five stations situated in the Amursky Bay (Sea of Japan) in areas having different degrees of contamination. The first station was chosen near the dump of contaminated ground excavated from the Golden Horn Bay; the second one in area of intensive discharge of domestic wastes; the third, in a comparatively normal zone of the urban seaside; the fourth and fifth were controls, situated accordingly in the Marine State Reserve (Island of Popov) and remoted from the waters of the Amursky Bay (Figure 20.2). The two latter places were at a considerable distance from anthropogenic sources of pollution. Animals were sampled during spawning period in August and September, 1984 and 1985. Sampling stations were almost identical in substratum (rock, coarse sand, shell-ground) and biocenosis composition. The algae Laminaria cichorioids end Zostera provided food for animals. The animals were sampled from depths of 3 to 5 m. While sampling, water temperature was about 27°C in August 1984, about 23°C in August 1985, and 17.5°C in September 1985. Only at Station 5 the August water temperature was about 3°C lower.

Morphological studies of gonads. Histological analysis of female gonads in sea urchin *Strongylocentrotus intermedius* has shown that most animals sampled at Stations 3,4, and 5 were at reproductive stages described as active gametogenesis, prespawning and spawning periods (Gnezdilova,1971). There was a big number of large wall oocytes and free lying ones in gonadial follicles, and some animals had eggs (Figures 20.3 a,b). No histopathological changes were found in gonads of sea urchins sampled at these stations in 1984 - 1985.

Figure 20.2: A layout of sampling stations in the Amursky Bay (Sea of Japan). 1, Egersheld Cape; 2, Skrebtsov Island; 3, Sportivnaya Harbour; 4, Alekseyev Bight (Popov Island).

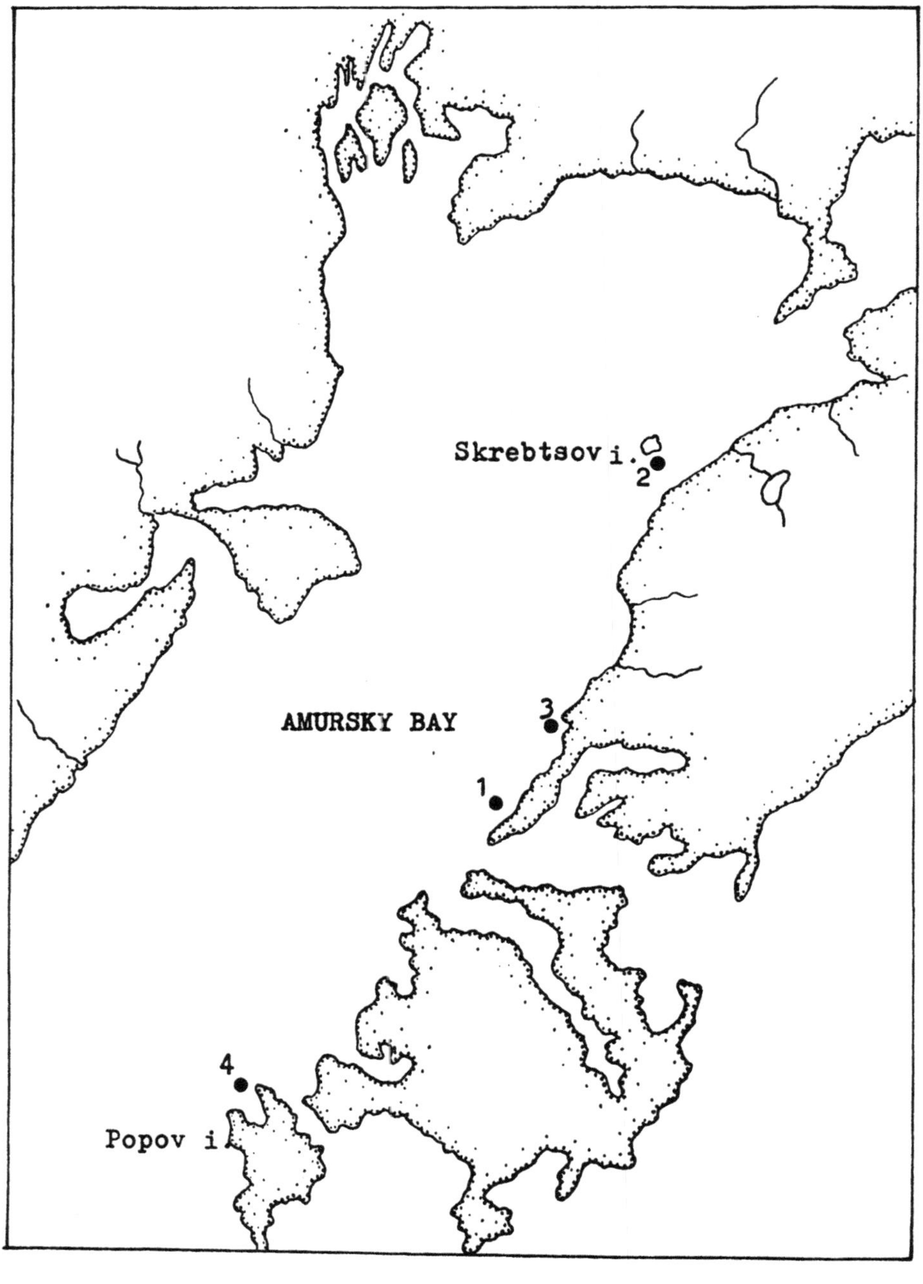

Figure 20.3: Female gonads of sea urchins S.intermedius sampled in the Amusky Bay in August 1984. (A,B, large oocytes (o) and eggs (e) in the follicles of sea urchins, sampled at Stations 3,4; (C-E) ovaries of sea urchins, sampled at Station 1: (C), oocytes absent and large black inclusions (i) occur in follicles; (D), strands (s) consisting of epithelium and connective tissue with lipofucsine granules: (E), necrotic oocytes (no) and remants of destroyed follicle (df); (F), resorption of eggs (re) in follicles of sea urchins sampled at Station 2. All sections are stained with haematoxylin and eosin. Scale bars are 100 m.

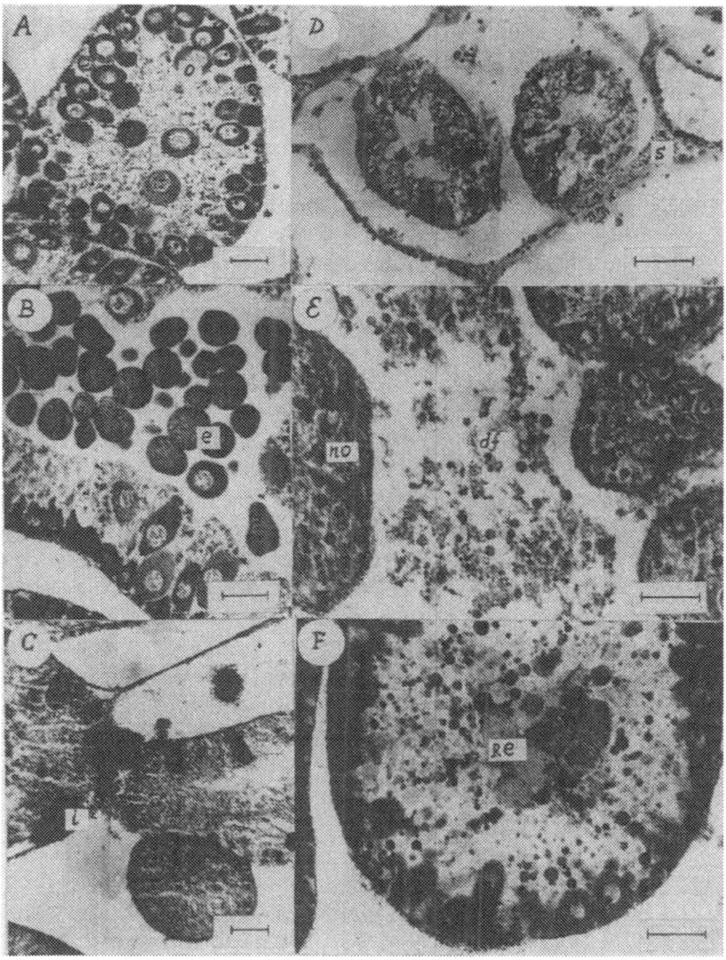

In 1984 histological picture of female gonads in sea urchins sampled at Station 1 differed from that described above. Sex cells were subjected to active decay processes (Figure 20.3,e). No ovocytes were found in gonadial follicles of some animals (Figure 20.3,c). The cytoplasm of accessory cells contained a lot of black inclusions and yellow-brown lipofuscine granules. The heavy growth of epithelial, muscular and connective tissues occurred in gonads of such animals. Damaged follicles were braided with strands consisting of epitheliums and connective tissue (Figure 20.3, d), Lipofuscine granules concentrated in these strands. Only a small number of minute wall oocytes (10-20 m in diameter) most of which were decaying, were observed in gonadial follicles of other specimens. The condition of these gonads corresponded to the early stage of the reproductive cycle.

Sea urchin gonads from Station 2 showed almost the same histopathological changes as from Station 1, though less prominent. We revealed here a decrease in the number of sex cells, of reduced size, hypertrophical accessory cells and resorption of oocytes and eggs (Figure 20.3,f). Histological analysis of gonads of sea urchins, sampled at the same stations in 1985 revealed no changes. Only sea urchins from Station 1 were characterized by necrosis of some oocytes, a heavy growth of muscular and connective tissue in Follicle wall and by the presence of large black inclusions.

Thus, morphological analysis revealed significant histopathological changes in gonads of sea urchins sampled from Stations 1 and 2. As all the populations studied inhabit similar ecological conditions, the serious derangements in gonads seem to be due to a long-term effect of pollution on animals in the Amursky Bay. Gonads of sea urchins, sampled at Stations 1 and 2 in August 1984,were even more suppressed than in summer of 1985. This may be due to a combined effect of pollution and abnormally high water temperature. It is a common knowledge that the effect of a toxic agent depends generally on a kind of toxicant, its concentration, temperature and time of exposure (Bayne, 1975; Myint, Tyler, 1982). Besides, the closure of the dumping-place for contaminated ground may have played a positive role in amelioration of ecological conditions in 1985 at Station 1.

The results of laboratory experiments revealed histopathological changes in gonads of marine animals under the action of high-concentrated toxicants. For example, the resorption of ovocytes and eggs was reported for sea urchins *S.intermedius* exposed to cadmium concentrations of 1 mg/l (Khristoforova et al.,1984). No histopathological changes have been noted in gonads of sea urchins exposed to hydrocarbons at concentrations of 0.04-0.3 mg/l (Vashchenko, 1985). In their experiments, the authors used one toxicant and exposure incomparable to many year pollutant effect on hydrobionts of the Amursky Bay. Here as it seems, lies the reason why the morphological changes which we revealed in sea urchin, gonads in the Amursky Bay are more profound than in animals exposed to experimental pollution.

<u>Investigations into the quality of offsprings</u>. As mentioned above, no changes were observed in the development of sexual glands in sea urchins sampled at Stations 3,4 and 5 in 1984 and at Stations 2,3,4, and 5 in 1985. However from experimental data it is known that the absence of morphological changes in gonads does not mean the full value of sex cells. The offspring quality is more sensitive criterium of the erect of toxicants on gametes (Vashchenko, 1985).

For the estimation of the quality of sexual products of sea urchins sampled in

September 1985, at all 5 stations, the animals were put at once in clean running water at the control station 5. The progeny of five parent couples sampled at each station was observed for development from fertilization to pluteus 1 in the same water. The results of these investigations have shown that the progeny of sea urchins taken from different stations developed in different ways. The biggest number of normally developing embryos were observed during fertilization, first cleavage and blastula stages in the offspring of Rex urchins from the control station 4,5 (Table 20.1).

Table 20.1: The development of the offspring of sea urchin S. intermedius secured from Amursky Bay (Sea of Japan).

Year/ Station	Ferti lized eggs	Double-cell embryos	Blast ula	Gastrula late	Gastrula middle	Gastrula early	Gastrula anor mal	Early plutei
	%	%	%	%	%	%	%	%
1984/1	14.9	17.7	30.8	-	-	-	-	-
1984/2				no eggs were obtained				
1984/3	69.8	26.8	70.3	-	-	-	-	-
1984/4	34.8	39.9	76.0	-	-	-	-	-
1984/5	84.8	63.6	72.2	-	-	-	-	-
1985/1	79.6	55.1	63.3	17.5	55.0	24.4	3.1	53.6
1985/2	50.0	66.3	50.8	46.0	39.0	5.5	9.5	91.0
1985/3	76.1	67.9	72.0	32.3	39.4	26.4	1.9	93.7
1985/4	93.1	85.2	92.9	65.0	29.5	4.6	0.9	92.8
1985/5	99.1	94.3	94.5	77.3	20.2	1.0	1.5	92.8

"-" not determined

The development of larvae from stations 1,2 and 3 was asynchronous, retarded and accompanied by a big number of abnormalities. At the above stations the larvae hatched two hours later than at control stations and was characterized by inhibited formation of the archenteron at the gastrula stage. As a result, 22 hours after fertilization the number of larvae of the 1985 progeny at the post-gastrula stage was 30 to 60% lower than the control. The most prominent deviations in embryo genesis were typical of the 1984 and 1985 progeny were typical of Station 1, though some retardation in the development of sea urchin progeny was also observed for the

Island of Popov (station 4). This may be associated with insufficient remoteness of the control station 4 from the sources of pollution, though one cannot reveal the environmental factors other than pollution. For example, ecological conditions of stations 4 and 5 were not identical. Nevertheless, the disturbances in the development of embryos and larvae at all pollution-affected stations 1, 2, and 3 testify to defective sex cells which may be due to disorder in biochemical processes as a result of a long-term impact of pollutants on the gametogenesis of parent specimens.

20.3 The influence of oil hydrocarbons on gametogenesis in sea urchins

The particular features of pollutants' effect on the reproductive function of sea animals may be studied more finely in laboratory experiments than in field observations. In this case eliminated the influence of scatter in ecological conditions between stations is eliminated.

This section of the present work is devoted to the long-term impact of pollutants upon sea urchin sampled at the control station 5 and exposed to the action of soluble fraction of diesel fuel oil throughout gametogenesis. In experimental aquarium the concentration of hydrocarbons varied within 0.04 to 0.3 mg/l and was controlled by mean of IR-spectroscopy. The detailed description of this method was done earlier (Vashchenko,1985; Vashchenko et al.,1988). The study of gonad morphology and the development of the offspring obtained from the experimental and control sea urchins was carried out in the game manner as during field observations. Also, we studied processes of protein synthesis in gonads, as well as leucine and Ca^{2+} transport through the membranes of unfertilized eggs.

Morphological studies. Histological analysis did not reveal any noticeable changes in gonads of sea urchins exposed to watersoluble hydrocarbons for 50 days.

The offspring quality investigations. Disturbances of embryogenesis and early larval development were caused by a long-term action of water-soluble hydrocarbons of diesel fuel oil upon the adult sea urchins *S.intermedius* (Table 20.2). These disturbances were similar to those in sea urchins from polluted stations in field experiments. The offspring of both the control and experimental sea urchins had developed up to the pluteus I stage. However, the number of fertilized eggs, obtained from experimental animals, was 15% lower and the number of two-cell embryos during the first cleavage was 3 times less, than in the control. This testified to the retarded fertilization and cleavage. The retarded and asynchronous development of experimental sea urchin offspring was observed at the gastrula stage as well: their archenteron formed much slower. During all developmental stages, embryonic abnormalities were higher in number as compared with the controls.

The study of protein-synthesis processes. When speaking about the problem of hydrocarbon's effect on marine organisms, many scientists noted that disorders in the reproductive function of a species (physiological reaction) may be caused by disturbances in biochemical processes related to gametogenesis (Davis, 1972; Bayne, 1981; Cowles, Remillard, 1983). At the same time, the data on the effect of hydrocarbons on biochemical processes in reproductive organs are practically absent. The results of autoradiographical experiment with labelled precursors of the RNA and protein synthesis, have shown that the intensity of the ^{3}H-uridine and ^{14}C-leucine uptake into the ovocytes and accessory cells of experimental sea urchins *S.nudes*, subjected to a prolonged impact of dissolved hydrocarbons a of diesel fuel oil, was reliably lower as compared with the control (Figures 20.4 a,b).

Figure 20.4: The effect of long-term exposure of sea urchins to water-soluble hydrocarbons of diesel fuel oil (0.04-0.3 mg/l) for biosynthesis of protein in gonads. (A), incorporations of ^{3}H-uridine into oocytes of *S.nudus*; (B), incorporation of ^{14}C-leucine into oocytes of *S.nudus*; (C), incorporation of ^{13}C-leucine into gonadial protein of *S.intermedius*. (A,B) Incubation for 1 hour. Abscissa: (A,B) (a) nucleulus, (b) nucleus, (c) cytoplasm, (C) incubation time in presence of ^{14}C-leucine, min. Ordinate: (A,B) number of silver grains per 100 m^{2}, (C) incorporation of ^{14}C-leucine, cpm x 10^{-3}/mg gonadial protein. Shaded columns are experiment; non-shaded, control. Vertical bars are standard deviations.

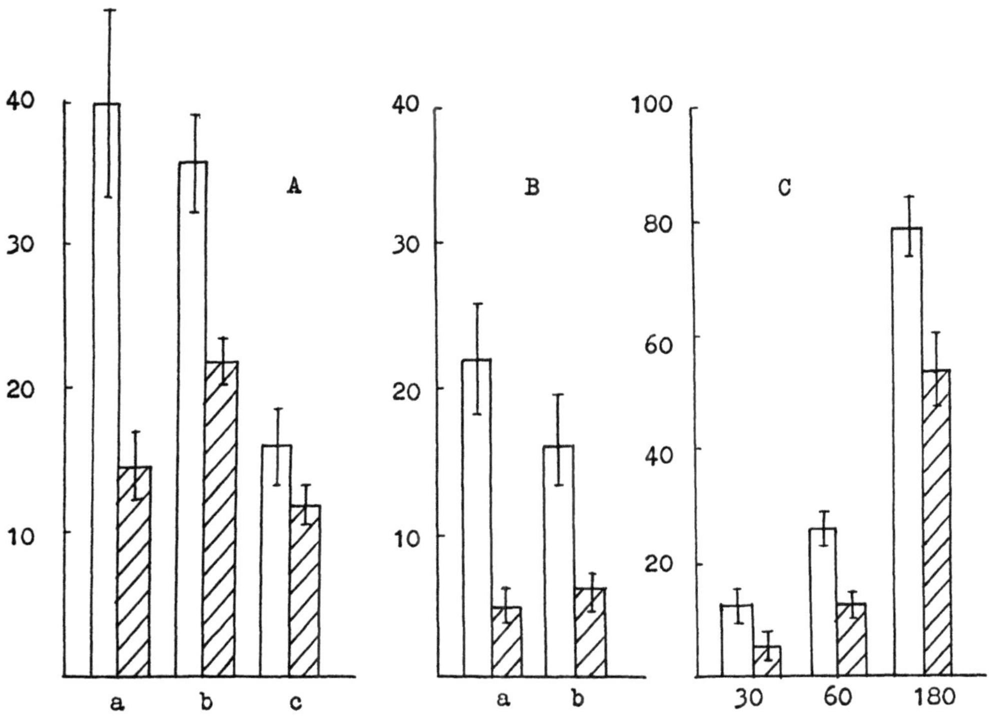

Moreover, biochemical investigations showed that the intensity of incorporation of ^{14}C-leucine into proteins of experimental animals were also reliably lower than in the control (Figure 20.4,e). These facts reveal the inhibition of processes of protein synthesis in sexual and accessory cells of experimental animals.

The investigation of ^{14}C-leucine transport. The inhibition of protein synthesis observed may be caused by a decrease in amino acid uptake through the cell membrane. It is also possible that oil hydrocarbons cause a reduction in metabolic rate of proteins which is associated with disturbances in some cell regulating mechanisms.

To investigate this problem we carried out some experiments on the uptake of C-leucine in unfertilized cells of sea urchins exposed to a prolonged effect of hydrocarbons. These experiments have shown the uptake of ^{14}C-leucine in egg-cells to be 60% higher than in the control (Figure 20.5).

Figure 20.5: The effect of long-term exposure of sea urchin *S.intermedius* to water-soluble hydrocarbons of diesel fuel oil (0.04-0.5 mg/l) on ^{14}C-leicine uptake in unfertilized egges. Abscissa: duration of egg incubation in presence of ^{14}C-leucine, min. Ordinate: ^{14}C-leucine uptake, cpm 10^{-3}/mg of egg protein. Shaded circles are experiment; unshaded, control. Vertical bars denote standard deviation.

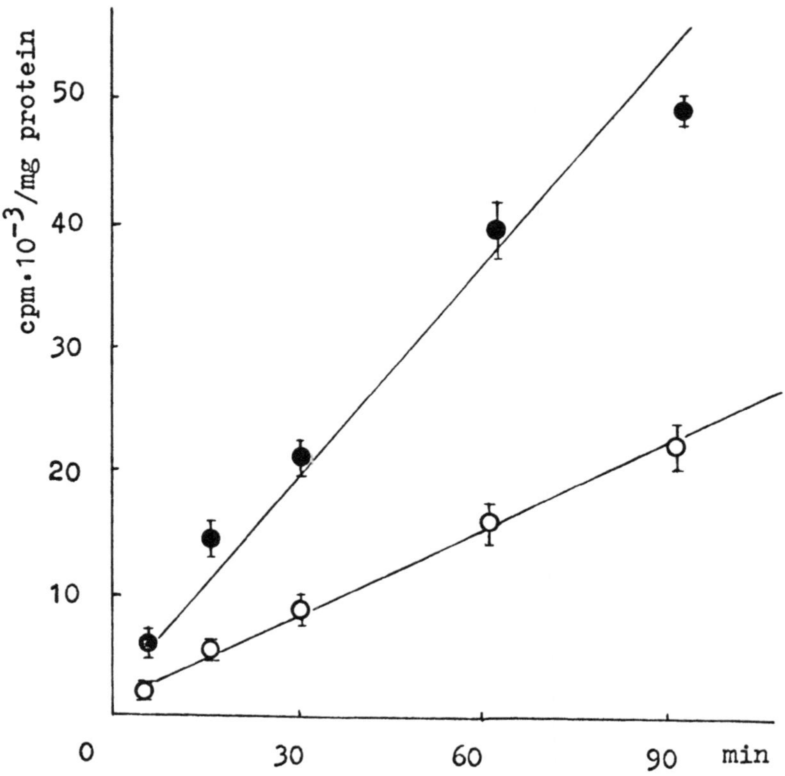

The investigation of $^{45}Ca^{2+}$ uptake in unfertilized eggs.

The amino acid transport through cell membranes is known to be closely related to that of inorganic ions (Stephens, 1988; Van Winkle,1988). Diesel fuel hydrocarbons are expected to change the rate of amino acid transport and to disturb the membrane permeability for inorganic ions in egg-cells. In this connection it would be extremely interesting to explore the transport of Ca^{2+} ions. First, the increase in the uptake of Ca^{2+} ions through plasmic membrane is known to be accompanied by an increase in amino acid uptake (particularly leucine) in some types of cells (Tokadera, Mohri, 1987). Second, Ca^{2+} ions are very important in regulation of gametogenesis and early embryo genesis (Isant,1983; Risen et al.,1984; Petzelt, 1984; Poenie et al., 1985). Besides, there is a lot of data indicating the connection between the alteration of the intracellular free-calcium content and the intensity of protein synthesis (Hesketh et al.,1985; Klaus,1986).

As was shown by our experiments, an increase in the rate of Ca^{2+} uptake in unfertilized eggs is caused by a chronic effect of hydrocarbons on sea urchins. After 90-minute incubation of eggs in the presence of $^{45}Ca^{2+}$, the radioactivity of eggs from urchins exposed to hydrocarbons was 30 to 60% higher than in the control (Figure 20.6 a). It is not without interest that a steep increase in the $^{45}Ca^{2+}$ uptake in egg-cells was caused by temperature rise in the range optimum for development (i.e. from 17°C to 21°C) (figure 20.6 b). The relative rise in the rate of Ca^{2+} uptake with increasing temperature was almost the same in experimental and control egg-cells.

Figure 20.6: The effect of long-term exposure of sea urchins *S.intermedius* to hydrocarbons of diesel fuel oil (0.04-0.3 mg/l) (A) and temperature (B) on $^{45}Ca^{2+}$ uptake into unfertilized eggs. Abscissa: time of egg incubation in presence of $^{45}Ca^{2+}$ uptake, cpm x 10^{-3}/mg egg protein. Unshaded circles are control, shaded ones are experiment. Temperature of incubation, (A), 19°C; (B): unshaded square 19°C, shaded square, 17°C; unshaded triangle, 21°C; shaded triangle, 20°C. Vertical bars are standard deviations.

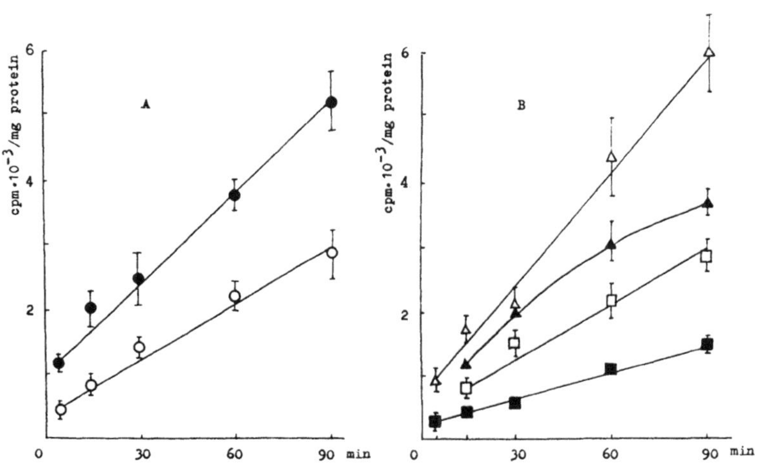

So, the changes in the process of protein synthesis, an increased transport of amino acids and inorganic ions through egg-cell membranes, deranged embryogenesis and larval development and, in some cases, considerable histopathological changes in gonads, testified gametogenetic disorders caused by long-term influence of marine pollution (field studies) or diesel oil hydrocarbons (laboratory studies) upon sea urchins. The comparison of laboratory and field experiments data revealed an apparent similarity in the effect of pollution on the development of the offspring in both cases. These are a decreased extent of fertilization, retarded and asynchronous cleavage as well as gastrula and early pluteus formation, and also an increase in developmental abnormalities. The similar disturbances in embryonic and larval development of sea urchins were also observed when parent specimens were exposed to some heavy metals, e.g. cadmium (Vashchenko et al., 1988), copper and zinc (unpublished data). This non-specific physiological response seems to be based upon the similar changes in biochemical processes. For example, the cations of heavy metals (such as hydrocarbons) also cause disturbances in protein synthesis and in transport of calcium ions (Verbose et al.,1987,a,b; Plishker, 1984; Berisha et al.,1983).

It should be noted that many researches also observed disturbances in embryonic development of sea urchins in acute experiments when toxicants directly affected sexual cells prior to and during fertilization. However, visible alterations in embryogenesis had occurred only when toxicants had a very high concentration, e.g. when hydrocarbon concentrations in the medium made up 1 to 10 mg/l and more (Lönning, Hagström, 1975; Nicol et al., 1977; Kobayashi, 1981). In the case of hydrocarbons effect on the gametogenesis of parent specimens, the disturbances in the offspring development occurred at much lower concentrations, 0.04 to 0.3 mg/l. Hence, gametogenesis in sea urchins is more sensitive to pollution than embryonic and larval development. It is obvious that this fact should be taken into account in developing marine biomonitoring methods.

20.4 The effect of heavy metals upon the embryonic and larval development of bivalve molluscs

The effect of heavy metals upon bivalves was studied on gametes of animals taken from non-polluted areas. Egg-cells and sperm were obtained by provoking spawning when serotonin solution in the concentration of 10^{-4} M was introduced into the gonads of adult specimens. Fertilized eggs were placed in solutions of the hollowing salts: $CuSO_4$ with copper concentrations 0.0004 mg/l, 0.004 mg/l and 0.04 mg/l; $ZnSO_4$ with zinc ion concentrations O.01 mg/l, 0.1 mg/l and 1 mg/l; and $CdSO_4$ with cadmium ion concentrations 0.05 mg/l, 0.5 mg/l and 5 mg/l. Metal salt solutions were changed daily. Embryos, developing in clean sea water from the natural mollusc habitats served as control.

Experiments were carried out with embryos of six mass species, inhabiting the Sea of Japan, i.e. *Mytilus edulis, Crassostrea gigas, Mizuhopecten yessoensis, Callista brevisiphonata, Spisula sakhalinensis and Mactra chinensis*. Heavy metal effect was registered at the stages of sterroblastula, conchostoma and early veliger sterroblastula is the first mobile stage in the development of bivalves, it it a large-cell embryo without cavities and is provided with three ciliate bands (Figure 20.7). Conchostoma is an actively floating larva, it has the invaginated shell gland rudiment and archenteron, a powerful locomotor ciliate band and sensitive ciliate apical tuft

(Figure 20.7 b). Early veliger has straight-hinge bivalve shell, vellum with locomotor cilia, the intestine dissected on stomodeum, stomach and small intestine, complicated muscles (Figure 20.7 c).

Figure 20.7 Stages of normal development of bivalve mollusc *Callista brevisiphonat*: (a), Sterroblastule; (b), Conchostoma; (c), Veliger. 1, Blastopore; 2, shell gland; 3, mouth; 4, velum; 5, shell; 6, adductor; 7, retractor muscles; 8, anus; 9, stomach.

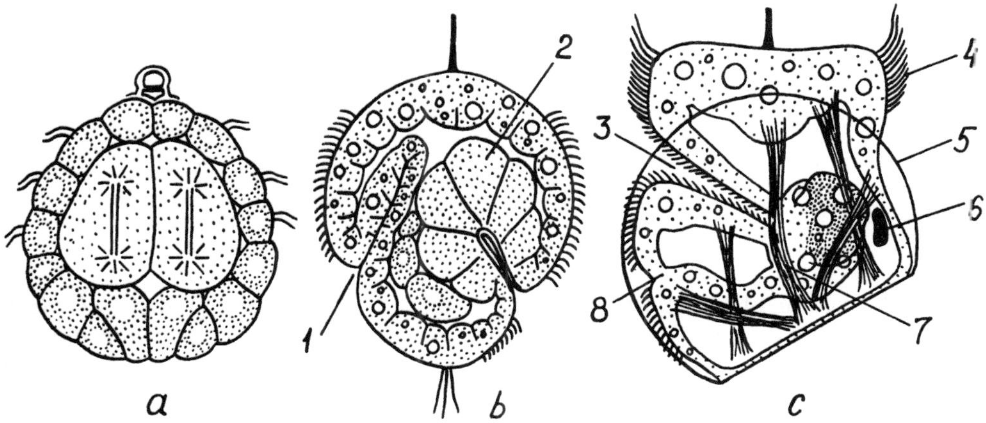

At the blastula stage the number of abnormal embryos was insignificant, their damages being inspecific with respect to the kind of toxicant, and different species had similar damages. At this stage, among the deformed embryos we found mobile blastulas of anomalous shape, blastulas with separate protruding cells, embryos, much retarded in development, with abnormal distribution of cells.

At the conchostoma stage all the species studied had similar abnormalities. Embryos which had the structure of abnormal celloblastula with voluminous internal cavity could be found. Embryos having clearly separated animal and vegetative halts occurred frequently. Al a rule, abnormal embryos had a relatively normal locomotor ciliate band in the animal part of the body, with apical ciliate tuft; whereas in the vegetative part shapeless cell conglomerates were seen.

Specific effect of heavy metals was conspicuous at the early veliger state. The maximum concentrations of copper (0.04 mg/l) rendered the shield gland uncapable of evagination, and calcium carbonate deposited inside the embryo in the form of two dense conglomerates (Figure 20.8 a). If the shell gland does evaginate, two fine-wall vehicles form instead of the valves and calcium carbonate accumulates in the form of two dark-coloured globules in the centres of these vehicles. In *Crassostrea gigas* the maximum copper concentrations lead to the formation of reduced valves with calcium carbonate plates of anomalous shape positioned in their centres.

Figure 20.8: Specific morphological abnormalities in *Callista brevisiphonata* at the veliger stage: 1, velum; 2, accumulation of calcium carbonate; 3, shell gland; 4, stomodium material; 5, damaged, wrinkled shell.

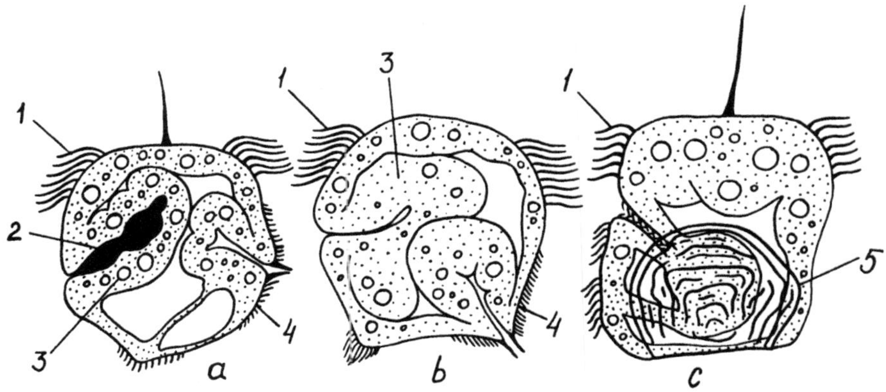

Under the action of the maximum used concentrations of zinc (1 mg/l), *Callista brevisiphonata* and *Crassostrea gigas* did not evaginate the shell gland and did not invaginate stomodeum material (Figure 20.8 b). Other species evaginated shell gland but instead of valves they had two thin-walled vehicles. None of the species studied revealed any signs of calcium carbonate deposition. Some species (e.g. *C.gigas*) showed hypertrophic ciliate cover.

The maximum concentration of calcium (5 mg/l) resulted in reduced shell with wide-open valves; the larval body came off the valves. Peculiar concentric lines emerged on the surface of the shell in C.brevisiphonata (Figure 20.8 c).

The maximum concentrations of metals resulted in underdeveloped musculature in all the mollusc species. The front abductor was wanting, that is why the valves of the shells were wide open (if any). The larvae could contract a little, but were unable to retract into the shell. The intestine was always underdeveloped. *C.brevisiphonata* and *C.gigas* sometimes had an uninvaginable stomodeum.

As to the vellum and other ciliate structures in larvae, subjected to the action of metals, no pathology was observed.

No specific metal action was found in solutions where concentrations were one to two orders lower. The shell of veligers developing in such solutions, used to have reduced unequal valves with uneven edges. The internal organs were underdeveloped to some extent. The reduced shell sizes and irregular form of valves in deformed larvae makes it easy distinguish them from the normal veligers.

There are general rules in the effect of heavy metals on the embryonic development of molluscs. At all developmental stages, the deleterious effect of heavy metals was the most heavy on the structures of the vegetative part of the embryo, i.e. on those developing from the first quartet of IA-ID macromeres, whereas the structures of the animal pole with the velum in the first turn, developed almost normally. Judging by the character of morphological abnormalities (insufficient development of the shell and the shell gland, of mesodermal derivatives, and, on the contrary, the practically normal development of the vellum), bivalve larvae, affected by heavy metals, are similar to those formed after the removal of polar lobes (e.g. Rattenbury, Berg, 1954).

Table 21.2: Long-term effect of diesel fuel hydrocarbons (0.03-0.4 mg/l) on the development of the offspring of sea urchin S. intermedius.

	Fertilized egg	First cleavage			Blastula	Gastrula				Plutei		
		Double-cell embryos	Undivided eggs	Abnor-mal		Late	Middle	Early	Abnormal	Normal	Abnormal	Other abnormalities
	%	%	%	%	%	%	%	%	%	%	%	%
Control	100	93.1	6.1	0.8	97.6	87.5	9.8	1.6	1.1	93.0	5.0	2.0
Experiment	84.7	29.2	61.8	9.0	80.9	63.4	25.1	3.1	8.4	77.2	21.7	1.1

Table 21.3: Heavy metal effect on the development of Callista brevisiphonata.

Stage and time of development	Proportion of normal larvae in metal solutions and control (%)									
	Cu (mg/l)			Zn (mg/l)			Cd (mg/l)			Control
Sterroblastula 6 hours	89.2±0.7	84.0±2.9	80.6±3.3	85.6±10.5	91.0±4.3	83.6±1.7	92.0±2.7	90.0±2.4	91.2±0.7	93.6±3.3
Conchostoma 12 hours	80.6±4.3	72.2±3.0	35.6±8.3	84.9±4.4	66.5±3.9	42.1±6.3	83.9±4.6	85.7±2.9	78.9±2.3	90.0±0.9
Veliger 18 hours	75.8±3.5	64.8±8.7	-	71.2±6.0	59.5±9.1	-	74.9±4.9	66.2±16.1	4.5±2.4	90.6±1.6

"-" not determined

Molluscs, annelids and other Spiralia are known to have polar lobes or polar plasma at the vegetative pole, which during cleavage is distributed mostly between the 2d and 2d blastomeres. The removal or damage of this plasma results in defects in the vegetative pole of the embryo: underdeveloped shell and shell gland, mesoderm, underdeveloped intestine. Among the structures of the animal pole only apical ciliary tuft is subjected to reduction (Crampton, 1896; Wilson, 1904). The resemblance of developmental aberrations arisen under the heavy metal stress, with those, emerged when some parts of the vegetative pole of the egg are damaged or removed, testify to the fact that metal ions affect biochemical processes occurring in the polar plasma, polar lobes and the homological structures of the vegetative pole of the embryo.

The second worth mentioning aspect is that there are processes which are extremely sensitive to the effect of heavy metals. To those undoubtedly gastrulation belongs, as well as formation of shell gland and the shell.

The first group of processes creates conditions for the formation of conchostoma. And the emergence of a noticeable amount of abnormal embryos with uninvaginated ectoderm and often with uninvaginated shell gland, protruding as conglomerates of large-scale cells, is indeed associated with the suppression of these processes.

The evagination of shell gland is the main process in the chain of morphogenetic phenomena resulting in the formation of veliger. If the shell gland won't evaginate, the veliger (even abnormal) is not formed and calcium carbonate is either not deposited at all or is deposited inside the unevaginated shell gland (Figure 20.8 a).

Quantitatively, to effect of heavy metals is similar in different species. As seen from Table 20.3, the number of normal embryos in metal solutions decreases as it develops. At the sterroblastula stage, even the maximum concentrations caused but a little decrease in the percentage of normal larvae, while at the veliger stage the maximum concentrations of copper, zinc and cadmium had resulted in 100% abnormal embryos (Table 20.4). A significant increase in deformed larvae proportion was observed also with other concentrations used.

The increase in the percentage of abnormal larvae towards the advanced stages of development is partly due to increased exposure time. In different species the sterroblastula stage comes in 3 to 6 hours, and that of veliger, in 24 to 72 hours. However, this is not the only cause. The toxic effect of metals appear first at the biochemical level and is hidden at the morphological one. At the advanced stages biochemical damages begin to show at the morphological level.

As follows from Table 20.4, the maximum concentrations of heavy metals used have caused the development of 100% abnormal larvae at the veliger stage. Comparison of our results with the data of literature (Calabrese et al.,1973; Martin et al., 1981; Akberali, True man, 1985) permits the conclusion that the emergence of 50% abnormal larvae at the veliger stage is to be expected under the action of solutions with copper ion concentration 0.004-0.006 mg/l, zinc, 0.01-0.1 mg/l, and cadmium, 0.5-1 mg/l.

Bivalve embryos are 2 to 3 orders more sensitive to heavy metal effect than adults. It means that water harmless for adults can be undoubtedly mortal for embryos. Even numerous physiological and behavioral reactions of molluscs, which are proposed as tests I or controlling the quality of marine environment, have proved inadequate to reveal concentrations harmful for developing embryos. For example, copper has lowered the filtration rate of *Mytilus edulis* by 50% at the concentration of 0.05 mg/l, absolutely mortal for embryos (Abel, 1976).

Table 21.4: Heavy metal effect on development of six bivalve species.

Mollusc species	Proportion of normal veligers in metal and control solutions (%)									
	Cu (mg/l)			Zn (mg/l			Cd (mg/l)			
	0.0004	0.004	0.04	0.01	0.1	1.0	0.05	0.5	5.0	
Mytilus edulis	47.9±2.9	17.3±3.3	-	84.3±2.4	60.6±5.3	-	93.4±3.0	87.9±2.7	12.4±5.1	87.6±3.9
Mizuhopecten yessoensis	21.6±4.8	21.7±2.1	-	32.0±4.5	-	-	33.6±2.9	31.7±2.8	-	67.4±3.1
Mactrachinensis	74.5±2.9	68.1±8.4	-	72.9±6.4	11.9±4.3	-	82.6±1.4	79.4±2.6	-	76.2±1.0
Spisula sakhalinensis	81.2±2.7	54.8±4.6	-	74.0±6.2	55.6±10.9	-	57.8±6.6	26.7±9.3	-	82.0±5.1
Callista brevisiphonata	75.8±3.5	64.8±8.7	-	71.2±6.0	59.5±9.1	-	74.9±4.9	66.2±16.3	4.4±2.4	90.6±1.6
Crassostrea gigas	50.0±2.2	51.8±1.1	-	36.9±4.3	5.8±3.1	-	61.5±5.9	47.0±3.9	-	78.8±2.1

"-" normal larvae absent

Significant increase in oxygen consumption rate after exposure to 0.05-0.1 mg/l of copper was reported for the adult oyster *C.virginica* (Engel, Fowled, 1979).
Simple behavioral reactions of bivalves such as the withdrawal of the siphons into the shell are often proposed as a test of environmental quality. Bivalve mollusc *Scrobicularia plana* (one of the species proposed for the use as a test-organism in biomonitoring) retracts the siphons and closes its valves after a six-hour exposure to copper concentrations from 0.1 to 0.5 mg/l and zinc concentrations from 5 to 10 mg/l (Akberali et al., 1981). These concentrations are two orders higher than those at which 50% embryos of bivalve molluscs become deformed and inviable larvae.
All this implies that using various behavioral and physiological indices of adult animals, one may claim a particular population normal in a given environment, whereas it is already uncapable of reproduction.
In conclusion, it is necessary to emphasize that the study of aberrations in gametogenesis, embryonic and larval development, caused by anthropogenic pollution and certain toxic substances, is an important aspect of monitoring. The development of special measures of embryological monitoring as part of the complex system of biomonitoring marine environment, seems to be tipical for modern toxicology. This complex system must comprise the definition of toxicant concentration, affecting gametogenesis, embryos and larvae of marine organisms; the correction of MPC (maximum permissible concentration) levels and the identification of real concentrations of toxicants in sea water; revealing biochemical alterations in gametes which result in abnormal embryonic and larval development and also particular abnormalities in structure due to the action of particular toxic substances in order to use these data for biological (in this case, embryological) water testing and to determine the state of populations of marine organisms in nature. This will enable forecasting of not only the survival of adult animals, but also their reproductive abilities.

References

Abel, P.D.: Effects of some pollutants on the filtration rate of Mytilus. Mar.Pollut. Bull. 7, 228-231 (1976).
Akberali, H.B; Wong, T.M.; Trueman, E.R.: Behavioral and siphonal tissue responses of Scrobicularia plana (Bivalvia) to zinc. Mar.Environ.Res. 5, 251-264 (1981).
Akberali, H.B.; Trueman, E.R.: Effects of stress on marine bivalve molluscs. Ad.Mar.Bill. 22, 163-198 (1985).
Bayne, B.L.; Abbott, P.A.; Windows,T.: Some effects of stress in the adult on the eggs and larvae of Mytilue edulis. J.Mar.Biol. Assoc.U.K. 55, 675-689 (1975).
Bayne, B.L.; Clarke, K.R.; Moore, M.N.: Some practical considerations in the measurement of pollution effects on bivalve molluscs, and some possible ecological consequences. Aqua. Toxicol., 1, 59-1 74 (1981).
Berisha, A.; Busnikov, G.A.; Malchenko, L.A.; Rakitch, L.: Influence of trace metal salts on the development of sea urchins embryos and protein synthesis in the mouse carcinoma cells. Ontogenez, 14, 173-179 (1983).(in Russian)
Calabrese, A.; Collier, R.S.; Nelson, D.A.; MacInnes, J.R.: The toxicity of heavy metals to embryos of the America oyster Crassotrea virginica. Mar.Biol. 18, 162-166 (1973).
Cowls, T.J.; Remillard, J.F.: Effects of exposure to sublethal concentrations of crude

oil on the copepod Centropages hamatus. 1. Feeding and egg production. Mar.Bill. 78, 45-51 (1983)

Crampton, H.E.: Experimental studies on gastropod development. Arch.Entw.Mach. 3, 1-18 (1896).

Davis, C.C.: The effects of pollutants on the reproduction of marine organisms. In: Ruivo, M. (Ed.), Marine pollution arid sea life, pp.305-311. Fishing News (Books) Ltd., London. 1972.

Dixon, D. R.: Methods for assessing the effects of chemicals on reproductive function on marine molluscs. In: V.B.Vouk, P.J.Sheehan (Eds.), Methods for assessing the effects of chemicals on reproductive functions. SCOPE, pp.439-457.1983.

Eisen, D.P.; Wieland,S.T.; Reynolds, G.T.: Temporal sequence and spatial distribution of early events of fertilization in single sea urchin eggs. J. Cell. Bill. 99, 1647-1654 (1984).

Engel, D.W.; Fowled, B.A.: Copper-and-cadmium-induced changes in the metabolism and structure of molluscan gill tissue. In: Vernberg, W.B,; Calabrese, A.; Thurnberg, F.P.; Vernberg, J. (Ed.), Marine Pollution Function Responses, pup.239256. Academic Press, New York, 1979.

Gnezdilova, S.M.: Morphological and cytochemical character of ovogenesis and reproductive cycles of sea urchins Strongylocentrotus nudes and S.intermedius. Candidate Thesis, Vladivostok, p.224,(1971).(in Russian)

Hesketh, T.R,; Moore, J.P.; Norris, J.D.H.; Taylor, M.V.; Rogers, J.; Smith.Y.A.; Metcaffe, J.C.: A common sequence of calcium and pH signals in the mitogenic stimulation of eukaryotic cells. Nature 313, 481-484 (1985).

Isant, J.G.: The role of calcium ions during mitosis - calcium participates in the anaphase trigger. Chromosome 88, 1-10 (1983).

Khristoforova, N.K.; Gnezdilova, S.M.; Vlasova, G.A.: The effect of cadmium on gametogenesis and the offspring of the sea urchin Strongylocentrotus intermedius. Mar.Ecol. (Prog.ser.) 17, 9-14 (1984).

Klaus, Y.Y.B.; O'Yarra, A.; Bijsterbosch, M.K.; Holman, M.: Activation and proliferation signals in mouse B cells. VIII. Induction of DNA synthesis in B cells by a combination of calcium ionophores and phorbol myristate acetate. Eur. J. Immunol. 16, 92-97 (1986).

Kobayashi, N.: Comparative toxicity of various chemicals, oil extracts and oil dispersant extracts to Canadian and Japanese sea urchin eggs. Publ. Seto Mar.Biol.Lab. 26, 123-133 (1981).

Kobayashi, N.: Marine ecotoxicological testing with echinoderms. In: Persoone, G.; Jaspers, E.; Claus, C. (Ed.), Ecotoxicological testing for the marine environment, vol.1, pp.341-405. State Univ. Gent and Inst.Mar.Sci.Res., Breden, Belgium, 1984.

Lönning, S.; Hafström, B.E.: The effects of crude oils and the dispersant corexit 8664 on sea urchin gametes and embryos. Nov. J.Zool. 23, 121-129 (1975).

Martin, J.M.; Osborne, K.E.; Billig, P.; Glickstein,N.: Toxicities of 10 metals to Crassostrea gig as and Mytilus edulis embryos and Cancer magister larvae. war.Pollute.Bull. 12, 305-308 (1981)

Mint, U.M.; Tyler, D.A.: Effects of temperature, nutritive and metal stressors on the reproductive biology of Mytilus edulis. Mar.Bill. 67, 209-223 (1982)

Nicol, J.A.C.; Donahue, W.N.; Wang, R.T.; Winters, K.: Chemical composition and effects of water extracts of petroleum on eggs of the sand dollar Melitta quinquies -

perforate. Mar.Bill. 40, 309-316 (1977)
Petzelt, Oh.; Wulroth, P.: Cell cycle specific variations in transport capacity of an isolated Ca^{2+}-transport system. Cell. Bill.Intern.Rep.8, 823-840 (1984)
Plishker, U.A.: Effect of cadmium and zinc on calcium uptake in human red blood cells. Am.J.Physiol. 247 C, 143-149 (1984)
Poenie, M.; Alderton, J.; Tsien, E.T.; Steinhordt, R.A.: Changes of free cadmium levels with stages of the cell division cycle. Nature 315, 147-149 (1985)
Rattenbury, J.C.; Berg, Ew.E.: Embryonic segregation during early development of Mytilus edulis. J.Morph. 95, 393-414 (1954)
Stephens, Y.C.: Epidermal amino acid transport in marine invertebrates. Biochem.Biophys.Acta 947, 113-138 (1988)
Tokadera, T.; Mohri, T.: Involvement of membrane potential in stimulation of glycine transport by calcium ion. J. PharmacoBio.-Dyne 10, 5-U 2 (1987)
Vashchenko, M.A.: Effect of water-soluble hydrocarbons of diesel oil on gametogenesis of sea urchin Strongylocentrotus nudes. Candidate Thesis, Vladivostok, p.176, 1985
(in Russian)
Vashchenko, M.A.; Durkina, V.B.; Gnezdilova, S.M.: The influence of diesel fuel hydrocarbons and cadmium on the development of sea urchin offspring. Ontogenez, 19, 82-88 (1988)
Verbost, P.M.; Flik, Y.; Lock, R.A.C.; Bonga, S.E.W.: Cadmium inhibition of Ca^{2+} uptake in rainbow trout gills. Ann.J. Physiol.253, R216-R221 (1987,a)
Verbost, P.M.; Senden, M.H.M.N.; Van Os, C.H.: Nanomolar concentrations of Cd^{2+} inhibit Ca^{2+} transport systems in plasma membranes and intracellular Ca^{2+} stores in intestinal epitheliums. Biochem.Biophys. Act a 902, 247-252 (1987,b)
Van Winkle,L.J.: Amino acid transport in developing oocytes and early concept uses. Biochem.Biophys.Acta 947, 173-208 (1988)
Wilson, E.B.: Experimental studies on germinal localization. J.Exp.Zoo. 1, 197-268 (1904).

Zhadan, P.M.
Pacific Oceanological Institute
USSR Academy of Sciences
Far East Branch
Vladivostok
USSR

Vashchenko, M.A.,
Institute of Marine Biology
USSR Academy of Sciences
Far East Branch
Vladivostok
USSR

21 EFFECTS OF TEMPERATURE AND CADMIUM ON THE ACCUMULATION OF COPPER BY MUSSEL TISSUES

V.P. Chelomin, O.N. Lukjanova and E.A. Bobkova

Abstract

A study was performed on the uptake of copper by edible mussel, Mytilus edulis, and influence of temperature and cadmium on this process. Laboratory experiments have demonstrated that mussels could accumulate dissolved copper from sea water and initial uptake of copper was linear with time and the exposure concentration. Significant differences in velocities of copper uptake were found for mussels, exposed to +10° and +20°C. After the exposure concentration of copper was higher in the hepatopancreas than in the gills to +10°C, but mussels exposed for 12 days to copper-containing water to +20°C accumulated copper in higher concentrations in gills than in the hepatopancreas. The uptake of cadmium by tissues of mussels was time dependent. Competitive interactions between cadmium and copper was shown for hepatopancreas of mussels with the result that uptake of these metals was reduced. The data obtained is discussed in view of using Mytilus edulis as biological indicator of heavy metal pollution.

21.1 Introduction

Bivalve molluscs have a unique capability to accumulate the metals within the concentration factor 10^3-10^6 depending on both the species and type of metals (Wright, 1978; Simkiss et al., 1982; Ray, 1984). In this connection, several laboratories have used bivalve molluscs for heavy metals monitoring in the aquatic environment (Talbot et al., 1976; Cunningham, 1979; Gordon et al.,1980; Popham et al.,1980;: Riisgard et al., 1987). However, many unsolved theoretical problems prevent from use of these animals for practical purposes. Although the literature on the relationship between accumulation process and several physiological and ecological factors is getting more numerous those factors sre weakly studied until now. Information is available showing that accumulation of heavy metals by mussels in influenced by intrinsic factors including age, size, sex (Schultz-Baldez, 1973; Boyden, 1974; De Wolf, 1975; Walting, Walting 1976), reproductive conditions (seasonal variations) (Boyden, 1974; Delhaye, Cornet, 1975; Phillips,1976; Cossa et al., 1980, Boalch et al., 1981) and by extrinsic environmental factors including hydroclimate, particularly temperature and salinity (Phillips, 1976; George et al., 1977; Cossa et al., 1980; Fischer et al., 1983), chemical forms of the metals in sea water (George, Coombs, 1977; 1977 a) and the influence of one metal on the uptake of another (Phillips, 1976; Jakim et al., 1977; Simkiss, Mason, 1984). Nevertheless, the attempts to systematize and make clear the resolution of these problems are restrained by uncertain and contradictory experimental data, which makes the model laboratory experiments nesessary. This paper deals with some aspects of copper and cadmium accumulation by tissues of Mytilus edulis: the uptake of metals by gills and hepatopancreas cells and effects of temperature and Cd presence on copper accumulation.

All of the adult mussels, Mytilus edulis Linne (dimensions of shell 4-5 cm) used in these experiments were collected in Peter the Great Bay of the Japan Sea from June through September. The mussels were collected in the area free from any industrial input of heavy metals, and having copper and cadmium concentration lower than 1.0

V. I. Ilyichev and V. V. Anikiev (eds.), Oceanic and Anthropogenic Controls of Life in the Pacific Ocean, 287–296.

ppb and 0.1 ppb, respectively. After 72 hours of acclimatzation mussels were exposed to sea water with either cadmium or copper (0.1 ppm). The animals were exposed in the laboratory to sea water for 14 days under air and temperature control. Sea water in exposure tanks was changed every day. Controls were treated in a similar way, but not exposed to metal. At certain time the mussels were removed from the aquarium and then the gills and hepatopancreas were taken out. Isolated tissues were dried to constant weight. Copper and cadmium concentrations in these tissues were determined using atomic absorbtion spectrophotometer "Shimadzu" AA-610 after acidic decomposition of samples. In each measurement 10 mussels per sampling time were used and each experiment was replicated twice.

21.2 Accumulation of copper in mussels

By now the ability of bivalve molluscs to accumulate heavy metals is studied more in Mytilidae. It is due to geographical distribution of these molluscs in moderate temperature waters and to their relatively studied physiology. Their ability to accumulate dissolved heavy metals proportionally to metal concentration in the environment made the researches to propose them as an effective indicator of Pb, Zn, Cd, Fe and Cu (Phillips, 1976; Phillips, 1977; De Silva, Kureishy, 1978; Ritz et al., 1982; Amiard-Triquet, et al. 1986). However, this view is not supported by Phillips (1976;1977), who found no correlation between copper concentration in mussels and pollution areas.

Phillips also found that accumulation kinetics of copper differs from one of the other metals, which on his behalf, is due to Cu higher toxicity. In our experiments (Figure 21.1) the mussels kept the ability to accumulate copper in a wide range of metal concentrations in the environment (from 0.05 ppm to 0.5 ppm). In addition, after 14 days exposure at +10°C the content of copper in gills and digestive gland was proportional to the external copper concentration growth without appreciable saturation, thus indicating to some extent, that copper passive diffusion through cell membrane takes place. Linear relation between the amount of metal uptake by mussels and external metal concentration was shown by other authors as well (De Silva, Kureishy,1978; Amiard-Triquet et al., 1986a). As shown in Figure 21.1, practically under all experimental conditions concentration of copper in hepatopancreas cells was four times higher than in gills. In addition, mussel tissues' copper accumulation has a linear function to exposure time and tissue specificity maintained.

Figure 21.1: The accumulation of copper by tissues of mussels, exposed to different concentration of copper for 5 days (+10°C).
Control copper concentrations in mussel tussues (gills -4.55 μg Cu/g dry weight tissue; hepatopancreas - 11.18 μg Cu/g dry weight tissue) remained constant throughout experiments.
Abscissa: ppm copper in environmental seawater.
Ordinatae: content of copper in gills (A) and hepatopancreas (B) - μg Cu/g dry weight tissue.
Symbols: shaded columns - hepatopancreas; unshaded columns - gills.

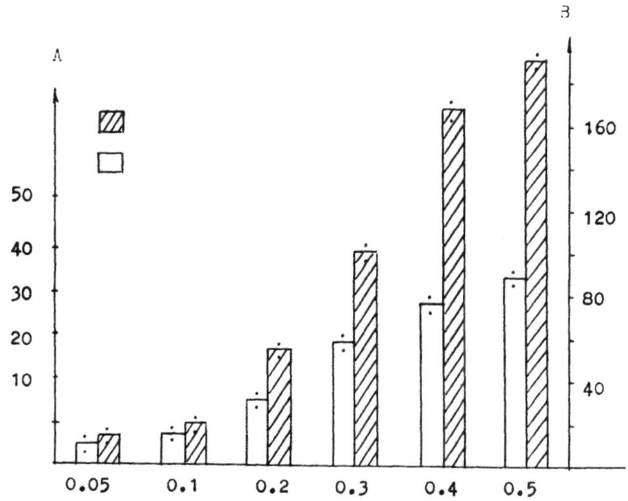

21.3 Effect of temperature on uptake of copper by mussels

Though the temperature is a significant factor in vital functions of aquatic organisms, the evidences of temperature influence on the process of heavy metals accumulation by mussel tissues is contradictory. Phillips (1976) showed a significant influence of temperature on Mytilus edulis uptake of Cd, Pb, Cu and no effect on the uptake of Zn. Later it was found that the accumulation of copper by this mussel increases at fluctuating temperature (Watkins, Simkiss, 1988). Jackim with colleges (Jackim et al., 1977) showed that Cd uptake by 3 species of bivalves increased with temperature. However, Mytilus edulis showed no significant differences in this respect (Fowler, Benayoun, 1974; Jackim et al., 1977). The attempts were made to trace quantitative variations in the contents of heavy metals during a year and to connect them to seasonal fluctuations of temperature (Fowler, Oregioni, 1976; Cossa et al. 1980; Boalch et al., 1981). However, obtained seasanol maxima of accumulation of some metals were hard to explain and the authors connected them to the reproductive conditions of animals only (Fowler, Oregioni, 1976; Cossa et al.,1980). Model experiments can spot the light on temperature influence on accumulation ability of mussels as we showed an increase of temperature results in increase of copper accumulation in mussel tissues. The kinetics of copper accumulation in cells of mussel hepatopancreas and gills in 12 days period at +20°C is shown in Figure 21.3. When the water temperature was changed from +10° to +20°C, the rate of copper uptake increased approximately 10 times by hepatopancreas cells (from 0.5-0.7 to 7.0 μg Cu/ g dry weight tissue/ day), and 60-70 times by cells of gills (from 0.20-0.25 to 15.0 μg Cu/ g dry weight tissue/ day). Besides the environment temperature heavy metals mixtures bring significant corrections into hepatopancreas and gills accumulating abilities. While at +10°C by the 12-14 day of the experiment copper content in hepatopancreas was 18-19 μg Cu/ g dry weight tissue (Figure 21.2), which was twice as high as in gills, at +20°C by the same period the results were the opposite: 340 μg Cu/ g of dry tissue weight in gills and 140 μg Cu/ g of dry tissue weight in hepatopancreas. Probably, this phenomena is impossible to explain only by distinctions in physical and chemical properties of the membrane lipid matrix, responsible for the process of passive diffusion (Simkiss, 1983), in cells of hepatopancreas and gills at various temperatures. Probably, the temperature also can exercise different influence both on the metabolic and penetration pathways of metals in hepatopancreas and gills. Gills of various bivalve molluscs participate in feeding by active absorption of organic and inorganic molecules dissolved in sea water and a number of heavy metals among them. Later transport systems redistribute heavy metals accumulated in the gill cells to other organs, in particular, to hepatopancreas and kidney. In this connection, differences in copper accumulation in tissues at +10° and +20°C can be explained by different temperature influence on the processes of uptake and transport of heavy metals.

As it is shown in Figure 21.4 the time kinetic of cadmium accumulation by mussel tissues is a linear function at 0.1 ppm of Cu in the environment. Hepatopancreas cells were twice more active in Cu accumulation than gill cells. The data obtained (Figure 21.3 and 4) indicate that there is tissue specificity in accumulation of both copper and cadmium. At +20°C gill cells preferably accumulate copper, while hepatopancreas cells accumulate cadmium. As field and laboratory studies showed there is tissue specificity in accumulation of Cu, Pb, Zn, Cu and Hg (Chow et al., 1976; George et al., 1976; Cunningham, 1979; George, 1980; King, Davies, 1987). In addition to the physico-chemical processes, the effect of biological mechanisms may also be significant in selective accumulation. Consequently, a differentiated approach in using mussels as heavy metals indicator is needed.

Figure 21.2: The accumulation of copper by tissues of mussels, exposed to 0.1 ppm Cu for 14 days (+10°C).
Abscissa: incubation days.
Ordinate: content of copper in tissues - μg Cu/g dry weight.
Symbols: shaded columns - hepatopancreas; unshaded columns - gills.

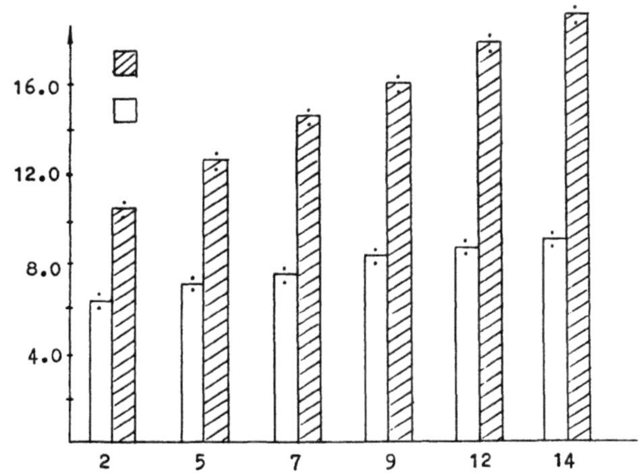

Figure 21.3: The accumulation of copper by tissues of mussels, exposed to 0.1 ppm Cu for 12 days at +20°C.
Abscissa: incubation days.
Ordinate: content of copper in tissues - μg Cu/g dry weight.
Symbols: shaded columns - hepatopancreas; unshaded columns - gills.

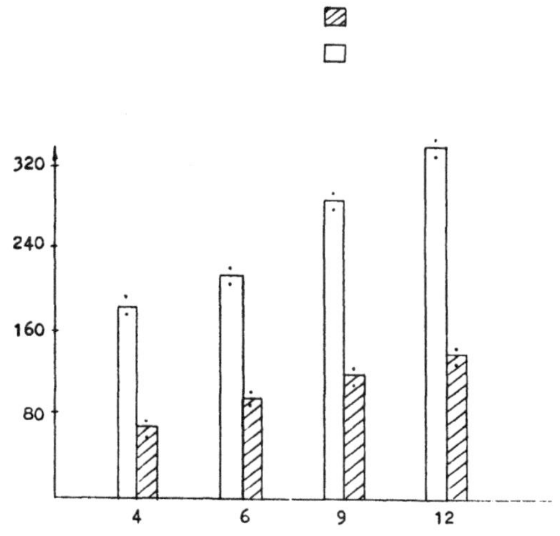

Figure 21.4: The accumulation of cadmium by tissues of mussels, exposed to 0.1 ppm Cd for 14 days at +20°C.
Control cadmium concentrations in mussel tissues (gills - trace; hepatopancreas - 1.65 μg Cd/g dry weight tissue) remained constant throughout experiments.
Abscissa: incubation days.
Ordinate: content of cadmium in tissues - μg/Cd/ g dry weight.
Symbols: shaded columns - hepatopancreas; unshaded columns - gills.

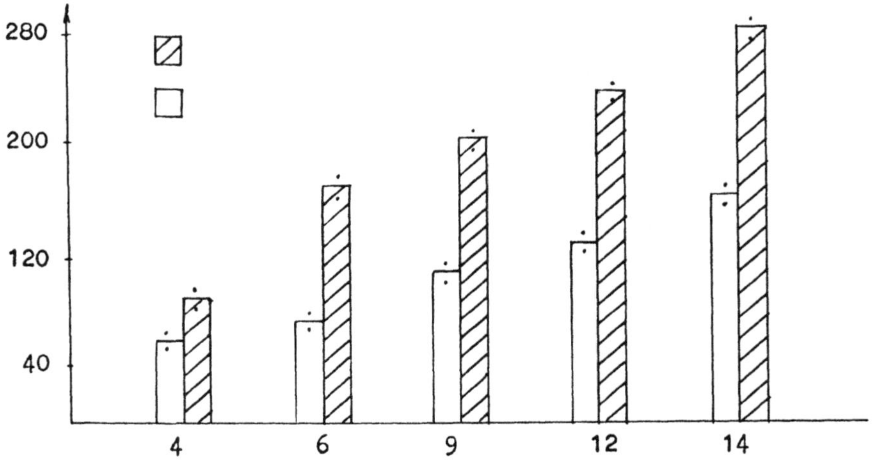

21.4 Effect of cadmium on uptake of copper by mussels

The ability to maintain accumulation rate of a single metal in combination with other ones allows to use a the animals for monitoring tasks. This property acquires a greater importance taking into account complex heavy metal composition of industrial inputs. Results of copper and cadmium accumulation by mussel tissues for 4 days in combinations of equal quantity of both metals in environment (0.1 ppm) are demonstrated in Figure 21.5. Copper did not effect the uptake of cadmium by gill cells. However, copper accumulation decreased by 20% under these conditions, thus indicating cadmium influence on copper uptake in gills. The combined concentration of cadmium and copper in sea water decreased to a great extent due to ability to accumulate these metals by hepatopancreas cells. The results of our experiments showed that the content of copper and cadmium of mussel hepatopancreas in 4 d period was approximately 1.8-2.0 times and 3.5 respectively decreased compared to individual kinetics of uptake of these metals. Obtained data suggest a competition between cadmium and copper for binding sites of hepatopancreas cell membranes and its absence in gill cell membranes.

Figure 21.5: Accumulation of copper and cadmium by tissues of mussels exposed to 0.1 ppm Cu + 0.1 ppm Cd for 4 days at +20°C.
Abscissa: A - mussels, exposed to single metal,
B - mussels, exposed to metal mixture.
Ordinate: content of copper or cadmium in tissues - μg Cu or Cd/g dry weight tissue.
Symbols: shaded columns - copper; unshaded columns - cadmium.

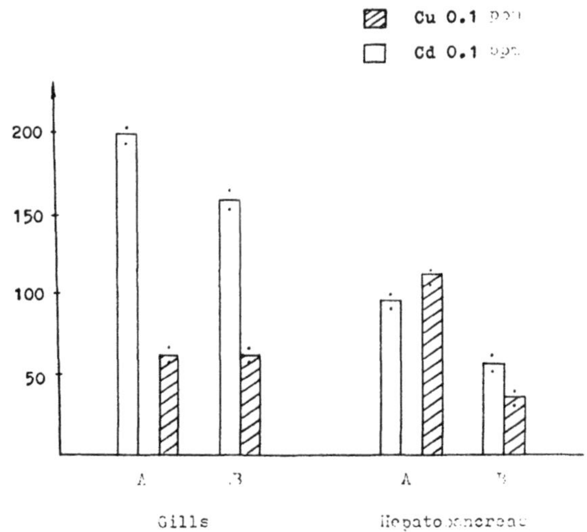

Several metals' combination influence on the uptake of a single metal by animal tissues is very complex because of a big number of combinations and possible biological responses. Phillips (1976) demonstrated that presence of Zn, Cd and Pb influenced copper uptake by M. edulis but at the same time, the uptake of these metals did not depend on the presence of other elements. Cadmium inhibited the uptake of zinc by mussel tissues (Simkiss, Mason, 1984), but at the same time zinc did not influence the uptake of cadmium (Fowler,Benayoun, 1974; George, 1980; Carpen, George, 1981). Competition in accumulation of Fe, Cd and Pb was not found (George, 1980).

21.5 Conclusions

Our experimental data testifies high sensitivity of heavy metals uptake by mussel tissues to various ecological factors, in particular, to temperature and presence of metal mixture. Using of mussels for practical biomonitoring programs requires development of theoretical questions related to processes of metal uptake in various environments. The data obtained is in agreement with other investigations where it was found that mussels are suitable for monitoring of not only cadmium but copper as well. Hepatopancreas of mussels can be also used as an indicator of copper in pollution areas at low temperature. At the same time, hepatopancreas of mussels collected from the area, contaminated by copper and cadmium cannot be used for the monitoring of these metals because of a mutual influence of the metals on accumulation process. Taking into account higher copper accumulation rate in gills and lower dependence of these process on Cd, we believe that using of mussel gills for copper monitoring is effective.

References

1. Amiard, J.C.; Amiard-Triquet, C.; Berthet, B.; Metayar, C.: Contribution to the ecotoxicological study of cadmium, lead, copper and zinc in the mussel Mytilus edulis. 1. Field study. Mar. Biol. 90, 425-431 (1986).
2. Amiard-Triquet, C.; Berthet, C.; Amiard, J.C.; Metayer, C.: Contribution to the ecotoxicological study of cadmium, lead, copper and zinc in the mussel Mytilus edulis. 2. Experimental study. Mar. Biol. 92, 7-13 (1986a).
3. Boalch, R.; Chan, S.; Taylor,D.: Seasonal variation in the trace metal content of Mytilus edulis. Mar. Pollut. Bull. 12, 276-280 (1981).
4. Boyden, C.R.: Trace element content and body size in molluscs. Nature, Lond. 251, 311-314 (1974).
5. Carpene, E.; George, S.G.: Absorption of cadmium by gills of Mytilus edulis. Mol. Physiol. 1, 23-34 (1981)
6. Chow, T.; Snyder, H.; Snyder, C.: Mussel (Mytilus sp.) as an indicator of lead pollution. Sci.Total Environ. 6, 55-63 (1976).
7. Cossa, D.; Bourget, E.; Pouliot, D.; Piuze, J.; Chanut, J.: Geographical and seasonal variations in the relationship between trace metal contents and body weight in Mytilus edulis. Mar. Biol. 58, 7-14 (1980).
8. Cunningham, P.: The use of bivalve molluscs in heavy metal pollution research. In: Vernberg, W.; Thueberg, F. et al. (eds), Marine Pollution: Functional Responses, pp.183-221, Acad. Press, N.Y., 1979.
9. Davenport, J.: A study of the effects of copper applied continuously and discontinuously to specimens of Mytilus edulis exposed to steady and fluctuating salinity levels. J. Mar. Biol. Assoc. U.K. 57, 63-74 (1977).
10. Delhaye, W.; Cornet, D.: Contribution to the study of the effect of copper on Mytilus edulis during the reproductive period. Comp.Biochem.Physiol. 50A, 511-518 (1975).
11. D'Silva, C.; Kureishy, T.: Experimental studies on the accumulation of coper and zinc in the green mussel. Mar. Pollut. Bull. 9, 187-190 (1978).
12. De Wolf, P.: Mercury content of mussels from West European coasts. Mar. Pollut. Bull. 6, 61-63 (1975).
13. Fischer, L.: Influence of temperature, salinity and oxygen on the cadmium balance of mussels Mytilus edulis. Mar. Ecol. Prog. Ser. 32, 265-278 (1986).
14. Fowler, S.; Benayoun, G.: Experimental studies on cadmium flux through marine biota. I.A.E.A. Proc.Symp. Nuclear Techniques in Comp, Studies of Food and Envir. Contam., Otaniemi, Finland, pp. 159-178. IAEA, Vienna. 1974.
15. Fowler, S.; Oregioni, B.: Trace metals in mussels from the NW Mediterranean. Mar. Pollut. Bull. 7, 26-29 (1976).
16. George, S.; Pirie, B.; Coombs, T.: The kinetics of accumulation and excretion of ferric hydroxide in Mytilus edulis and its distribution in the tissues. J.Exp.Mar.Biol.Ecol. 23, 71-84 (1976).
17. George, S.; Coombs, T.: Effects of chelating agents on the uptake and accumulation of cadmium by Mytilus edulis. Mar. Biol. 39, 261-268 (1977).

18. George, S.; Coombs, T.: Effects of high stability iron-complexes on the kinetics of iron accumulation and excretion in Mytilus edulis. J.Exp.Mar.Biol.Ecol. 28, 133-140 (1977a).
19. George, S..; Carpene, E.; Coombs, T.: The effect of salinity on the uptake of cadmium by the common mussel Mytilus edulis. In: Physiology and Behaviour of Marine Organisms. McCluskey, D.; Berry, A. (eds), pp 189-193, Oxford, Pergamon Press, 1977.
20. George, S.: Correlation of metal accumulation in mussels with mechanisms of uptake, metabolism and detoxification: a review. Thalassia Yugoslavia 16, 347-365 (1980).
21. Gordon, M.; Knauer, G.; Martin, J.: Mytilus californianus as a bioindicator of trace metal pollution: variability and statistical considerations. Mar. Pollut. Bull. 11, 195-198 (1980).
22. Jackim, E.; Morrison, G.; Steele, R.: Effects of environmental factors on radio cadmium uptake by four species of marine bivalves. Mar. Biol. 40, 303-308 (1977).
23. King, D.; Davies, I.: Laboratory and field studies of the accumulation of inorganic mercury by the mussel Mytilus edulis. Mar. Pollut. Bull. 18, 40-45 (1987).
24. Phillips, D.J.: The common mussel, Mytilus edulis, as an indicator of pollution by zinc, cadmium, lead and copper. 1. Effects of environmental variables on uptake of metals. Mar. Biol. 38, 59-69 (1976).
25. Phillips, D.J.: The use of biological indicator organisms to monitor trace metal pollution in marine and estuarine environments. Envir. Pollut. 13, 281-317 (1977).
26. Popham, J.; Johnson, D.; D'Auria, J.: Mussels (Mytilus edulis) "point source" indicators of trace metal pollution. Mar. Pollut. Bull. 11, 261-263 (1980).
27. Ray. S.: Bioaccumulation of cadmium in marine organisms. Experientia 40, 14-23 (1984).
28. Riisgard, H.; Bjonrestad, E.; Mohlenberg, F.: Accumulation of cadmium in the mussel Mytilus edulis: kinetics and importance of uptake via food and sea water. Mar.Biol. 96, 349-353 (1987).
29. Ritz, D.; Swain, R.; Elliot, T.: Use or the mussel Mytilus edulis planatus in monitoring heavy metals levels in sea water. Aust. J. Mar. Freshwat. Res. 33, 491-506 (1982).
30. Schulz-Baldes, M.: Die miesmuschel Mytilus edulis als indikator fur die bleikonzentration om weserrastuar und in der Deutschen bucht. Mar. Biol. 21, 98-102 (1973).
31. Simkiss, K.; Taylor, M.; Mason, A.: Metal detoxification and bioaccumulation in mollusks. Mar.Biol.Lett. 3, 187-201 (1982).
32. Simkiss, K.: Lipid solubility of heavy metals in saline solutions. J.Mar.Biol.Ass. UK 63, 1-7 (1983).
33. Simkiss, K.; Mason, A.: Cellular responses of molluscan tissues to environmental metals. Mar.Envir.Res. 14, 103-118 (1984).
34. Talbot, V.: Relationship between cadmium concentrations it seawater and those in the mussel Mytilus edulis. Mar.Biol. 35, 51-54 (1985).
35. Talbot, V.; Magee, R.; Hussain, M.: Cadmium in Port Phillip Bay mussels. Mar. Pollut. Bull. 7, 84-86 (1976).
36. Watkins, B.; Simkiss, K.: The effect of oscillating temperatures on the metal

ion metabolism of Mytilus edulis. J.Mar.Biol.Ass. UK 68, 93-100 (1988).
37. Watling, H.; Watling, R.: Trace metals in Choromytilus-meridionalis. Mar.Pollut.Bull. 7, 91-94 (1976).
38. Wright, D.: Heavy metal accumulation by aquatic invertebrates. In: Coak, T.H. (ed), Applied Biology, Vol.3, pp.331-394. Academic Press, London, 1978.

Chelomin V.P. and Lukjanova, O.N.
Pacific Oceanological Institute
Institute of Marine Biology
USSR Academy of Sciences
Far Eastern Branch
Vladivostok
USSR

PART V
METHODS OF OBSERVATIONS AND CALCULATIONS

22 LASER ANALYTICAL PHOTOIONIZATION SPECTROSCOPY - A NOVEL POTENTIAL IN OCEAN RESEARCH

G.I. Bekov and A.S. Egorov

22.1 Introduction

Investigation of the distribution of the rare elements in tune Earth's crust and in the ocean is of importance for many fields of present-day natural science, including oceanology, geology, geochemistry and cosmochemistry. Noble elements are among those elements. Studies of their geochemical behaviour, apart from its scientific importance, is of practical use in search of the new promising resources on land and sea. If, however, geological investigations on land can be fully based on analysis of the samples collected during expeditions, similar oceanological tasks can be solved in complete measurements directly on board a ship only. This is due to incidental availability of subjects of inquiry and to the properties of sea water, one of the basic objects in the ocean. Sea water is often called a living organism because of its complex and unstable composition, impossibility of long storage and some other factors. Since the concentration of noble metals in sea water is very low (ppt level and less) their determination by classical spectral methods is impossible without their multi-stage concentration. The realization of this approache entails such difficulties as little information available on concentration of noble metals in sea water. It applies equally to determination of noble metals' concentration in the Earth's crust where they are one or two orders higher. Nevertheless, for some platinum metals no data is available on their content in many important geological rocks.
What limits the sensitivity of traditional methods of analysis in determining the content of noble metals in natural objects? The point is that in the spectral methods of determination of element' traces in a substance commonly used due to their relative simplicity, some of the basic stages, atomization and detection of atoms, can be realized with an efficiency much less than unity. Such methods of optical detection as emission, absorption and fluorescence do not enable attaining the ultimate detection level, that is, a single atom. Besides, in ultimate atomization of the substance in a buffer gas the atoms being detected may undergo losses as a result of their high temperature reactions with the atoms and molecules formed by the atomization of the basic substance and with the impurities in the gas. The analysis becomes more complicated due to the so called "matrix effects" when the analytical signal depends not only on the concentration of the element to be detected, but also on the composition and form of the substance under study (or the matrix) and on the amount of the third elements. Suppression of the matrix effects is a primary problem in traditional analysis now. New concentration techniques enabled in some cases to increase the sensitivity of analyses by tens and even hundreds of times. But in most cases the sensitivity gain does not exceed an order of magnitude.
It is obvious that, to solve the problem of detecting traces of noble metals in natural objects, it is necessary to apply methods with ultimate detection sensitivity and free from the above shortcomings. One of such methods developed in the last decade is laser stepwise photoionization that under certain conditions provides registration of single atoms in a laser beam [1,2]. In combination with thermal atomization of samples in vacuum we used this method for direct determination of impurities in high-purity materials, traces of elements in natural and biological objects [3]. Atomization in vacuum was shown to enable the matrix effects to be suppressed to

V. I. Ilyichev and V. V. Anikiev (eds.), Oceanic and Anthropogenic Controls of Life in the Pacific Ocean, 299–313.

a considerable degree. Detection limits of elements obtained in direct analysis (without preliminary chemical preparation of the sample) rang from 1 ppb to 10 ppt and in case of reconcentration are 1 ppt and less [3].

22.2 Experimental technique of laser photoionization analysis

In the method of laser stepwise photoionization the atoms are excited by laser radiation to an intermediate high-lying state through one or several steps and then only the excited atoms are subjected to photoionization. Three approaches in this method may be distinquished, differing from each other in the way of ionization of the atoms from an intermediate state. They are: 1) nonresonant photoionization of the excited atoms to the continuum; 2) resonant excitation of the atoms to a Rydberg state with subsequent ionization by a pulsed electric field; 3) resonant photoionization of excited atoms via narrow autoionization state lying in the continuum.

As far as the use of laser stepwise photoionization for analytical purposes is concerned, the most suitable ionization approach for atoms detection is multistep excitation of atoms by laser radiation to a Rydberg state and subsequent ionization of Rydberg atoms by electric field pulse. In this case the maximum elemental selectivity can be reached with minimal nonselective ion background [3].

Figure 22.1 schematically shows a laser analytical atomicphotoionization spectrometer (LAPhIS). The laser part comprises a pumping laser (N2-laser excimer laser or Nd:YAG laser) that excites simultaneously two or three tunable dye lasers. This laser system enables the multistep excitation of atomic Rydberg states for more than 80% of elements of the periodic system.

Figure 22.1: Scheme of laser photoionization analytical spectrometer.

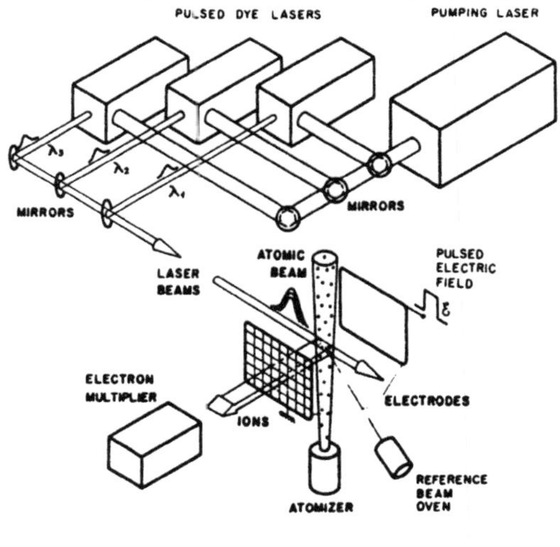

During the analysis (Figure 22.1) the sample was placed into a graphite, tantalum or tungsten crucible electrically heated to about 3000 K. The system of atomization and ion registration was located in a vacuum chamber with a residual pressure of 10 6 tour. During the analysis of solutions moisture it was firstly evaporated from

the crucible. The strong background caused by thermal ions and electrons was removed by special ion suppression system installed above the heater.
The evaporating substance was formed by the crucible walls into an atomic-molecular beam which was simultaneously irradiated at 90° angle by all the laser beams between two electrodes. For precise tuning of the laser radiation to the excited transitions of the elements an additional furnace in the chamber took place which formed a reference beam of this element. The basic and reference atomic beams crossed the laser beams at the same region, in front of the hole in one electrode covered by a wire mesh. After the atoms were excited by the laser pulses to a Rydberg state, an ionizing pulse of electric field with its duration of 10 nm and amplitude of 10 kV (the distance between the electrodes was 1 cm) was supplied to the electrodes with a delay of 25 no. The ions formed by field ionization of Rydberg atoms were pushed by the same ionizing pulse through the mesh in the % Eros potential electrode into a time-of-flight mass filter and after mass separation were registered with an electron multiplier. The signal from the multiplier was amplified, supplied to the boxcar averager and then to the recorder.
This method was applied to measure the concentration of aluminium, ruthenium, gold, platinum and rhodium in oceanic objects. Both direct analyses of the substance and chemical sample decomposition followed by reconcentration of the elements under study were used in these studies. The experiments were performed with a laboratory laser photoionization spectrometer and with its expeditionary version specially designed for being installed an board a ship.

22.3 Dissolved aluminium in natural waters

Aluminum, one of the most important elements in the lithosphere, is usually used in geochemistry as a terrigenous runoff indicator. Besides, some data shows that aluminum in sea water is associated with biological processes [4,5]. Correct qualitative estimation of the role of aluminum in the processes of the ocean demands a universal method which would allow determining the content of aluminum in waters with different mineralization and variable content of organic matter which is able to form soluble organometallic complexes with aluminium.
At present there is no direct method and universally adopted technique to determine dissolved aluminum in natural waters. There are a lot of methods [6] which differ from one another by the application of various chemical reagents contaminating the samples and producing an undue hirable background. Those methods are complicated and unsatisfactory, are not responsible for the absence of reliable data on dissolved aluminum in sea and ocean waters and for the presence of just scarce data for pore waters. We have developed a technique for direct determination of the aluminum concentration in natural waters. A 40 μl water sample was introduced into a tantalum crucible and evaporated to a dry residue. Then the crucible was placed in a vacuum chamber evacuated to 10^{-6} Torr. The crucible with the sample was heated in a stepped mode to 1000°C. The dye laser wavelengths were tuned for two-step excitation of aluminum atoms to a Rydberg state by the scheme

$$3^2P_{3/2} \xrightarrow{396.15\ \text{nm}} 4^2S_{1/2} \xrightarrow{447.4\ \text{nm}} 15\ P$$

We investigated the content of dissolved aluminum in the Amazon and its tributaries, the zone of mixing of the Amazon with the Atlantic Ocean, the pore waters from the Indian Ocean and the Sea of Okhotsk.

Dissolved aluminum in the Amazon basin. Water samples were obtained during the 9th cruise of the scientific-research vessel "Professor Shtokman", February-May 1983. Figure 22.2 shows the concentrations of Al in the waters of the Amazon and its main tributaries. For the Amazon river they range from 25 to 85 μg/l (on the average, 58 μl). The Al concentration range for the main tributaries is wider and lies between 14 μg/l (Xingu) and 182 μg/l (Rio Negro). As for the content of Al, there are three basic types of Amazonian waters, white, clear and black. The clear waters of such tributaries as the Xingu Tapajos and Trombetas are characterized by the lowest concentration of Al (14 to 55 μg/l, 35 μg/l on the average), the white waters of the Amazon itself and its tributary Madeira contain 25 to 90 μg/l (71 μg/l on the averages), the black waters of the Rio Negro are the most rich in Al, 140 to 182 μg/l (161 μg/l on the average). This is probably explained by the relationship of the dissolved Al with the dissolved organic matter which is abundant in black waters. A direct relationship between dissolved Al and dissolved organic carbon was observed [7].

Figure 22.2: Concentration of dissolved aluminium (μg/l) in the Amazon and its tributaries.

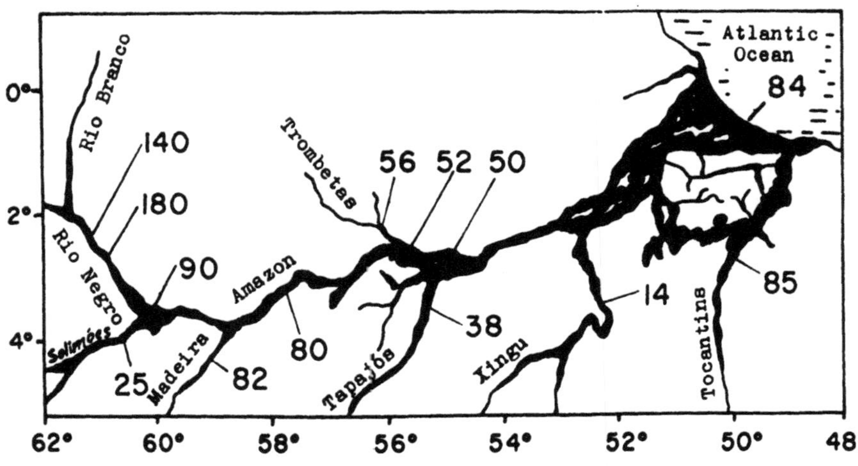

The behaviour of chemical elements in the zones of mixing of river and oceanic waters determines the process of river runoff transformation. The analysis of the dependence C=f(S), where C denotes the element concentration and S is salinity, enables to estimate in first approximation whether the element is involved into chemical or biochemical reactions in the estuary. Figure 22.3 shows the dissolved aluminum concentration as a function of the salinity of waters in the Amazonian estuary. The deviation of this dependence from the direct one indicates that the behaviour of Al is nonconservative and degree of deviation characterizes the losses at the river-sea boundary. The quantitative estimation of Al losses in the Amazonian estuary carried out with the model suggested in [8] is 70%.

Aluminum in pore waters. Little information on the Al content in pore waters impedes the calculation of its global cycle in the ocean. Therefore, it is a pressing problem now to determine Al concentrations in pore waters.

Figure 22.3: Dependence of the dissolved aluminium concentration on the water salinity in the Amazonian estuary.

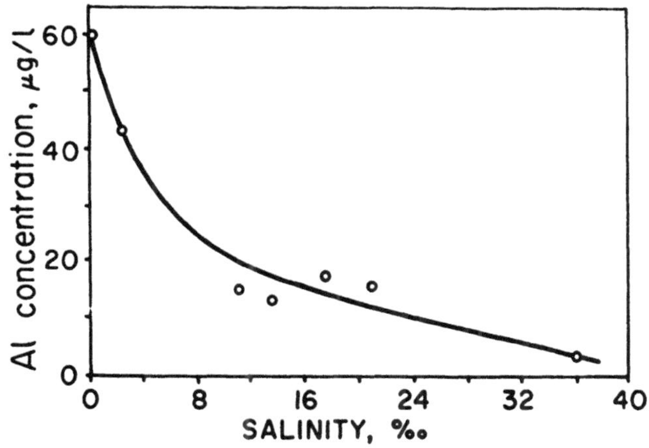

Table 22.1 contains the results of determination of the Al content in pore water samples taken from the Indian Ocean. It makes 14 to 54 μg/l (30 μg/l on the average) for pelagic and 100 to 220 μg/l (160 μg/l on the average) for metalliferous sediments. With the use of this data and the Fick law [9] the flux of Al through the surface of the sediment water boundary was found to be $2*10^{-7}$ g/cm^2 for pelagic and $12.10*^{-7}$ g/cm^2 per year for metalliferous sediments.

Table 22.2 presents the results of determination of the Al content in the pore and bottom waters of the sea of Okhotsk near the Kuril Islands. The water samples were obtained in the vicinity of fumarole activity and contemporary volcanic deposits. The Al concentration in the pore waters in fumarole activity areas, 19 to 30 μg/l (26 μg/l on the average), is lower than in the region of contemporary volcanic deposits, 45 to 102 μg/l (74 μg/l on the average). The aluminum flux through the surface of the sediment, the water boundary in the region of fumarole activity in the Sea of Okhotsk, is 10 7g/cm^2 per year.

22.4 Noble metals in the ocean

Behaviour of noble metals in the ocean is the least investigated field of geochemistry. This is due to the fact that the content of these elements in sea water is extremely low and their analytical determination is rather difficult. The required sensitivity of analysis must be from 10 pp to 0.1 ppt. The conventional methods enable rather easy analysis of sea ores and sediments for the content of Au, Pt and As. As for Ru, Rh and Os, their content in sea objects has not been really studied. The most complicated problem, however, is to determine the content of noble metals in sea waters. The LAPhIS technique was applied to analyze a great number

Table 22.1: Content of dissolved aluminium in the pore waters of the Indian Ocean.

Station Lat.	coordinates Long.	Ocean depth (m)	Sediment type	Al concentration (μg/l)
8°45'S	102°19'E	5300	Pelagic slime	17±2
2°63'N	80°10'E	4300	Clay slime	14±2
0°04'N	80°02'E	4600	Clay slime	30±3
10°10'S	67°31'E	4060	Carbonaceous slime	34±4
9°08'S	67°16'E	6015	Siliceous slime	29±3
26°38'S	67°31'E	4780	Metalliferous sediment	220±25
23°54'S	73°56'E	3780	-"-	54±6
23°00'S	75°21'E	4320	-"-	100±11

Table 22.2: Content of dissolved aluminium in the pore waters of the Sea of Okhotsk (Kuril Islands Region).

Station Lat.	coordinates Long.	Depth (m)	Sample	Al concentration (μg/l)
50°30'N	155°18'E	820	Bottom water (I)	11±2
-"-		-"-	Pore water (I)	19±3
		-"-	Bottom water (II)	16±3
		-"-	Pore water (II)	28±4
44°37'N	146°59'E	960	Bottom water	15±3
-"-		-"-	Pore water	30±6
50°46'N	155°29'E	490	Pore water (volcanic)	102±10
50°46'N	155°41'E	455	-"-	45±10

of sea objects for Ru and Rh as well as to study the behaviour of Au, Pt and Rh in hydrothermal springs of the Pacific.

Solid samples were prepared for analysis in two ways: 1) acid-alkaline decomposition of samples, 1 to 5 g in mass, with subsequent sorption of noble metals on polymer thioether [10] or chelates sorbent POLYORG XI-N [11] and 2) fire assay enrichment of samples, 5 to 25 g in mass, into a lead alloy [12]. Water samples were analyzed in three ways: reconcentration of 0.1 l samples on polymer thioether with subsequent thermal decomposition of the sorbent and analysis of the ash residue; reconcentration of samples (0.05 to 1 l) on chelates sorbent POLYOXG XI-N with subsequent thermal decomposition of the sorbent and analysis of the ash residue; evaporation of 0.3 to 1.5 l of sea water to salts with subsequent fire assay enrichment of the salts into a lead alloy, its currelation and analysis of the Pb bead obtained. The application of the different methods was an attempt to check the correctness of analysis and to find an optimal method for noble metal concentration. The technique of reconcentration on thioether was used for mass determination of Rh in ocean sediments and ores and for single determinations in natural waters. Concentration on POLYORGS XI-N was used in mass analysis of natural waters and single analysis of ores for Au, Pt and Rh. Fire assay enrichment was employed in determining the Ru content in natural objects.

Ruthenium in the ocean. Table 22.3 contains short descriptions of some subjects of investigation and the results obtained [13]. The Ru content (input) in biogenic siliceous clay sediments of the Pacific and Indian Oceans amounts to 150-250. In the metalliferous sediments enriched in iron and some heavy metals the Ru content is much higher: from the Pacific is 1700, from the Red Sea is 7600.

Another group of oceanic formations enriched with iron, manganese and heavy metals are ferromanganese nodules and crusts. The Ru content in the nodules studied varies between 1600 and 3500. The minimal content was found in the nodules from the S Pacific. The Ru content in the nodule from the radiolarion zone of the NE area of the Pacific is two times higher, 320O. A higher content, 3500, was found in the nodule from the Central part of the Pacific. The maximum content, 3800, was found in the ore crust from a sub-sea mountains of the N part of the ocean.

In the samples of phosphorites which we studied, the Ru content ranges from 23 to 66O ppt. It is only 23 ppt in the phosphate rocks from a reef in the Sea of Japan. In the phosphorities from the mid-pacific mountains the content of ruthenium is much higher (170-260 put) and reaches its maximum in the lithified Holocene nodules from the inner shelf of Namibia, 660 ppt.

In the sea water sample from the antarctic sector of the Indian Ocean the Ru content is 1.1 pp. This value is much lower than the platinum content from [14] but more than order of magnitude higher than the palladium content determined in [15]. Thus, the Ru concentration in sea water is two orders of magnitude lower than in pelagic sediments. For most of other heavy metals the situation is quite different: their content in sea water is usually 3 or 4 orders lower than in sediments [16].

In the pore water of siliceous, clayey ooze from the Indian Ocean, the Ru content is 6.9 ppt. This shows that in the process of pelagic oceanic diagenesis ruthenium is able only to undergo some redistribution between the solid and liquid phases of sediments. Therefore, the enrichment of ferromanganese nodules with Ru may depend on both its supply from sea water and pore water.

Of particular interest is the high Ru content (910 ppt) in an aerosol sample taken in the central part of N Atlantic of Sahara. This concentration is about three orders of magnitude higher than in sea water and 4-5 times higher than in pelagic bottoms sediments. If such concentration with Rut is characteristic of oceanic aerosols as a whole, this may be considered as an argument for the hypothesis on cosmogenic origin of the main part of Rut in oceanic bottom sediments.

Table 22.3: Ruthenium content in oceanic objects (analytical precisions (15) are 15 and 25% for solid and water samples respectively).

Sample	Location Lat.,Long.,depth (m)	Ru concentration (ppt)
Clayey-siliceous ooze	Northern Pacific 14°23'N, 117°18'W, 4125	250
-"-	Southern Indian Ocean 8°45'S, 102°19'E, 5300	150
Metalliferous sediment	East Pacific Rise 38°56'S, 99°58'W, 4330	1700
-"-	Red Sea, Discovery Deep	7600
Ferromanganese nodule	Southern Pacific 22°45'S, 160°50'W, 4850	1600
-"-	Northern Pacific 14°23'N, 117°18'W, 4125	3200
-"-	Northern Pacific 10°58'N, 153°22'W, 4980	3500
Ferromanganese crust	Mid-Pacific Mountains 20°41'N, 170°32'W, 1930	3800
Phosphorite (Pliocene)	Sea of Japan 38°'43'N, 130°12'E	23
Phosporite (Pleistocene)	Mid-Pacific Mountains 32°16'N, 172°51'E, 370	170
Phosphorite (Cretaceuos)	Mid-Pacific Mountains 18°31'N, 172°06'W, 1050	260
Phosphorite (Holocene)	Namibian shelf 22°08'S, 13°15'E, 85	660
Ocean water	Southern Indian Ocean 62°12'S, 76°26'E, surface	1.1
Pore water of siliceous-clayey ooze	Southern Indian Ocean 8°45'S, 102°19'E, 5300	6.9
Atmosphere aerosol	Northern Atlantic 20 March 1983, 5°43'N, 29°13'W- 21 March 1983, 10°37'N, 27°59'W	910

Rhodium in the ocean. Table 22.4 contains the results of Rh content determination in oceanic samples [17].

Table 22.4: Rhodium content in oceanic ore formations.

Sample	Location	Rh concentration (ng/g)
	Pacific Ocean	
Ferromanganse nodule	Klarion-Klipperton area	6.6±4.1(11)
-"-	Southern hollow	11.4±8.1(4)
-"-	Antarctic zone	9.9±3.4(5)
Ore crust	Mid-Pacific Mountains	12.0±3.6(3)
-"-	Southern zone, Gayota Ita Mountain	3.8±34
Crayey-radiolarian ooze	Klarion-Klipperton area	0.78
Massive sulphide ore	East-Pacific Rise	1.75±0.7(2)
	Atlantic Ocean	
Ferromanganese nodule	Brasil hollow	10.0
	Indian Ocean	
-"-	South-eastern zone	7.6
Ferrous ooze	Discovery Deep	3.4±1.2(2)

Value in brackets - number of measurements (different aliquots of the same sample).

The nodules from the Clarion-Klipperton ore province contain 6.6 ng/g, from the S hollow of the Pacific - 10.5 ng/g, from the Central hollow - 1.3 ng/g, from the Antarctic sector 9.9 ng/g, from the Atlantic and Indian Oceans - 10.0 and 7.6 ng/g respectively. Ferromangenese ore crusts from the Mid Pacific Mountains contain 1.9 to 12 ng/g, from Gayota Ita Mountain -3.8 to 34 ng/g of rhodium. In the rest of the samples the Rh concentrations are lower. Clayey-Radiolarian ooze contains 0.78 ng/g, fossilized whale auditory bone - 3.2 ng/g, massive sulphide ore - 1.7 ng/g, ferrous ooze - 3.4 ng/g of rhodium.
The contents of Rh and basic ore components in the nodules were found not to correlate significantly. There are still no results on the Rh concentration in sea and oceanic sediments. Its average content in the Earth's crust is taken to be 0.4 ng/g [18]. The Rh concentration determined in a clayey radiolarian ooze sample (0.78 ng/g) is twice as high compared to the average value for the Earth's crust. This allows approximately evaluating the coefficient of Rh accumulation in nodules

compared to sediments that comes to 10. The Rh concentration in ocean water samples increases as follows: surface water - 3.3 ng/l (7 determinations), bottom water - 4.3 ng/l pore water - 12 ng/l.
The only analyzed sample of biogenic material is represented by a whale auditory bone. Relatively high content of Rh in this sample (3.3 ng/g), compared to bottom sediments (0.78 ng/g), seems to be explained by its long exposition on the bottom surface and Rh accumulation on hydroxides.

Noble metals in hydrothermal waters of the Pacific. This section contains data on distribution of Au, Pt and Rh in regions of the Pacific characterized by hydrothermal activity: in the caldera of Axial Sea Mount, Juan de Fuca Ridge (45 N, 130°W) and in the Guaymas basin, the Gulf of California (27°N, 111°W). The subjects of investigation were sea and pore waters, water suspension and solid phase of hydrothermal fluid. A unique feature of this work was that, for the first time, analyses were carried out on board a ship with the use of a laser analytical photoionization spectrometer during the 12th cruise of research vessel "Akademik Mstislav Keldysh", August - December 1986.
Figure 22.4 shows a histogram and an averaged profile for dissolved gold concentrations measured in the water samples from the caldera of Axial Sea Mount. Averaging was carried out over both the caldera area (8.5*3.3 km^2) and the time (5.09.8629.09.86). Samples were taken from depths of 1450 to 1580 m. Gold concentration ranges from < 0.4 to 9.0 ng/l, its average value being 1.7 ng/l. The concentration histogram (Figure 22. 4a) shows that in the range less than 0.4-4.5 ng/l there are 95% of values and in the range 5.6-9.0 ng/l - only 5%. Concentrations from < 1 to 5 ng/l are recommended to be considered as background ones [19]. By this criterium 5% of analyzed samples from the caldera region contain anomalous gold concentrations.
In the averaged gold concentration profile (Figure 22.4b) a well-defined maximum at the horizon from 400 m to the bottom can be seen. Analysis of the area distribution of gold concentrations in the caldera shows that anomalous values are characteristic of the S part. The maximum value (9 ng/l) was detected in the water samples at the southernmost station (St. 1486). This station (Table 22.5) is characterized by a decrease of concentrations towards the bottom and a correlation

Table 5. Concentrations of dissolved gold and suspended aluminium in seawater samples from Axial Seamount caldera, Juan de Fuca Ridge.

Distance from the bottom, m	Al_{susp} concentration, μg/l	Au_{diss} concentration, ng/l
10	0.74	1.0
150	0.41	3.4
400	0.13	9.0

of gold with suspended aluminum. Such a correlation is typical of all the samples taken in the south of the caldera. This is not observed for the N part of the caldera, where there are no anomalous gold concentrations either. Since suspended aluminum characterizes terrigenous deliveries of material, a decrease in its content can be explained perhaps by the hydrothermal nature of waters. Then this assumption enables to explain the anomalous gold concentrations in the S by a hydrothermal activity. On the other hand, the anomalous gold concentrations seem to be conditioned by a titanic outburst of unknown character 70 km SW from Axial Seamount (17-23.08.1986), a month before the beginning of work on the caldera area [20].

Figure 22.4: Dissolved gold concentration in the waters from Axial Seamount caldera, Juan de Fuca Ridge: a - histogram for gold concentration; b - averaged profile of dissolved Au concentrations.

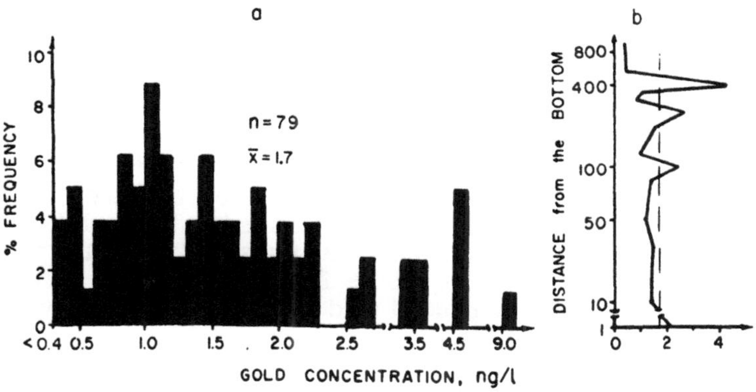

Figure 22.5: Dissolved gold concentration in the waters of Guaymas basin: a - histogram for gold concentrations; b - averaged profile of dissolved Au concentrations.

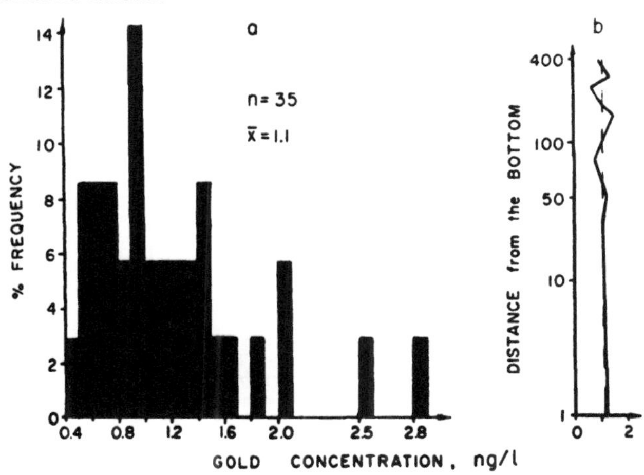

Figure 22.5 shows a histogram and an averaged profile for dissolved gold concentrations measured in the water samples from Guaymas basin. Averaging was done over both the basin area (12.5*3 km^2) and the time (15.10-3.11.86). Samples were taken from depths of 1880 to 2115 m. Gold concentrations range from 0.4 to 2.8 ng/lo the average value being 1.1 ng/l. The concentration histogram (Figure 22.5a) shows that all the concentrations lie within the limits of background content. In the averaged concentration profile (Figure 22.5b) there are no well-defined maxima observed.

The results on dissolved gold concentrations in Guaymas almost agree with the gold distribution in the open part of the Pacific, not far from California that ranges from 0.5 to 2.1 ng/l, the average value being 1.0 ng/l [21]. All these facts enable assuming that the hydrothermal activity of the region under investigation does not affect the concentration of dissolved gold.
Table 22.6 contains averaged gold concentrations in sea water, water suspension and solid hydrothermal fluid from Guaymas. This region is characterized by the fact that suspended gold fraction is three times higher than the dissolved fraction (Table 22.6). In the open part of the Pacific the dissolved fraction is ten times higher than the suspended one [21]. The relative concentration of the gold in suspension of Guaymas waters, 17 ppm (Table 22.6), is much higher than that in the open part of the Pacific, ≤ 0.4 ppm [21]. Such gold concentration in water suspension is unlikely to be caused by the hydrothermal activity of the region because the solid phase of hydrothermal fluid contains just 0.85 ppm of gold (Table 22.6). High gold concentrations in Guaymas water suspensions can be explained by the specific properties of this region. The basin of Guaymas is fed by the runoff of rivers which drain a vast belt of Cretaceous granitoidal batholiths [21] where the gold content is higher than the average one [23]. These facts enable us to assume that the anomalous concentrations of suspended gold in the waters of Guaymas basin are conditioned by terrigenous runoffs.

Figure 22.6: Dissolved platinum concentration in the waters of Guaymas basin: a-histogram for platinum concentrations; b - averaged profile of dissolved Pt concentrations.

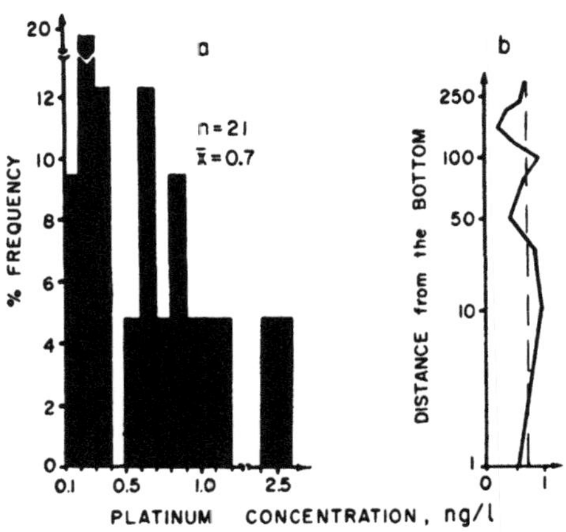

Figure 22.6 presents a histogram and an averaged profile for dissolved platinum concentrations in the waters of the Guaymas basin. sampling was performed at depths from 1880 to 2115 m. The platinum concentrations range from 0.1 to 2.5 ng/l the average being 0.7 ng/l (Figure 22.6a). Little information from literature says that in sea water the concentration is 0.2 ng/l [24]. The averaged profile (Figure 22.6b) does not have well defined maxima relative to the average level. The comparison between dissolved (0.7 ng/l) and suspended (2.4 ng/l) platinum concentrations (Table 22.6) shows that the latter prevails. The relative concentration of platinum in suspension (9 ppm) exceeds considerably its concentration in the solid phase of hydrothermal fluid (0.52 ppm) (Table 22.6). This points to the fact that, as in the case of gold, the supply of platinum into the basin of Guaymas is terrigenous by nature. The platinum concentration in pore water (3.4 ng/l) is nearly 5 times higher than its average concentration in sea water (0.7 ng/l) (Table 22.6).

Table 22.6: Averaged concentrations of gold, platinum and rhodium in sea and pore waters, water suspension and solid phase of hydrothermal fluid (Guaymas basin).

Element	Seawater, ng/l		Seawater	Hydrothermal fluid	Pore
	Dissolved form	Suspended form	suspension	(solid phase)*) ppm	water, ng/l
Au	1.1	3.5	17	0.85	-
Pt	0.7	2.4	9	0.52	3.4
Rh	0.029	0.041	0.015	0.008	1.2

*) Content of solid phase in hydrothermal fluid is about 0.1 g/l.

Figure 22.7 shows histograms for dissolved Rh concentration in the waters of Juan be Fuca Ridge and Guaymas basin. It can be seen that the rhodium concentration lies in the range ≤ 0.003 to 0.15 ng/l for Juan de Fuca and 0.003 to 0.22 ng/l for Guaymas. For Juan de Fuca 54% of concentration values are below the detection limit (3 pg/l). For Guaymas 23% of concentration values are below the detection limit. The average concentration for Juan de Fuca is 0.012 ng/l and for Guaymas 0.028 ng/l. The considerable spread of Rh concentration values in the samples analyzed does not have a reasonable explanation yet. The average concentrations of rhodium in dissolved and suspended formes for waters of Guaymas are nearly equal. The values of Rh concentrations in sea water suspension and solid phase of hydrothermal fluid are close too. But Rh content in pore water (1.2 ng/) is much higher than the second one for sea water (0.028 ng/l). The data obtained show than the rhodium geochemical behaviour differs substantially from the behaviour of gold and platinum in the investigated regions.

Figure 22.7: Histograms for dissolved rhodium concentrations in the waters of Juan de Fuca Ridge (a) and Guaymas basin (b).

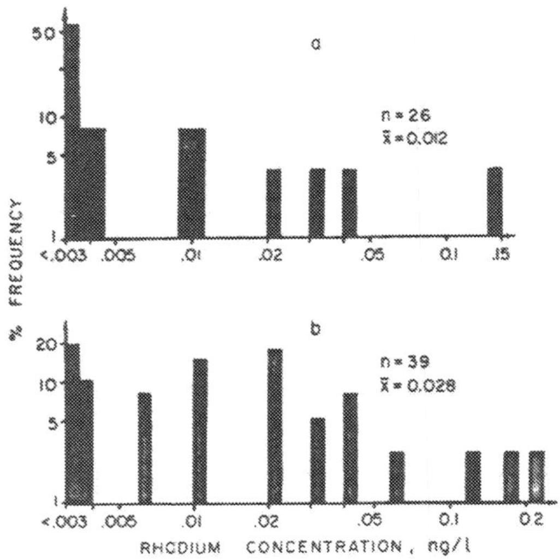

22.5 Conclusion

The measurements performed have demonstrates the universal character of laser photoionization spectroscopy in combination with thermal atomization of a substance in vacuum. The results obtained stimulate now the wide use of the LAPHIS-technique as a new analytical approach both in laboratories and on board a ship, particular for the purposes of ultrasensitive determination of noble metal traces in oceanic objects. The sensitivity levels attained are not ultimate. They can be improved by one or two orders of magnitude by perfecting the laser analytical equipment.

Acknowledgements

The authors gratefully acknowledge the supports of V.S.Letokhov and A.P.Lisitsyn and the contribution of V.N.Radaev.

References

1. Bekov G.I., Letokhov V., Matveev O.I., Mishin V.I. Optics Letters **3**, 159-161 (1978).
2. Hurst G.S., Nayfeh M.H. and Young J.P. Phys.Rev. A. 15, 2283-2292 (1977).
3. Bekov G.I. and Letokhov V.S. In: Laser Analytical Spectrochemistry. Ed. V.S.Letoknov, Adam Hilger, Bristol (1986), pp. 98-151.
4. Mackenzie F.T., Wollast R. and Stoffyn M. Science **199**, 680-682 (1978).
5. Caschetto S. and Wollast R. Geochim. Cosmochim. Acta **43**, 412-428 (1979).

6. Dickinson Burrows W. Critical Reviews in Environmental control. CRC, 167-216, (1977).
7. Monin A.S. and Gordeev V.V. Amazoniya. Moscow, Nauka, p.104 (1988) (in Russian)
8. Boyle E., Collier R., Dangler A.T., Edmund J.M., Ng A.C. and Stallard R.F. Geochim. Cosmochim. Acta 38, 1719-1728 (1974).
9. Stoffyn M., Mackenzie F.T. Marine Chemistry 11, 105-127 (1982).
10. Petrukhin O.M., Malofeeva G.I., Nefedov V.I., Salon O.V. et al. Zh.Anal.Khim. 38, 250-254 (1983).(in Russian)
11. Shvoeva O.P., Kuchava G.P., Myasoyedova G.V., Savvin S.B. et al. Zh.Anal.Khim. 40, 1606-1610 (1985).(in Russian)
12. Haffty J., Riley L.B. and Gross W.D. Bull.Geol.Surv.G.B. 1445, 1-58 (1977).
13. Bekov G.I., Letokhov V.S., Radaev V.N., Baturin G.N., Egorov A.S., Kursky A.N., Narseyev V.A. Nature 312, 748-750(1984).
14. Popov N.I., Fedorov K.N. and Orlov V.M. Seawater (in Russian), Nauka, Moscow (1979).
15. Lee D.S. Nature 305, 47-48 (1983).
16. Bruland K.W. In : Chemical Oceanography 8, 157-225, (Academic, New York, 1969).
17. Baturin G.N., Narseyev V.A., Bekov G.I., Kursky A.N., Pakhomov D.Yu., Radayev V.N. Geokhimiya, No 3, 376-385 (1988).(in Russian)
18. Goldschmidt V.M. Geochemistry. Oxford: Clarendon Press, 730 (1958).
19. McHugh J.B. Journal of Geochemical Exploration, 30, 85-94, (1988).
20. Bekker E.T., Massoth G.Y., Feely R.A., Nature 329, 149-151 (1987).
21. Krendelev F.P., Pogrebnyak Yu.F., Melnikova R.D. Geokhimiya, No 3, 439-445 (1977).(in Russian)
22. Khain V.E. Regionalnaja geotektonika. Severnaia i Yuzhnaia America. Antarktida i Africa Moscow, Nedra (1971), 548p.(in Russian)
23. Vinogradov A.P. Geokhimiya, No 7, 555-571 (1962).(in Russian)
24. Hodge V., Stallard M., Koide M. and Goldberg E.D. Anal. Chemistry 58, 616-620 (1986).

G.I. Bekov
Institute of Spectroscopy
USSR Academy of Sciences
Troitsk
Moscow Region 142092
USSR

A.S. Egorov
Shirsnov Institute of Oceanology
USSR Academy of Sciences
Moscow 117218
USSR

23 HIGH-RESOLUTION MEASUREMENTS OF pH, FLUORESCENCE AND TURBIDITY IN SEA WATER

D.J. Mackey, E.C.V. Butler, P.D. Nichols and H.W. Higgins

Abstract

In coastal waters, vertical profiles of temperature, salinity, *in situ* fluorescence (chlorophyll) and depth were recorded at 2 sec intervals by coupling a Variosens II fluorometer to a submersible data logger. For measurements down to 800 m, a Variosens III was interfaced with a Neil Brown CTD. In addition to *in situ* fluorescence, the Variosens III enabled us to measure profiles of turbidity.

Underway surface measurements of pH, *in situ* fluorescence, temperature and salinity were made by diverting some of the water from a thermosalinograph through a Turner Fluorometer and a pH cell. All four parameters were logged 75 times a minute and the data was stored on a VEX computer. The partial pressure of carbon dioxide (p_{CO2}) in the surface waters was calculated from the pH. Simultaneous measurements of a range of parameters enabled us interpret changes in p_{CO2} on the basis of physical and biological processes.

23.1 Introduction

The past decade has seen impressive advances in the acquisition and processing of oceanographic data. Digital computers can simultaneously collect and store information from many instruments, at speeds that are limited only by the response times of the sensors. This has enabled physical oceanographers to compare the information from diverse instruments at a spatial and temporal resolution that was not previously attainable. CTD data is now routinely reported at 2 debar intervals.

Chemical oceanographers have few techniques available for making *in situ* measurements of chemical and biochemical parameters. This is mainly because most sensors do not have a sufficiently fast response time and sensitivity to measure chemical and biological parameters at environmentally significant concentrations. The development of such sensors is important, for in addition to the obvious advantage of obtaining real-time data, *in situ* techniques can reduce or eliminate problems associated with collection, storage and contamination.

In situ fluorescence (due to chlorophyll) and pH can be measured in oligotrophic oceanic waters with sufficient accuracy and precision to provide useful information on the global carbon cycle. The two parameters are related, since the pH of sea water is determined primarily by the concentration of carbon dioxide species. These species are taken up by phytoplankton during photosynthesis, and are released by respiration and the remineralisation of organic detritus. In surface waters, these processes occur at time scales of days to weeks, while the exchange of carbon dioxide between the atmosphere and the surface mixed-layer of the ocean occurs on a time scale of about one year.
Small-scale changes in the pH (carbon dioxide species) and *in situ* fluorescence of surface waters provide information on recent biological processes, while large-scale changes in pH are important in assessing the ability of the oceans to take up carbon dioxide. These changes must be taken into account if we are to make accurate assessments of the productivity of the oceans and the ability of the oceans to ameliorate the greenhouse effect.

V. I. Ilyichev and V. V. Anikiev (eds.), Oceanic and Anthropogenic Controls of Life in the Pacific Ocean, 315–324.

Continuous vertical profiles have been made of pH (Disteche and Dubuisson, 1960; Ben-Yaakov and Ruth, 1974; Zotov et al., 1984) and chlorophyll fluorescence (Derenbach et al., 1979), while Berman (1972) used a surface-based fluorometer and pump to obtain vertical profiles of fluorescence. Ishimaru et al. (1984) developed a system whereby an oxygen sensor, *in situ* fluorometer, *in situ* transmissometer and *in situ* quanta meter were added to a CTD system. Continuous surface measurements have been made of pH (Zirino et al. 1983; Zirino and Lie Berman, 1985; Zirino et al., 1986; Fuhrmann and Zirino, 1988; Zirino et al. 1988) and fluorescence (Zirino et al., 1983; Takahashi et al., 1986). While these and other measurements have established the variability of pH and fluorescence, they have seldom been logged simultaneously with temperature, salinity (and depth), which limits their value. In addition, there are advantages when vertical and horizontal profiles are determined concurrently.

The pH and fluorescence of surface waters are intimately related to productivity and biomass, although there may be no direct correlation at any given time, since preexisting pools of dissolved and particulate carbon can be uncoupled by dynamic processes. Surface measurements of these properties can provide ground-truth data for the calibration of satellite measurements. However, before satellite data can be used to assess oceanic biomass, the surface biomass data (limited by the penetration depth of the sensors on the satellite) must be extrapolated to total biomass integrated throughout the water column. A comparison of the results obtained from continuous vertical and horizontal profiles of pH and fluorescence (chlorophyll) is of great importance if satellite imagery is to provide quantitative data on biological processes. In this paper, we describe some preliminary data from three systems we have used to obtain *in situ* vertical and horizontal profiles of pH and fluorescence and vertical profiles of turbidity. In all the systems, salinity, temperature, and depth (where appropriate) were logged simultaneously, and all parameters were digitally stored for subsequent processing and analysis.

23.2 Experimental

Vertical Profiles

For near-shore measurements down to 300 m, we have interfaced a Variosens II (IMPULSPHYSIK GmbH) *in situ* fluorometer with a YEO-KAL submersible data logger (SDL). The Variosens was powered by a submersible, rechargeable battery pack. The system records pressure (depth), conductivity (salinity), temperature and chlorophyll fluorescence, and can be readily deployed from a small winch. An inductively coupled interface is used to charge the internal batteries, program the SOL and retrieve the data via an IBM-compatible personal computer. The output of the Variosens II spans three orders of magnitude and is approximately log-linear in fluorescence intensity (Parker and Audrey, 1980). The output is reported as percentage of full-scale response. The complete system has been calibrated as a function of chlorophyll *a* from a range of laboratory cultures. A response of (26 ± 5)% corresponded to 1 μg l^{-1} of chlorophyll *a*.

For open-ocean measurements, we interfaced a Variosens III to a Neil Brown conductivity-temperature-depth (CTD) probe. The Variosens III is similar to the Variosens II but has an additional detector unit that measures scattered radiation (turbidity). No signal was recorded on this channel when the Variosens III was placed in a phytoplankton culture medium containing chlorophyll concentrations of

100 lug I-I. The system can therefore discriminate between living phytoplankton and zooplankton or detritus. The Variosens III is powered by a submersible, rechargeable battery pack and is pressure rated to 800 m. This is sufficient to investigate the distribution of phytoplankton within the euphoric zone and the resuspension of bottom particles on the continental shelf. The output of the Variosens III is log-linear over four decades but is very noisy over the lower decade (0-25%).

Surface pH and Fluorescence

For shipboard measurement of surface pH and chlorophyll fluorescence, we used a system similar to that described by Zirino and co-workers (Zirino et al., 1983; Zirino and Lie Berman, 1985; Zirino et al., 1986; Fuhrmann and Zirino, 1988; Zirino et al., 1988). A T-connection was inserted into the water supply to the thermosalinograph on the RV "Franklin" to provide a constant head of water to a Turner Fluorometer and a pH electrode system. A constant flow (500 ml min^{-1}) was maintained through the pH system by a peristaltic pump, and a pneumatic pulse dampener was inserted to minimize pulsations in the flow rate. The sea water stream was earthed via a platinum wire.

The pH was calibrated by circulating freshly prepared TRIP [(tris)hydroxymethyl)amino methane] buffers of nominal pH 7.3 and 8.8 (Smith and Hood, 1964) through the cell until the pH remained constant. One of the buffers was also circulated through the cell at a temperature at least 10°C different from that of the ambient seawater. The algorithms supplied with the Radiometer pH meter were used to determine the pH at the *in situ* temperature. The temperature probe, supplied by Radiometer, is accurate to 0.2°C, which introduces an uncertainty of 0.006 pH units in the calibration with the TRIP buffers (Smith and Hood, 1964). We estimate that our pH values are accurate to + 0.008 pH units. Measurements of the short-term stability and long-term drift of the pH system indicate that the precision of the system is about 0.001 pH units.

The surface temperature, salinity, pH and fluorescence were logged by a PUP-11 computer at about one second intervals and periodically copied to the VEX computer. One-minute averages were transferred to a desktop computer for processing. During cruise FR10/87, the signal from the Turner Fluorometer (TUG) was calibrated against chlorophyll a determined by the spectrophotometric method of Jeffrey and Humphrey (1975). The system was calibrated regularly, on samples collected at night and during the day, over a two week period, and on water masses spanning a distance of more than 2000 km. A least squares regression of the data showed that they were related by the equation

$$\text{Fluorescence (TU)} = 278\ (\text{chl } a) - 7.6 \quad (n=14, r^2=0.969)$$

where chl a was the concentration of chlorophyll a in $\mu g\ l^{-1}$.

Calculations

The pH scale used for calibrating the electrode system must be the same as that used for determining all equilibrium constants relevant to the calculations of carbon dioxide speciation. In this paper, all calculations of the carbon dioxide speciation in sea water are based on the Hansson sea water pH scale (Smith and Hood, 1964;

Hansson, 1973a) as modified by Almgren et al. (1975). The acidity constants for carbonic acid and boric acid were obtained from Hansson (1973b, 1973c). The appropriate equations are summarized in Skirrow (1975). Full details are given in Mickey et al. (1989).

23.3 Results and discussion

Vertical profiles

The SOL/Variosens II system has been deployed at weekly internals in Storm Bay, Tasmania since the middle of 1987. During late (astral) winter and early spring, the water column is often stratified due to the runoff of cold, fresh water from the Huon and Derwent Rivers and the surface waters are enriched in phytoplankton. Figure 23.1 shows vertical profiles of temperature, salinity and fluorescence that are typical under these conditions. The sample was collected from the middle of Storm Bay (43°09'S, 147°31'E) on July 6 1988. The fluorescence profile shows that the distribution of phytoplankton is very inhomogeneous which would not be apparent from discrete sampling, even at depth intervals of 5 m. It is clear that, in cases such as this, continuous measurements give more realistic values of phytoplankton biomass integrated throughout the water column.

Figure 23.1: Vertical profiles of (a) fluorescence, (b) temperature and (c) salinity in Storm Bay (43°09'S, 147°31'E), Tasmania, on 6 July 1988.

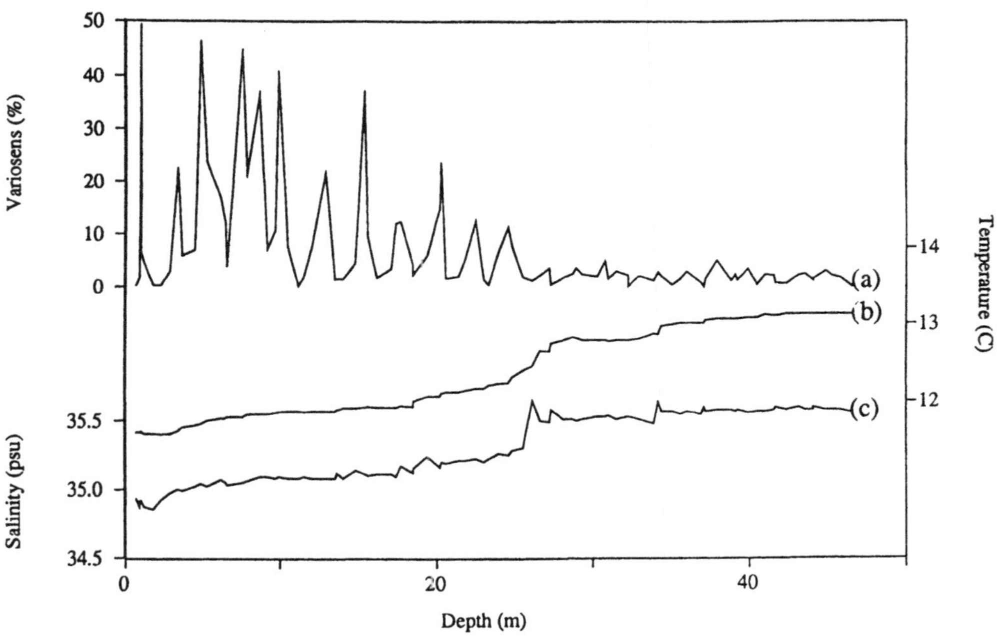

Typical profiles of data obtained from the CTD/Variosens III system deployed from the REV "Franklin" are shown in Figure 23.2.

Figure 23.2: Vertical profiles of turbidity (top) and fluorescence (bottom) at Stn. 63 on FR10/87. The Station was located off the north west coast of Australia at 15°21'S, 122°43'E.

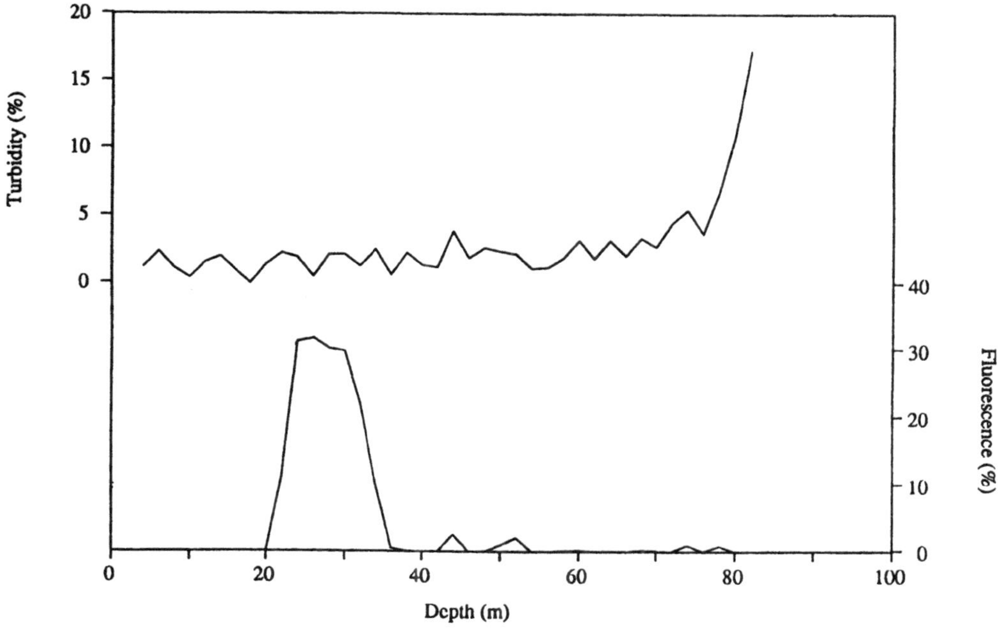

The data have been processed to provide values at two-metre intervals. The station was off the entrance to King Sound on the NW coast of Australia (15°21'S, 122°42'E) and the water depth was 87 m. The area is subject to strong tidal currents; at the time of sampling, the tidal range was approximately 8 m. The small but significant signal from the turbidity sensor throughout the water column increases rapidly about 10 m from the bottom. The chlorophyll maximum occurs at 25 m, which is just below discontinuities in the vertical profiles of temperature and salinity (Figure 23.3). Since there is no correlation between the fluorescence and turbidity signals, it is apparent that the turbidity in the water column is due to the resuspension of bottom sediments.

Figure 23.3: Vertical profiles of temperature (top) and salinity (bottom) at Stn. 63 on FR10/87. The Station was located off the north west coast of Australia at 15°21'S, 122°42'E.

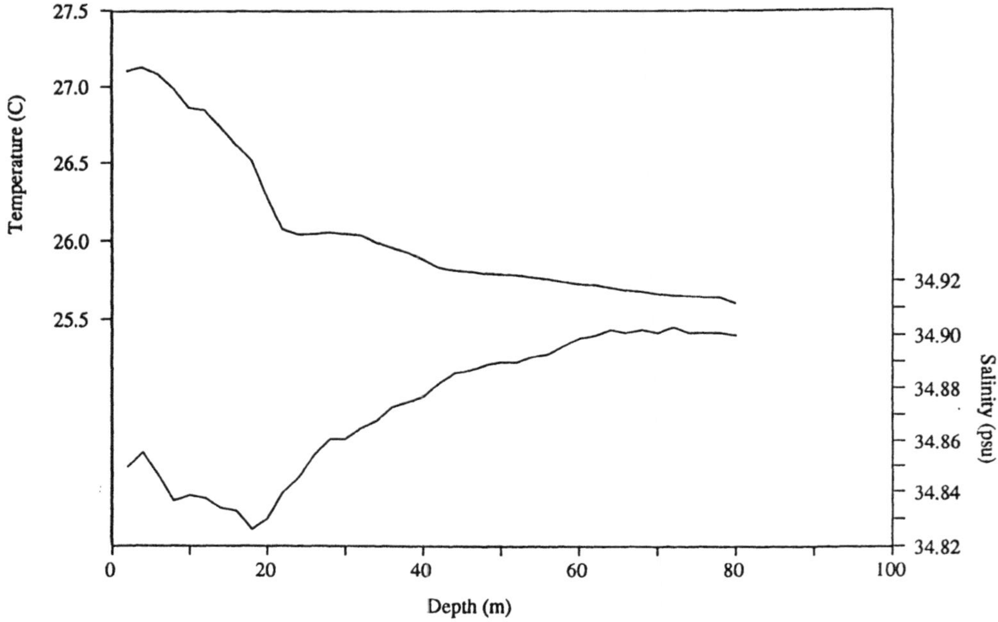

Surface pH and fluorescence

In regions where photosynthesis is the dominant mechanism controlling the concentration of carbon dioxide species in sea water, a strong positive correlation should exist between pH and fluorescence. Such a correlation was observed for extended periods during cruise FR02/88 in the Gulf of Carpentaria. If respiration and/or remineralisation are more important than photosynthesis, pH and fluorescence should be inversely correlated. If the pH is controlled by the mixing of water masses of different temperatures and carbon dioxide concentrations, then there will be no correlation between pH and fluorescence. Examples of all three patterns were observed during a 24 h period in the N of the Gulf of Carpentaria at about 12° 30 S, 141° 30 E (Figure 23.4).

For the interpretation of changes in surface pH at temporal scales on the order of days and spatial scales on the order of hundreds of kilometres, it is essential that simultaneous measurements are made of fluorescence, salinity and temperature. These pH changes are superimposed on a background pH that is determined by the long-term integration of the effects of mixing, biological processes and exchange with the atmosphere. In productive waters, where uptake of carbon dioxide by photosynthesis is the dominant process, the surface waters become under saturated with carbon dioxide.

Figure 23.4: Variation of surface pH (top) and fluorescence (bottom) as a function of time (UTC) on 21 February 1988 during FR02/88. The data were collected in the Gulf of Carpentaria at about 12°30'S, 141°30'E.

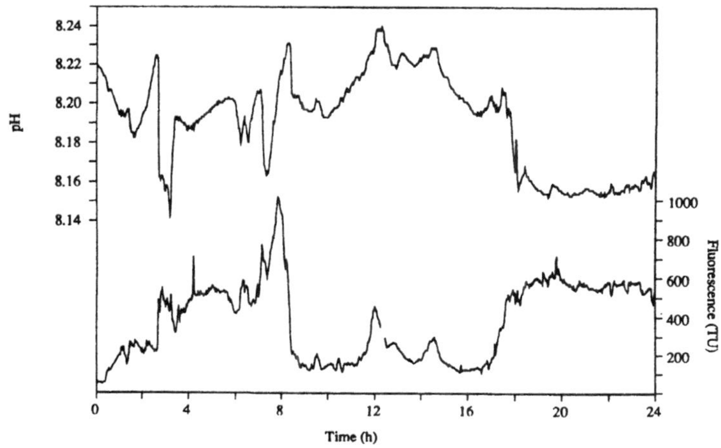

Assuming a value of 0.1197 for the specific alkalinity (Broecker et al., 1982), and using the equations of Hansson as given in Skirrow (1975) it is possible to calculate the partial pressure of carbon dioxide (pCO_2) as a function of temperature, salinity and pH. The accuracy in the determination of pCO_2 is about 10 ppm, while the precision is better by about an order of magnitude (Mickey et al., 1989). In Figure 23.5 we have plotted pCO_2 as a function of time for the same 24 h period shown in Figure 23.3. The waters are under saturated with carbon dioxide by 50 - 100 ppm.

Figure 23.5: Calculated values of pCO_2 found on 21 February 1988 (UTC) during FR02/88. A value of 0.1197 was assumed for the specific alkalinity. The data were collected in the Gulf of Carpentaria at about 12°30'S, 141°30'E.

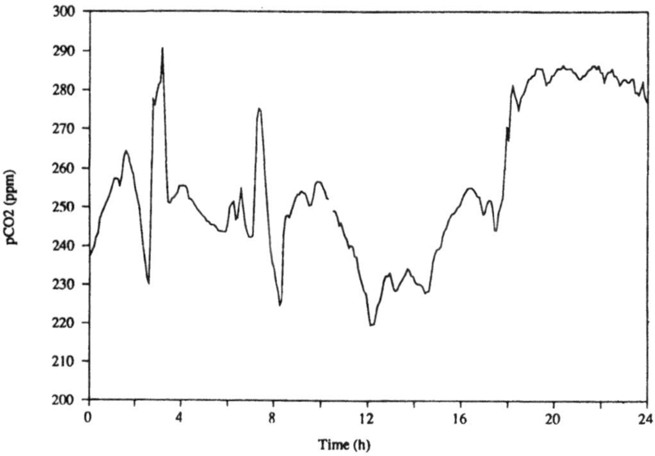

23.4 Conclusions

The exchange of carbon dioxide between the atmosphere and the oceans has a powerful influence on the atmospheric concentration of carbon dioxide and the resultant greenhouse effect. Direct determinations of pCO_2 can be made with an accuracy of about 1%, but the determinations are tedious, time-consuming, and difficult to automate for continuous measurements. pCO_2 can be calculated from pH if the alkalinity is known, but although the precision of the calculated value is about 1%, the accuracy is worse by an order of magnitude. The combination of continuous surface measurements of pH and a few direct measurements of pCO_2 should enable us to obtain continuous, accurate values of pCO_2. Such measurements would be invaluable in studying the spatial and temporal variations in the pCO_2 values of the oceans.

Algorithms are continually being developed for estimating primary productivity and biomass from satellite imagery. These algorithms are of limited value in coastal waters subject to interference from suspended matter and dissolved organic matter (Gelbstoff). However, these productive areas are readily accessible by ship and continuous surface measurements of fluorescence (chlorophyll) combined with frequent calibration can provide good estimates of the biomass of phytoplankton in these surface waters. Such measurements will not only complement satellite imagery, but may also lead to the development of suitable algorithms for coastal waters.

It is important that surface data be combined with continuous vertical profiles of fluorescence (chlorophyll), since the distribution of phytoplankton throughout the water column is in homogeneous. At a given location, there may be no correlation between surface fluorescence and phytoplankton biomass integrated throughout the water column. However, the combination of continuous surface measurements and regular vertical profiles of fluorescence will enable us to determine the temporal and spatial scales over which such a correlation is valid.

Continuous measurements of pH, fluorescence, temperature and salinity enable us to differentiate between the effects of gas exchange, respiration, photosynthesis and mixing processes. When these data are combined with profiles of other biologically significant parameters (dissolved oxygen, turbidity, light transmission and intensity), we will be able to study the interactions of physical, chemical, and biological processes at spatial and temporal scales that would not have been possible a decade ago. The development of new sensors for continuous, *in situ* measurements should have high priority in oceanographic research.

Acknowledgements

We thank the master and crew of the RV "Franklin" for their help in collecting samples, and L. Clementson, J. Atack and F.B. Griffith for collecting samples from Storm Bay. D. Crooks, R. Beattie, E. Madsen, D. Edwards and A. Poole provided valuable assistance in computing and electronics support. M. Rayner, J. Atack and P. Deprez helped nurse the surface pH system through its formative stages.

References

Almgren, T., Dyrssen, D. and Strandberg, M.: Determination of pH on the moles per kg sea water scale (Mw). Deep-Sea Res. 22, pp.635-646 (1975)
Ben-Yaakov, S. and Ruth, E.: An improved *in situ* pH sensor for oceanographic and limnological applications. Limnol. Oceanogr. 19, pp.144-151(1974)
Berman, T.: Profiles of chlorophyll concentrations by in vivo fluorescence: some limnological applications. Limnol. Oceanogr. 17, pp.616-618 (1972)
Broecker, W.S., Spencer, D.W. and Craig, H.: GEOSECS Pacific Expedition. U.S. Vow 3, p7. Government Printing Office, Washington DC (1982)
Derenbach, J.B., Astheimer, H., Hansen, H.P. and Leach, H.: Vertical micro scale distribution of phytoplankton in relation to the thermocline. Mar. Ecol. Pros. Ser. 1,pp.187-193 (1979)
Disteche, A. and Dubuisson, M.: Measures directs de pH aux grands profondeurs sous-marines. Bull. Inst. Oceanogr. Monaco 1174, pp.3-8 (1960)
Fuhrmann, R. and Zirino, A.: High-resolution determination of the pH of sea water with a flow-through system. Deep-Sea Reg. 35, pp.197-208 (1988)
Hansson, I.: A new set of pH-scales and standard buffers for sea water. Deep-Sea Reg. 20, pp.479-491 (1973a)
Hansson, I.: Determination of the acidity constant of boric acid in synthetic sea water media. Act a Chum. Stand. 27, pp.924-930 (1973b)
Hansson, I.: A new set of acidity constants for carbonic acid and boric acid in seawater. Deep-Sea Reg. 20, pp.461-478 (1973c)
Ishimaru, T., Otobe, H., Saint, T., Hasumoto, H. and Nakai, T.: OCTOPUS, an octo parameter underwater sensor, for use in biological oceanography studies. J. Oceanogr. Sock Japan 40, pp.207-212 (1984)
Jeffrey, S.W. and Humphrey, G.F.: New spectrophotometric equations for determining chlorophylls a, b, c1 and c2 in higher plants, algae and natural phytoplankton. Biochem. Physiol. Pflanzen 167,pp. 191-194 (1975)
Mickey, D.J., Butler, E.C.V., Nichols, P.D. and Higgins, H.W.: Continuous shipboard and in situ measurements of pH and fluorescence in sea water. Mar. chum.,(1989)
Parker, R.R. and Audrey, D.J.: A proposed reference standard for in vivo chlorophyll a fluorometry. CSIRO, Australia, Division of Fisheries and Oceanography, Report 125 (1980)
Skirrow, G.: The dissolved gases - carbon dioxide. In: Riley, J.P. and Skirrow, G. (Ed.), Chemical Oceanography, Academic Press, London, pp. 1-192. 1975
Smith, W.H., Jr. and Hood, D.W.: pH measurement in the ocean: a sea water secondary buffer system. In: Miyake, Y. and Koyama, T. (Ed.), Recent Researches in the Fields of Hydrosphere, Atmosphere and Nuclear Geochemistry, Maruzen Co. Ltd., Tokyo, pp. 185-202. 1964
Takahashi, M., Ishizaka, J., Ishimaru, T., Atkinson, L.P., Lee, T.N., Yamaguchi, Y., Fujita, Y. and Ichimura, S.: Temporal change in nutrient concentrations and phytoplankton biomass in short time scale local upwelling around the Izu Peninsula, Japan. J. Plankton Reg. 8, pp.1039-1049 (1986)
Zirino, A., Clavell, C., Seligman, P.F. and Barber, R.T.: Copper and pH in the surface waters of the eastern tropical Pacific Ocean and the Peruvian upwelling system. Mar. chum. 12, pp.25-42 (1983)
Zirino, A. and Lie Berman, S.H.: pH-temperature relationships in the Gulf of California. In: Zirino, A. (en.), Mapping Strategies in Chemical Oceanography, pp. 393-408. American Chemical Society, Washington 1985
Zirino, A., Fuhrmann, R.A., Oksanen-Gooden, D., Lie Berman, S.H., Clavell, C., Seligman, P.F., Mathewson, J.H., Jones, W.D., Kogelschatz, J. and Barber, R.T.: pH-

temperature-nutrient relationships in the eastern tropical Pacific Ocean. ski. Tot. Environ. 58,pp.117-137 (1986)
Zirino, A., Fiddler, P.C. and Kerr, R.S.: Surface pH, satellite imagery and vertical models in the tropical ocean. ski. Total Environ., in press (1988)
Zotov, A.V., Kryukov, P.A., Malov, V.S ., Obzhirov, A.I. and Prikhod'ko, V.A.: Measurement of sea water pH under natural conditions by potentiometric sounding. Doklady, Earth Sci. Sect. 277, pp.214-216 (1984).(in Russian)

D.J. Mackey and E.C.V. Butler
CSIRO Division of Oceanography
Marine Laboratories
GPO Box 1538
Hobart 7001
Tasmania
Australia

24 ESTIMATION OF PLANKTON ROLE IN MIGRATION OF TRACE ELEMENTS IN MARINE ESTUARIES

V.V. Anikiev, E.N. Shumilin, E.G. Starodubtsev, A.P. Kassatkina, A.A. Lobanov and N.A. Gorbachev

24.1 Introduction

Anthropogenic pollution of marine environments is dangerous for people due to using of contaminated hydrobionts for food. Increasing pollution of the World Ocean is especially manifested in the regions with the greatest protein supplies, i.e. shelf zones and estuaries. People are the final stage of the trophic chain which starts in the sea with plankton (Morozov, 1983). Therefore, the study of trace elements migration in plankton trophic cycle of estuary zones is an urgent problem which should be solved as soon as possible.

This study was carried out to estimate plankton impact on migration of some elements, in two areas of shelf zones, the Amursky Bay (in the estuary of the Razdolnaya River) in the Sea of Japan, and a shelf area near the estuary of the Mekong River, in the South China Sea.

24.2 Materials and Methods

Plankton was sampled with Judy's net (a small model) by a vertical method from the surface layer (10 to 0 m) in November December, 1981, in the Amursky Bay, and in August,1982, in the South China Sea (Figure 24.1). Two equal aliquots of plankton samples were used to study species composition of plankton by optical microscopy and to determine trace element contents by atomic absorption and neutron-activation methods (Guidance ..., 1980; Methods ...,1978). Additional studies, made in October, 1987, in the marine estuary of the Razdolnaya River included total plankton analysis as well as trace element concentration, determination in mass species of zooplankton, belonging to different trophic levels:

1. Fine filtrators, nannophagans (nauplial stages of all calanoids, mature stages of small calanoids, larval stages of mollusks (gastropods);
2. Coarse filtrators - euryphagans (euphasiids);
3. Predators (baby fish and chaetognaths).

24.3 Results and Discussion

Results of the study are given in Tables (24.1-24.4). Data for the Amursky Bay (Table 24.2) show trace elements concentration in plankton (with respect to dissolved fraction):

$$Fe > Zn > Cu > Ni > Cd \approx Mn$$

The corresponding sequence for the estuary zone of the Mekong River is:

$$Cr > Fe > Co > Zn \approx Sb$$

V. I. Ilyichev and V. V. Anikiev (eds.), Oceanic and Anthropogenic Controls of Life in the Pacific Ocean, 325–331

Figure 24.1: Scheme of station for plankton sampling:
a) in the Amursky Bay, Sea of Japan, near the estuary of the Razdolnaya River.
b) shelf of the South China Sea.

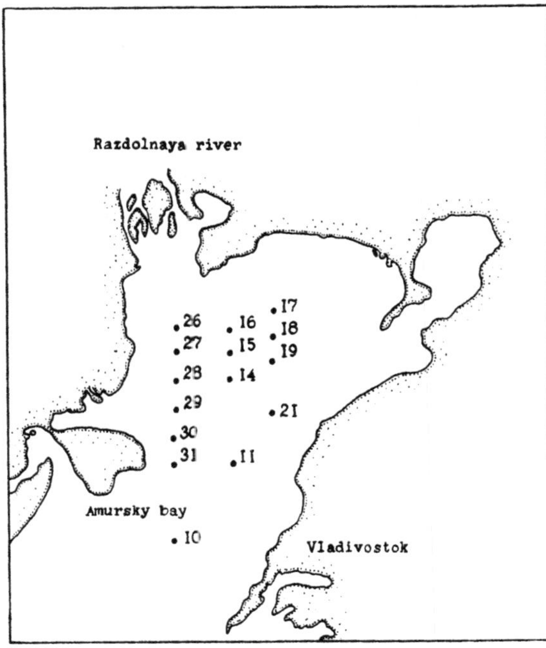

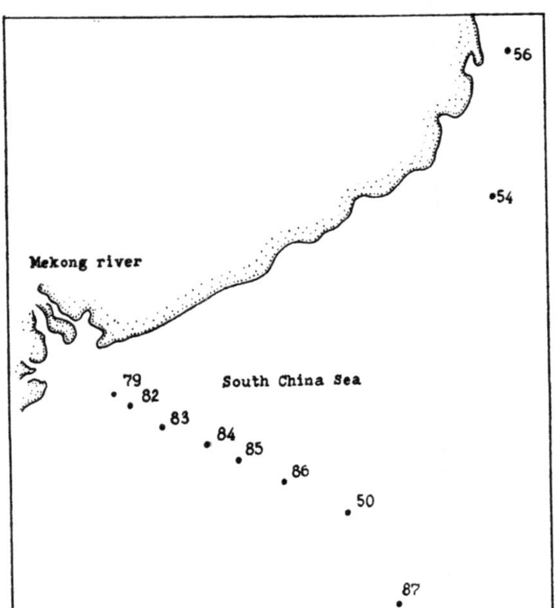

Table 24.1: The number of plankton organisms in the marine estuary of Razdolnaya River - Amursky Bay (Nov. - Dec., 1981).

Station No. Organisms	26	27	28	29	30	31	10
Oithona	65	50	50	175	480	460	100
Oncaea	50	160	-	25	30	25	-
Pseudocalanus	12	33	25	16	3	2	2
Sagitta	100	490	266	325	25	25	25
Bivalvia	716	100	-	75	30	50	25
Doliabaria	33	133	66	50	50	50	-
Tantanus	50	75	-	16	3	5	2
Polychaeta	33	-	-	-	-	-	-
Nauplius Copopeda	50	183	12	16	20	20	25
Acartia	165	360	75	50	50	35	75
Euterpina	12.5	25	12	16	12	14	25
Noctiluca	900	26000	12000	14000	10500	9700	11500
Calanus	-	33	-	16	33	66	-
Phytoplankton	6	22	8	2	11	9	2

Note: Zooplankton - specimens/m^3, phytoplankton - 10^6 cells/l

Table 24.2: Concentration of trace elements in plankton of the Amursky Bay in the profile river-sea (g/g dry substance).

Station No.	Fe	Mn	Zn	Cu	Ni	Pb	Cd
26	940	1.7	66	61	7.5	24.5	1.25
27	2000	12.9	26	12.1	3.5	4.6	0.34
29	290	3.8	40	15.2	9.5	-	2.00
11	480	6.5	34	22.0	8.5	9.0	1.76
14	290	7.0	21	4.9	7.0	10.4	1.23
15	360	8.6	26	5.8	6.3	7.3	1.62
16	320	3.9	23	5.8	12.2	-	1.90
17	1800	10.1	27	5.1	5.3	3.5	0.74
18	360	30.0	54	42.0	2.6	44.0	0.51
19	240	2.9	33	4.2	9.1	21.0	1.51
21	260	17.0	42	16.0	4.9	-	0.91
Average (n=11)	670	9.5	37	18	7.0	16.0	1.3

Results of estimation of trace elements frequency in plankton during a year in the marine estuaries was based on the calculations using experimentally obtained values of the primary production, 0.6 mg C/l per day or 200 gC/l per year (in the Amursky Bay), and 55 mg C/l per year (near estuary zone of the South China Sea), by means of trace element contents in total plankton and water of these areas taken into account (Table 24.3). As seen from Table 24.2, trace elements can be divided into two groups: 1. Fe,Cu,Zn, a complete trap of dissolved forms during 6-10 months; 2. Mn, Ni, Cd, during 2-3 years (for the Amursky Bay). Of greater importance is biogenic factor (3-10 times) when Fe and Mn are transformed in the estuary of the Mekong River which may result from a higher productivity of a greater number of plankton reproduction cycles per year; some species reproduce themselves 3-4 times per year due to unaltered water conditions, which is impossible for plankton in N areas (Alvarino, 1967). Specific features of elements accumulation in plankton are species heterogeneity and, which is more important, a different number of species in different estuary zones. This follows from comparison of Table 24.1 and 24.2. Evidently, mass development of phytoplankton in the Amursky Bay facilitates selective concentration of Fe and Mn via food chain with phyto- and detritophages involved. This conclusion is based on symbathic changes of Fe and Mn concentration

Table 24.3: Trace element contents (g/g) in plankton specimens from South China Sea (the 1st cruise of R/V "Akademik Nesmeayanov", August, 1982).

Element Samples	Fe $x10^2$	Zn $x10^2$	Co	Cr	Ce	La	Se	Sc $x10^{-2}$
%°50	14.2	0.8	0.34	6.1	0.5	3.7	2.9	1.0
54	7.9	0.7	0.46	8.6	0.8	8.9	1.9	26.0
56	19.3	5.1	0.66	15.3	2.9	2.0	2.6	66.0
79	50.0	12.4	1.56	108	10.6	11.8	3.9	170
82-630	23.0	5.1	1.50	28.8	13.3	10.8	3.8	137
83	13.2	18.2	1.65	81.3	5.3	8.6	3.5	157
84	27.7	4.6	0.98	12.1	0.5	4.6	5.8	1.0
85	39.9	5.1	1.59	33.8	6.4	59.2	3.9	150
86	52.7	18.0	2.37	62.9	6.8	6.3	2.8	152
87	15.6	9.4	1.08	48.2	1.7	2.7	4.1	4.3
Limit of determination	1.0	0.06	0.1	1.5	0.5	2.0	0.6	1.0

in sexton as compared to a small number of zooplankton in the total biomass. Similar correlation is observed between Zn, Cu and Pb trace element contents and Bivalvia larvae, whereas, generally, distribution pattern of Cd and Ni in plankton in the profile from the estuary to its marine part has correlations with homology Oithonia. Results of analysis of trace elements concentration in certain species of zooplankton are presented in Table 24.4. Mass planton belonging to fine filtrators, nannophagans, a mature stage of Sincalanus tenellus (Kikuchi,1928) was found to have the highest trace elements concentration. It is interesting that relative content of some metals (Mn, Cd) is bigger even in larval stages of calanoids than in all other plankton, including the top of the trophic plankton chain, baby fish and chaetognaths. Hoarse filtrators, euphausiids, which are in part predators, have a greater content of Fe and Zn than calanoid larval stages, with concentration of other metals (Mn, Cu, Cd, Ni, Pb) being somewhat lower. This can indicate higher capacities of euphasiids to release the above metals. This is possibly due to the fact that they have a more developed excretory system and relatively little chitin. Euphasiids are larger in size and their total chitin covers occupy less surface than those of copepods nauplius of the same weight. Evidently, it is a more developed excretory system and lack of chitin in fish larvae that account for the lowest contents of the above heavy metals in baby fish, although baby fish is at the top of plankton tropic chain consuming food with a high concentration of heavy metals (HM).

Table 24.4: Averaged coefficients of distribution and frequency of trace element dissolved fraction in the Amursky Bay and marine estuary of the Mekong River, South China Sea.

Element Parameter	Fe	Mn	Zn	Cu	Ni	Cd
Distribution coefficient 10^{-3}	6.3/185	1.6	6.6/80	5.6	2.1	1.7
Metal turnover frequency per year	1.4/10	0.34	1.5/4.4	1.23	0.46	0.38

Numerator - data for marine estuary of the Razdolnaya River, the Amursky Bay; denominator - data for marine estuary of the Mekong River, South China Sea.

Table 24.5: HM content in plankton and seston of the Amursky Bay (μg/g dry weight, October, 1987).

Nos.	HM	Fe	Mn	Cu	Zn	Cd	Ni	Pb	m,mg
	1	4400	222	111	667	111	222	444	0.09
1	2	1390	139	27.8	222	41.7	55.6	83.4	0.36
	3	80	40	20	180	40	40	80	0.5
11	4	840	37.4	37.4	178	14	28	46.7	1.07
	5	417	16.7	10.4	54.2	6.3	20.8	20.8	2.4
111	6	1080	81	40.5	216	40.5	81	135	0.37

1 - fine filtrators - nannophagans (1, mature small copopods; 2,- nauplical copedpod stages; 3, larval gastropod mollusks) 11 - coarse filtrators - euryphagans (4, euphausiids) 11 - coarse filtrators - euryphagans (4, euphausiids) 111 - predators (5, baby fish: 6, chaetognaths)

However, chitin presence does not always indicate a high concentration of heavy metals. Thus, for example mollusks' larvae were found to accumulate Cd and Ni, and, possibly, Mn, Cu, Zn, Pb, in greater amounts than euphasiids. The comparison here is made between nature euphasiids, whose life period ranges from several months to several years, and mollusks' larvae, which live in plankton for several weeks at most. Heavy metals specific content is bigger in another representative of the third trophic level (Aidanosagitta macilenta, Kassatkina,1971) than in a typical representative of fine filtrators, nannophagans, mature copepods, syncalanus. This cannot be attributed to different secretory systems, since both do not have any, in spite of different evolutions. This suggests that generally heavy metals accumulate in the chitin skeleton of crustaceans. Chitin-like substance (keratin) is contained in chaetognaths in small amounts, only in spines and teeth, surrounding the mouth. Ni correlation was found when different trophic groups were compared for heavy metals content. A direct correlation can be seen with the type of planters' tissues, chitin in the first place. Since chitin is not digested by predators, skeletons of small crayfish transport large amounts of heavy metals from the water mass and bury them in the bottom sediments.

The data obtained indicate that there is a great need to monitor element contents in plankton, control species composition of the plankton community, and study trace metal concentration in certain species and tissues of zooplankton, so that their role in biogeochemical element circulation in the marine estuaries of ecosystems could be understood.

References

Alvarino, A. In: Chaetognaths, Naga Report, part 2, pp.1-195, 1967.

Anikiev, V.V.: Short-term geochemical processes and ocean pollution. p.192, Moscow, Nauka, USSR, 1987.

Brügman L.: The influence of coastal zone processes on the balance for trace metals in the Baltic Sea. Rapp.p.-v. Reun.Cons.int.Explor.Mer., 186, pp. 329-342, 1986.

Guidance for methods of biological analysis of sea water and bottom deposits, p.191, Moscow, Gidrometeoizdat, USSR, 1980.
(in Russian)

Methods of hydrochemical studies of the ocean. Ivanenko V. N. and Bordovsky D.K, (Ed.), p. 260 Moscow, Nauka, USSR, 1978.
(in Ruissian)

Morozov, N. P. In: Biogeochemistry of the ocean. Ilonin A. S., Lisitsin A.P. (eds.), p.345, Moscow, Nauka, USSR, 1983.
(in Russian)

Trefry, F.N.; Presley, B. J. In: Marine pollutant transport. Wisdom H.L.; Duce R.A. (Eds.), p.350, N.Y.-L., 1975.

V.V.Anikiev and E.N.Shumilin
Pacific Oceanological Institute
USSR Academy of Sciences
Far East Branch
Vladivostok
USSR

25 NUMERICAL MODELLING OF NONCONSERVATIVE POLLUTANT TRANSPORT IN THE COASTAL ZONE

O.V. Zaitsev, V.V. Anikiev and A.V.Voitko

Abstract

At present two ways of simulating the processes of diffusion in marine environment exist, i.e. the solution of semi-empirical equation of pollutant advection-diffusion on Eulerian grid and/or the use of the random-walk method with Lagrangian statistics. The present work is devoted to comparison of the results of these alternative methods exemplified by a simulation of nonconservative pollutant distribution in a bay of the Sea of Japan. Current velocity field was specified on the basis of hydrodynamical modelling of wind-induced currents. Evaluation of turbulent processes' intensity was based on experimental measurements of currents. Calculations took into account the nonconservative feature of pollutant. It is shown that modelling in Lagrangian terms is preferable for short-term forecasting as it better describes the initial stages of the evolution of pollutant field and concentration fronts.

25.1 Introduction

At present, the coastal band of the ocean has been considerably contaminated. Heavy metals, detergents, herbicides, oil products and other pollutants get into the sea water mainly with river runoff and sewage disposal. As natural agents affect but slowly on their decomposition, the pollutants can be transported for long distances by currents. The development of complex measures to combat the negative aftermaths of contamination and establishment of pollution monitoring must be based on the forecast of space-and-time variations of pollutant fields in coastal zones. This problem can be solved with the help of mathematical modelling of pollutant transport. Turbulent diffusion of pollutants is one of the main agents controlling its distribution, so the main attention is to be devoted to problems of description of diffusion processes.

The advection-diffusion equation (ADE) is one of the most commonly used for simulation of pollutant transport. In natural water basins with a complex velocity field this equation is solved numerically on Eulerian computation grid using-the finite-difference method. A detailed example of numerical ADE solution can be found, e.g. in Holly and Usseglio-Polatera (1984). At the same time, solving the ADE in Eulerian terms causes some difficulties. In addition to the fact that numerical solution of the ADE is always connected with complexities of determining the tensor of turbulent diffusion ratios, Usseglio-Polatera and Lunge (1985) emphasize the complexity of description of the first stages of pollutant field formation when the size of the area polluted is less than that of the computational grid. The ADE is inadequate also in cases of large pollutant concentration gradients. Thus the solution of the ADE (either by finite-difference method or by finite-element one) is inadequate for description of physical processes with space-and-time scales minor to the computational grid sizes and time steps. Also worth mentioning is the need of taking into account restrictions, imposed on the choice of computational grid and algorithms of calculation to prevent computational instability of solution.

Calculation of pollutant transport in Lagrangian variables is devoid of the above drawbacks. In this case modelling diffusion processes is provided by the random-walk method (RWM). It is based on probability-statistical concept of the motion of elementary particles in the environmental velocity field. The solution of diffusion

V. I. Ilyichev and V. V. Anikiev (eds.), Oceanic and Anthropogenic Controls of Life in the Pacific Ocean, 333–345.

task is reduced to finding the function describing the probability of pollutant transport from one point of space to another during the term of forecast. An attractive side of the RWM is the opportunity to directly use statistical characteristics of turbulent current component rather than turbulent diffusion coefficients, and also to obtain solutions with any space-and-time averaging periods, corresponding to physical parameters of the environment. The foundations of the method and detailed examples of pollutant transfer calculations can be found in for example the works of Galkin (1975) and Bullin et al. (1974). On the other hand, the use of the RWM for practical and engineering purposes requires a large computational work and hence powerful computers. Actually, as technology progresses speed up, this shortcoming must vanish in a short time.
The present work describes the results of modelling of the propagation of nonconservative pollutant in one of the bays of the Sea of Japan using both methods. The area modelled is shallow, practically nonstratified, so the task may be considered as two-dimensional. Preliminary flow data analysis has shown the wind-induced currents in the area to be more significant than tidal ones, so for the current field velocity we assigned the field of wind-induced currents, obtained as a result of numerical solution of the equation of total stream for the shallow sea (Felzenbaum,1960).

$$\frac{\partial}{\partial x}\left(\frac{A}{H^3}\frac{\partial\psi}{\partial x}\right)+\frac{\partial}{\partial y}\left(\frac{A}{H^3}\frac{\partial\psi}{\partial y}\right)=\frac{1}{2}\operatorname{rot}\frac{T}{H} \tag{1}$$

where x,y are horizontal Cartesian coordinates;

Ψ, the total stream function;
H, the water depth;
A, the turbulent viscosity coefficient;
T, the vector of tangential wind stress at water surface.

For boundary conditions,
1. At the liquid boundary of the region to be calculated, G_1

$$\frac{\partial\psi}{\partial n}\Big|_{G_1} = Q/H$$

2. At the solid boundary, G_s

$$\psi(x,y)\,|_{G_s} = 0$$

where n is normal to the liquid boundary.
Q, the water exchange through the liquid boundary.
Though simple, this model yields satisfactory results for practical purposes.

Also, the simulation of turbulent diffusion in Lagrangian coordinates requires data on statistical characteristics of a turbulent component of current velocity in the water area studied. Up to this step we used experimental results of measurements of current velocity in the bay using autonomous current meters. From 1977 through 1981 in the area of the bay about 80 buoy stations were set equipped with current meters at different levels. After statistical processing of the time series obtained, including solution of the wind and tide components, we obtained spatial statistical characteristics of a turbulent component of current. Figure 25.1 illustrates the distribution of turbulent current component field variance. The distribution is built for characteristic invariant with respect to the turn of coordinate axis, i.e. the sum of variances of velocity components $\sigma^2 = \sigma_x^2 + \sigma_y^2$, characterizing turbulence intensity. In the N part of the bay, turbulent energy is at maximum, hence turbulent mixing is more intensive. On the S in the narrower part of the bay, turbulence intensity decreases owing to increasing bay's depth. Farther S turbulent energy increases and is distributed unevenly. Thus, the intensity of turbulent current velocity in the bay is non-uniform and anisotropic.

Figure 25.1: Variances of turbulent compnent velocity dispersion $\delta^2 = \delta_x^2 + \delta_y^2$ $[10^4\ m^2/s^2]$.

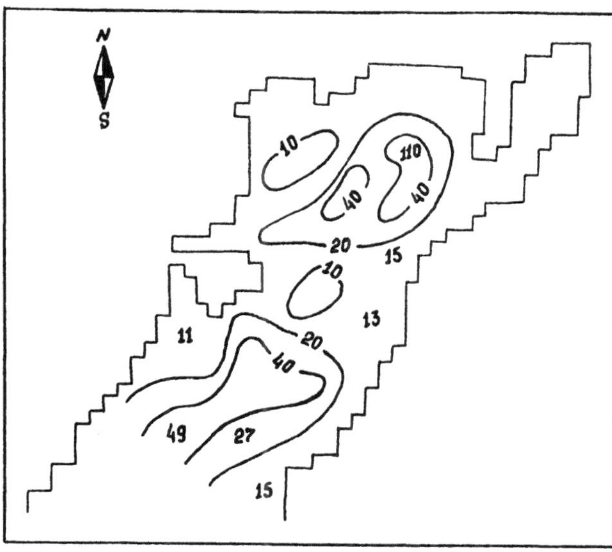

25.2 Advection-diffusion equation

The following equation was used to describe the propagation of distributed nonconservative pollutant:

$$\frac{\partial c}{\partial t} + U\frac{\partial c}{\partial x} + V\frac{\partial c}{\partial y} = \frac{\partial}{\partial x}\left(K_x\frac{\partial c}{\partial x}\right) + \frac{\partial}{\partial y}\left(K_y\frac{\partial c}{\partial y}\right) + Q - \alpha C \qquad (2)$$

where x and y are horizontal Cartesian coordinates;
t the time.
$C(x,y,t)$, the pollutant concentration;
$U(x,y)$, $V(x,y)$, the current velocities along x- and y-axes, respectively;
K_x, K_y, the turbulent diffusion tensor components;
$Q(x,y)$, the pollutant sources;
α, the pollution destruction constant.

The boundary conditions of Eq.(2):

1. The condition of full transport of pollutant across the liquid boundary G_1

$$C|_{G_1} = 0$$

2. The condition of non-propagation of pollutant on the solid boundary G_s

$$\frac{\partial c}{\partial n}|_{G_s} = 0$$

where n is normal to the solid boundary line.

3. Assignment of pollutant sources

$$Q(x_i, y_i) = Q_i$$

where i is source number. For initial condition we put the lack of pollutant in the entire region under calculation.

For numerical approximation of Eq.(2) we used the explicit scheme of splitting by physical processes, according to which at the first stage we calculated pollutant advection, and at the second stage pollution diffusion was calculated taking into account the calculated advection.

The choice of time step of modelling and computational grid sizes was done in according to criterion of computational stability of difference algorithm:

$$\frac{\Delta t(|u| + |v|)}{\Delta h} \leq 1$$

$$\frac{4\Delta t K}{\Delta h^2} < 1$$

where Δh is the size of computational grid cell;
Δt, the time step of modelling.

The calculations were carried out on the rectangular grid, approximating the bay with cell size 1000 x 1000 m. The time step of modelling was 1800 s. Turbulent diffusion coefficient was specified from the results of full-scale experimental investigation of fluorescent dye diffusion in the sea with the aid of air survey and was taken equal to $K=K_x=K_y=10m^2/s$.

25.3 Lagrangrian random-walk model

In a two-dimensional problem, the depth-averaged concentration of pollutant from a single point source of Q_0 capacity, situated in the point (x_0, y_0) can be put down as follows:

$$c(x,y,t) = \frac{1}{\Delta V} Q_0 e^{-\alpha t} P(x,y,t) \mid x_0, y_0, t_0)$$

where $P(x,y,t \mid x_0, y_0, t_0)$ is a conditional probability of particle transition from point (x_0, y_0) to point (x,y) during time $t - t_0$;
$\Delta V = h_x{*}h_y{*}H$, the elementary liquid volume;
H, sea depth;
h_x and h_y are horizontal cell sizes.
Denoting by μ the mass of one conventional pollutant particle, by N_0, the number of particles, released at time moment t_0 and also assuming as a result of modelling that n particles entered the elementary volume ΔV, we obtain

$$c(x,y,t) = Q_0 \frac{n}{\Delta V N_0}$$

Then, according to probability theory, we may proceed to the limit

$$c(x,y,t) = \frac{Q_0}{\Delta V} P(x,y,t) \mid x_0, y_0, t_0) = \frac{Q_0}{\Delta V} \lim_{N_0 \to \infty} \left(\frac{n}{N_0} \right)$$

Thus, pollutant concentration may be expressed in terms of conditional probability which is defined for the entire region calculated by means of calculation of trajectories of statistically meaningful set of conventional particles in current velocity field with a time step Δt. The distribution of terminal points of trajectories over the modelling domain permits calculation of the probability for particles to get into the cell (i,j) of calculational grid

$$P_{i,j}(t) = \frac{1}{h_x h_y H_{i,j}} \lim_{N_0 \to \infty} \left(\frac{n_{i,j}}{N_0} \right)$$

Because of linearity of diffusion model for L sources of power one may put down:

where $Q_l(t)$ is the amount of pollutant, ejected by each source;
N_l, the number of conventional particles, released by each source;

$$C_{i,j}(t) = \frac{1}{h_x h_y H_{i,j}} \sum_{l=1}^{L} Q_l(t) \frac{n_{i,j,l}}{N_l} = \frac{1}{h_x h_y H_{i,j}} \sum_{l=1}^{L} \mu n_{i,j,l}$$

$n_{i,j,l}$, the number of particles got into the volume ΔV.

According to Galkin (1975), time step of modelling Δt and cell sizes of calculational grid h_x and h_y must meet the following requirements providing the concentration of pollutant is arch as not to affect water motion velocity:

$$\Delta t \gg 2\pi\mu / A_c , \quad h_x \gg 2\pi d , \quad h_y \gg 2\pi d$$

where A_0 is particle friction coefficient in the environment;
d, the characteristic size of a particle.

Since the preliminary current velocity data processing for the area investigated has shown that wind-induced current and to a lecher extent tidal current are basic dynamic factors affecting the pollutant transfer the equations of particle motion can be written as follows

$$\frac{dx}{dt} = U_w(x, y, t) + U_t(x, y, t) + U'(x, y, t) \qquad (3)$$

$$\frac{dy}{dt} = V_w(x, y, t) + V_t(x, y, t) + V'(x, y, t) \qquad (3)$$

where U_w and V_w are velocity components of a wind-induced current;
U_t; V_t, the tidal current velocity components;
U'; V', the random velocity components due to turbulent diffusion.

The boundary conditions for pollutant will be put down in terms of particle transition probability:

1. Boundary conditions for the solid boundary G_s

$$P(x_i, y_i, t_i \mid x_{i-1}, y_{i-1}, t_{i-1}) \mid_{x_{i-1}, y_{i-1} \in G_s} = 1 - P_1$$

where P_1 is the probability of pollutant settling at the boundary.

2. Boundary conditions for the liquid boundary of the region, G_1

$$P(x_i, y_i, t_i \mid x_{i-1}, y_{i-1}, t_{i-1}) \mid_{x_{i-1} y_{i-1} \in G_1} = 0$$

The absence of pollutants in the region to be calculated at time moment t_0 was the initial condition

$$C(x, y, t_0) = 0.$$

To reconstruct the tidal current velocity field from the experimental data, we used an approach, based on the isolation of amplitudes F and phases φ of two most essential tidal harmonics, isolated from the time series of current velocities with the help of isolating algorithm of cyclic trends with nonmultiple frequency (Anderson, 1976). Later on the amplitudes and phases for the use in forecasting the tidal components of velocity field were isolated.

$$U_t(x, y, t) = \sum_{i}^{2} F_i(x, y) \cos\left[\frac{2\pi t}{T_i} + \varphi_i(x, y, t_0)\right]$$

where T_i are periods of the tide-induced waves isolated.
While analyzing the residual time series after the subduction of synoptic trends and two major tidal components, it turned out that in genes periodic components with the periods of the rest tidal waves, inertial oscillations and combinational tidal effects are still present. To account them we used the autoregressive process as

$$U'(x, y, t) = \sum_{j}^{J} a_j(x, y) U'(x, y, t_{i-j}) + \gamma\sigma(x, y)$$

where a_j are autoregression coefficients;
γ, the random number with normal distribution and variance $\sigma_2(x,y)$.
The components of wind-induced current velocity U_w and V_w were calculated with the help of a hydrodynamical model, described in Introduction.
Spatial distributions of all dynamic factors were interpolated on equidimensional grid, covering the calculated area and were entered into computer memory.
The area under calculation was approximated by a rectangular calculational grid with cell sized h_x = 375 m, h_y = 500 m. The time step of modelling for calculations, shown in this article, made up 1800 s.

25.4 Account of nonconservative pollutant behaviour

Here the nonconservative feature of pollutant is understood as its destruction due to hydrochemical and biological processes in sea water. The destruction of pollutant is commonly thought to be described by the equation

where q is amount of pollutant in the sea;
Q(t), pollutant supply to sea;

$$\frac{dq}{dt} = -\alpha q + Q(t) \qquad (4)$$

α, the pollutant destruction constant.

In the turbulent diffusion equation, pollutant destruction is accounted for by a special parameter and is included in the finite-difference calculational algorithm. It is different with implementation of Lagrangian algorithm for modelling pollutant transfer. Assuming $q(0)=0$ and α and $Q(t)$ constant, the solution (4) may be put down as follows:

$$Q(t) = \frac{Q}{\alpha}(1-e^{-\alpha t_p})$$

where t_p is time of forecast.

Galkin (1975) noted that the expression in brackets is a limit probability of an individual particle to survive until the end of the forecast

$$P_\alpha \geq 1-e^{-\alpha t_p}$$

Resolving this expression with respect to t_p, we obtain:

$$t_p \geq \frac{\ln(1-P_\alpha)}{\alpha} \qquad (5)$$

From this inequality an algorithm specifying the destruction rate, derives. At the beginning of each new particle trajectory a random number, distributed equally over the range [0,1] is selected which is then substituted in (5) for P_α. If the inequality (5) holds, the particle trajectory is not calculated. Another way to specify the nonconservative feature of pollutant is also feasible. By specifying the number of particles, released by the source at each particle step N_0, we determine the number of particles which have survived until the end of the forecast $N = N_0 e^{-\alpha(tp-t0)}$. In this case there is no sense in calculating trajectories of the rest particles.
More complex is another case when α has the known space-and-time distribution $\alpha(x,y,t)$. In this case the probability of particle conservation will depend on particle trajectory. The introduction of the concept of variable particle mass may provide a way out of the situation. By specifying the initial composition of pollutant μ_0 for the particle, we will calculate the current pollutant content at each step of calculation in parallel to the calculation of coordinates:

$$\mu_i = \mu_{i-1} \cdot e^{-\alpha(x_i, y_i, t_i)\Delta t}$$

where i is the number of time step of a trajectory.
When $\mu = 0$, the calculation of trajectory is suspended. This method requires an additional calculational time and memory expenses. Often, for convenience sake, the pollutant destruction constant is expressed in terms of half-life $T_{1/2}$.

25.5 Results and discussion

In order to demonstrate the results of the simulation of pollutant transfer in Eulerian and Lagrangian terms, we calculated the distribution of pollutants released from several sources for the similar fields of wind-induced current velocity. Figure 25.2 shows wind-induced current field, obtained with the help of numerical solution of the equation (1) for the northerly wind at 10 m/s.

Figure 25.2: North wind current field (wind velocity 10 m/s [m/s].

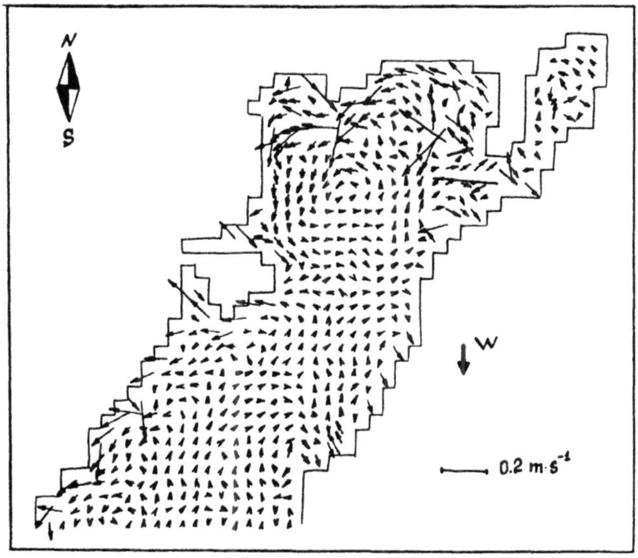

Figures 25.3a and 25.3b show the distribution of the pollutant, calculated with the help of the ADE solution and numerical simulation with the aid of the RWM, respectively. We took the similar amounts of conventional pollutants, erected by the sources, for both cases. The time of forecast was 240 h, the time step, 0.5 h. The half-life of pollutants was specified different: when solving the ADE, $T_{1/2}$ was 720 h, whereas with the RWM it was 240 h. Concentration distribution was assigned in the same relative units. The arrows indicate the position of pollutant sources, the quantity of pollutant, ejected by source 1, being two times as large as that ejected by source 2. In the calculation shown in Figure 25.3b, we specified an additional

source of pollution, the capacity of which was four times lest than that of source 1. Figure 25.4 depicts the calculated wind-induced current field for the southerly wind at 10 m/s, and in Figures 25.5a and 25.5b, the corresponding distributions of concentrations, calculated with the help of both methods. The distribution of capacities of sources as well as other parameters of calculation and the output of results were assigned in a way similar to calculations represented in Figures 25.3a and 25.3b.

Figure 25.3: Pollutant distibutions: a - (ADE), b - (RWM).
Forecast time - 480 h.

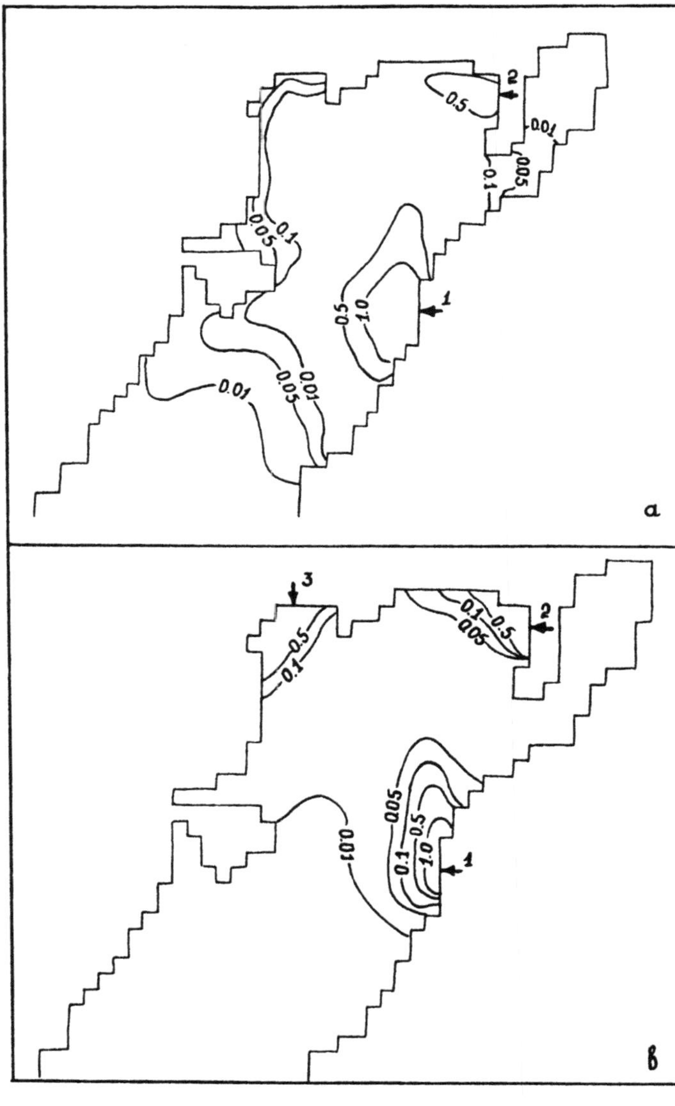

The comparison of Figures 25.2 and 25.4 shows that the reversal of wind direction to the opposite caused the rearrangement of bay currents. A cyclonic water circulation formed with northerly wind in the N part of the bay, and it changed to the opposite with the wind change. The maximum current velocities attained 25 cm/s, and in some shallow and narrow areas even more. At the same time, the maximum velocities of tidal currents, isolated from experimental observations, did not practically exceed 4 to 5 cm/s. This permitted a conclusion that wind-induced currents are dominating in the vast majority of cases.

Figure 25.4: Southely wind current field (wind velocity 10 m/s) [m/s].

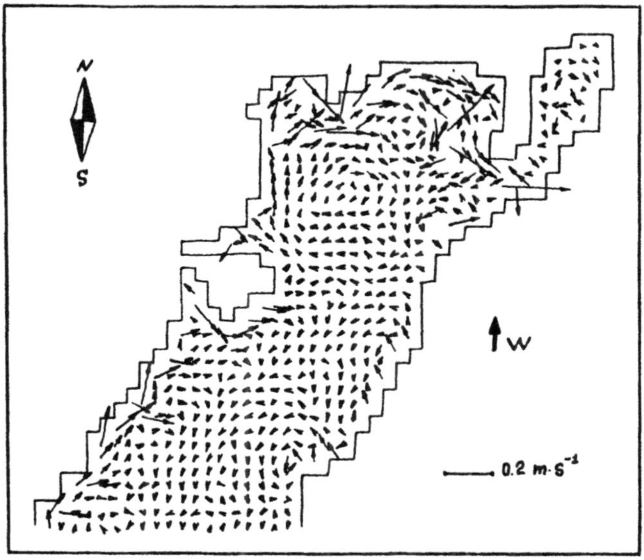

It is difficult to draw a direct comparison of the results of numerical ADE solution and of RWM modelling in our calculations even when the similar initial conditions are specified. The thing is that in the equation of turbulent diffusion, due to deficiency in data, we had to use the turbulent diffusion coefficient K constant in time and uniform in space, whereas while modelling in Lagrangian terms we had an opportunity, based on the experimental data, to specify space-variable and an isotropic variances $\sigma^2_x(x,y)$ and $\sigma^2_y(x,y)$ of turbulent velocity components U' and V'. Owing to this, in calculations using both methods we ventured to specify different values of pollutant destruction constants and also somewhat different sets of sources. The value $T_{1/2}$ = 720 h, employed in the ADE solution, corresponds approximately to the sum destruction speed and the rate of removal of total dissolved organic substance from water for the given bay, while the value $T_{1/2}$ = 240 h, assigned when modelling in Lagrangian coordinates, mirrors the rate of destruction of dissolved and emulsified fractions of oil products under a combined effect of hydrochemical, photochemical and biological processes.

The comparison of Figures 25.3 and 25.5 shows that both methods are efficient in isolating the regions of pollutant accumulation over the bay surface. The difference lies in that numerical solution of the ADE is smoother and is worse in describing the areas of high pollutant concentration gradients which is seen, e.g. from comparison of Figures 25.5a and 25.5b. In this case with decreasing time of forecast the RWM is preferred, while with longer forecasts the ADE solution has greater advantages including savings computer time.

Figure 25.5: Pollutant distributions: a - (ADE), b- (RWM).
Forecast time - 480 h.

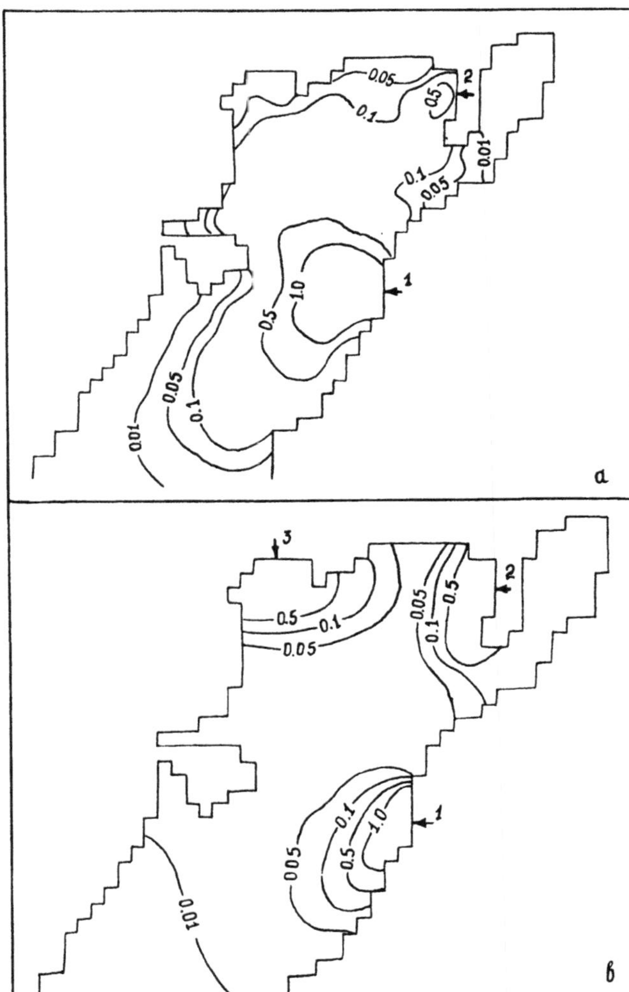

To conclude, it is necessary to present volumes of calculating efforts spent to solve the tasks. The programme of numerical ADE solution consumed 150 kBt of computer memory, and the solution of the examples cited lasted for about 20 min. The programme implementing the RWM required about 260 kBt, while the solution of the appropriate examples using the same computer took 3.25 h with total issued particles 860,000.

References

Anderson, T.: Statistics analysis of time series. Moscow. Nauka, 750p,1976.(in Russian)

Bullin, J.A.; Dukler, A.E.; Stochastic modelling of turbulent diffusion with hybrid computer. Environmental Science and Technology, 8, No.2, 156-164 (1974)

Felzenbaum, A.I.:Theoretical basses and calculation methods of stationary currents. Moscow. Nauka, 127p,1960.(in Russian)

Galkin, L.M.:Solution of diffusion tasks by Monte Carlo method. Moscow. Nauka, 1975.(in Russian)

Honey, F.M.Jr.; Usgeglio-Polatera, J.M: Dispersion simulation in two-dimensional tidal flow. Journal of Hydraulic Engineering, ASCE, vol.110, No.7 (1984)

Usseglio-Polatera, J.M.; Lunge, J.A.: Modelling of pollutant and suspended sediment transport with ARGOS modelling system. Proceedings of International Conference on Numerical and Hydraulic Modelling of Ports and Harbours, Birmingham, England, 83-92, 1985.

O.V.Zaitsev and V.V.Anikiev
Pacific Oceanological Institute
USSR Academy of Sciences
Far East Branch
Vladivostok
USSR

The GeoJournal Library

1. B. Currey and G. Hugo (eds.): *Famine as Geographical Phenomenon.* 1984 ISBN 90-277-1762-1
2. S. H. U. Bowie, F.R.S. and I. Thornton (eds.): *Environmental Geochemistry and Health.* Report of the Royal Society's British National Committee for Problems of the Environment. 1985 ISBN 90-277-1879-2
3. L. A. Kosiński and K. M. Elahi (eds.): *Population Redistribution and Development in South Asia.* 1985 ISBN 90-277-1938-1
4. Y. Gradus (ed.): *Desert Development.* Man and Technology in Sparselands. 1985 ISBN 90-277-2043-6
5. F. J. Calzonetti and B. D. Solomon (eds.): *Geographical Dimensions of Energy.* 1985 ISBN 90-277-2061-4
6. J. Lundqvist, U. Lohm and M. Falkenmark (eds.): *Strategies for River Basin Management.* Environmental Integration of Land and Water in River Basin. 1985 ISBN 90-277-2111-4
7. A. Rogers and F. J. Willekens (eds.): *Migration and Settlement.* A Multiregional Comparative Study. 1986 ISBN 90-277-2119-X
8. R. Laulajainen: *Spatial Strategies in Retailing.* 1987 ISBN 90-277-2595-0
9. T. H. Lee, H. R. Linden, D. A. Dreyfus and T. Vasko (eds.): *The Methane Age.* 1988 ISBN 90-277-2745-7
10. H. J. Walker (ed.): *Artificial Structures and Shorelines.* 1988 ISBN 90-277-2746-5
11. A. Kellerman: *Time, Space, and Society.* Geographical Societal Perspectives. 1989 ISBN 0-7923-0123-4
12. P. Fabbri (ed.): *Recreational Uses of Coastal Areas.* A Research Project of the Commission on the Coastal Environment, International Geographical Union. 1990 ISBN 0-7923-0279-6
13. L. M. Brush, M. G. Wolman and Huang Bing-Wei (eds.): *Taming the Yellow River: Silt and Floods.* Proceedings of a Bilateral Seminar on Problems in the Lower Reaches of the Yellow River, China. 1989 ISBN 0-7923-0416-0
14. J. Stillwell and H. J. Scholten (eds.): *Contemporary Research in Population Geography.* A Comparison of the United Kingdom and the Netherlands. 1990 ISBN 0-7923-0431-4
15. M. S. Kenzer (ed.): *Applied Geography.* Issues, Questions, and Concerns. 1989 ISBN 0-7923-0438-1
16. D. Nir: *Region as a Socio-environmental System.* An Introduction to a Systemic Regional Geography. 1990 ISBN 0-7923-0516-7
17. H. J. Scholten and J. C. H. Stillwell (eds.): *Geographical Information Systems for Urban and Regional Planning.* 1990 ISBN 0-7923-0793-3
18. F. M. Brouwer, A. J. Thomas and M. J. Chadwick (eds.): *Land Use Changes in Europe.* Processes of Change, Environmental Transformations and Future Patterns. 1991 ISBN 0-7923-1099-3

The GeoJournal Library

19. C. J. Campbell: *The Golden Century of Oil 1950–2050*. The Depletion of a Resource. 1991 ISBN 0-7923-1442-5
20. F. M. Dieleman and S. Musterd (eds.): *The Randstad: A Research and Policy Laboratory*. 1992 ISBN 0-7923-1649-5
21. V. I. Ilyichev and V. V. Anikiev (eds.): *Oceanic and Anthropogenic Controls of Life in the Pacific Ocean*. 1992 ISBN 0-7923-1854-4

KLUWER ACADEMIC PUBLISHERS – DORDRECHT / BOSTON / LONDON

GPSR Compliance
The European Union's (EU) General Product Safety Regulation (GPSR) is a set of rules that requires consumer products to be safe and our obligations to ensure this.

If you have any concerns about our products, you can contact us on

ProductSafety@springernature.com

In case Publisher is established outside the EU, the EU authorized representative is:

Springer Nature Customer Service Center GmbH
Europaplatz 3
69115 Heidelberg, Germany

www.ingramcontent.com/pod-product-compliance
Ingram Content Group UK Ltd.
Pitfield, Milton Keynes, MK11 3LW, UK
UKHW020235290926
15470UKWH00002B/551
9780792318545